Residential *Wiring*

Third Edition

AMERICAN TECHNICAL PUBLISHERS
ORLAND PARK, ILLINOIS 60467-5756

Gary J. Rockis
Suzanne M. Rockis
Thomas E. Proctor

Residential Wiring contains procedures commonly practiced in industry and the trade. Specific procedures vary with each task and must be performed by a qualified person. For maximum safety, always refer to specific manufacturer recommendations, insurance regulations, specific job site and plant procedures, applicable federal, state, and local regulations, and any authority having jurisdiction. The material contained is intended to be an educational resource for the user. American Technical Publishers, Inc. assumes no responsibility or liability in connection with this material or its use by any individual or organization.

American Technical Publishers, Inc., Editorial Staff

Editor in Chief:
Jonathan F. Gosse
Vice President—Production:
Peter A. Zurlis
Art Manager:
James M. Clarke
Technical Editor:
James T. Gresens
Copy Editor:
Talia J. Lambarki
Cover Design:
Jennifer M. Hines
Illustration/Layout:
Jennifer M. Hines
James M. Clarke
Melanie G. Doornbos

3 4 5 6 7 8 9 – 11 – 9 8 7 6 5 4 3 2

Printed in the United States of America

ISBN 978-0-8269-1656-3

 This book is printed on recycled paper.

Acknowledgments

The author and publisher are grateful to the following companies and organizations for providing technical information and assistance:

Advantage Drills Inc.
AEMC® Instruments
American Beauty Soldering Tools
Amprobe/Advanced Test Products
Ansul/Tyco Safety Products
Baldor Electric Company
BernzOmatic Corp.
Bowers, Division of Norris Industries
Briggs & Stratton Corporation
Broan-NuTone LLC
Bussman Mfg., a McGraw-Edison Co. Division
Carlon
Carrier Corporation
Coleman Cable Systems, Inc.
Cooper Wiring Devices
DeWALT Industrial Tool Co.
Fluke Corporation
General Electric Co.
Greenlee Textron Inc.
Harvey Hubbell, Inc.
ITT Holub Industries
Kennedy Mfg. Co.
Kidde

Klein Tools, Inc.
Ideal Industries, Inc.
Leviton Manufacturing Co., Inc.
Lew Electric Fittings Co.
Midland Ross Corp., Electrical Products Div.
Milwaukee Electric Tool Corp.
M & W electric Manufacturing Co., Inc.
NuTone, Division of Scovill
Panduit Corp.
Raco Inc.
Republic Steel Corp.
Ruud Lighting, Inc.
Sharp Electronics Corp.
Siemens
Square D Company
The Stanley Works
The Wiremold Co.
Trus Joist, A Weyerhaeuser Business
TRW Crescent Wire and Cable
Vaco Products, Inc.
The Wadsworth Electric Mfg. Co., Inc.
Wilden Enterprises, Inc.
Zircon Corporation

Contents

Introduction

Residential Wiring, 3rd Edition, provides an overview of typical residential wiring systems, electrical device and component installation, and basic troubleshooting procedures. This book also covers residential electrical system principles, tools, and safety. The operation and installation of common electrical devices (switches, receptacles, and boxes) and components (light fixtures, fans, appliances, and low-voltage lighting) are covered in detail. Structured cabling is covered with the latest technology in voice-data-video, security, and fire alarm systems. Wiring procedures are presented with references to personal protective equipment based on requirements specified in the National Fire Protection Association standard NFPA 70E®, *Standard for Electrical Safety in the Workplace*, and are emphasized throughout the book.

This edition has been revised to include the latest information on switch termination procedures, nonmetallic conduit, permanent standby generators, generator safety, and whole-house fan installation. Chapter 12 has been revised to cover residential control systems.

In addition, detailed illustrations are used to supplement concise text. Test instrument procedures are detailed with sequenced steps. Factoids, Technical Tips, and Safety Tips complement the content presented. An extensive Glossary and comprehensive Appendix offer additional reference material.

Residential Wiring is one of many high-quality training products available from American Technical Publishers, Inc. To obtain information about related training products, visit the American Technical Publishers, Inc. website at www.go2atp.com.

The Publisher

Features

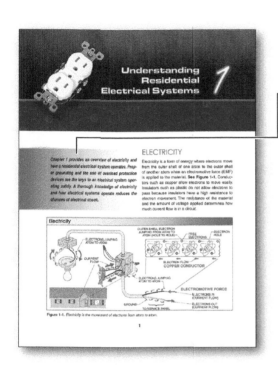

Chapter introductions provide an overview of key content in the chapter

Safety information is included throughout the book

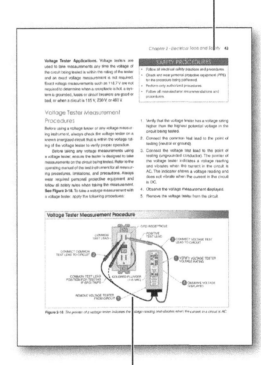

Illustrated step-by-step procedures depict common test instrument procedures

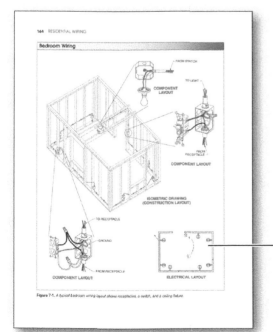

Large, detailed wiring layouts show common component connections

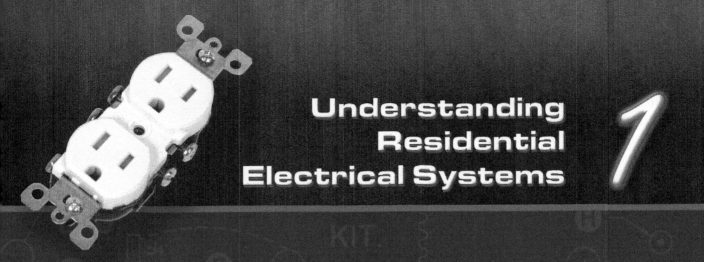

Understanding Residential Electrical Systems

1

Chapter 1 provides an overview of electricity and how a residential electrical system operates. Proper grounding and the use of overload protection devices are the keys to an electrical system operating safely. A thorough knowledge of electricity and how electrical systems operate reduces the chances of electrical shock.

ELECTRICITY

Electricity is a form of energy where electrons move from the outer shell of one atom to the outer shell of another atom when an electromotive force (EMF) is applied to the material. **See Figure 1-1.** Conductors such as copper allow electrons to move easily. Insulators such as plastic do not allow electrons to pass because insulators have a high resistance to electron movement. The resistance of the material and the amount of voltage applied determines how much current flow is in a circuit.

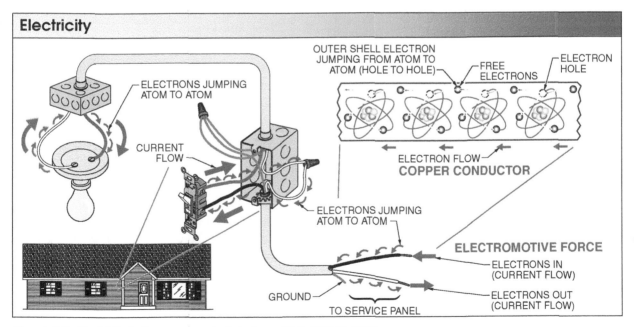

Figure 1-1. *Electricity is the movement of electrons from atom to atom.*

Ohm's law is the relationship between voltage, current, and resistance properties in an electrical circuit. The relationship between voltage, current, and resistance is typically visualized by using Ohm's law in pie chart form. **See Figure 1-2. See Appendix.**

Voltage, Current, and Resistance Relationship

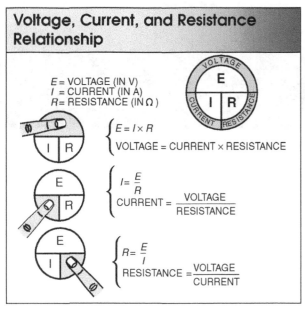

E = VOLTAGE (IN V)
I = CURRENT (IN A)
R = RESISTANCE (IN Ω)

$$E = I \times R$$
VOLTAGE = CURRENT $\times$ RESISTANCE

$$I = \frac{E}{R}$$
CURRENT = $\dfrac{\text{VOLTAGE}}{\text{RESISTANCE}}$

$$R = \frac{E}{I}$$
RESISTANCE = $\dfrac{\text{VOLTAGE}}{\text{CURRENT}}$

Figure 1-2. *Any missing value in Ohm's law can be found when the other two values are known.*

Generated Electricity

Generated electricity is the alternating current (AC) created by power plant generators. Power plant generators create three-phase alternating current electricity. Generators produce electricity when magnetic lines of force are cut by a rotating wire coil (armature). **See Figure 1-3.** The stronger the magnetic field and/or the faster the rotation, the higher the voltage produced.

The voltage output of commercial generators is several thousand volts (typically 22,000 V). The AC electrical output of commercial (power plant) generators travels to transformers for voltage and current control prior to distribution. A *transformer* is an electric device that uses electromagnetism to change (step-up or step-down) AC voltage from one level to another. **See Figure 1-4.** Electricity is used in circuits specifically designed to carry voltage and current through controlled paths (conductors) to operate specific loads.

Generators

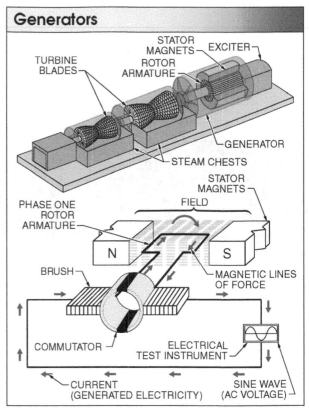

Figure 1-3. *Typical power company steam turbine generators have three armatures as part of the rotor to create 3ϕ power.*

Electricity Distribution

The transformers at a power plant step-up the voltage from the generators (22,000 V) to approximately 240,000 V. **See Figure 1-5.** The high- voltage electricity is distributed across high-voltage power lines to step-down transformer substations. The substations step-down the electrical voltage from 240,000 V to 12,000 V for local distribution. The local voltage distribution level of 12,000 V is step-downed further by transformers to voltages of 600 V, 480 V, 240 V, or 120 V for consumer use.

Residential Use

Local underground vaults and pole transformers step-down electricity to 120 V/240 V to service apartments, condominiums, townhomes, and homes. **See Figure 1-6.** Service entrances (overhead or underground) provide the connection for the distributed power to the power company meter of the residence. Electric meters record the amount of electricity used for billing.

Transformers

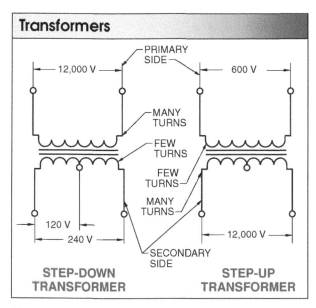

Figure 1-4. *Voltage and current change from the primary to secondary sides (windings) of transformers.*

Service Panels

A *service panel* is an electrical device containing fuses or circuit breakers for protecting the individual circuits of a residence and is a means of disconnecting the entire residence from the distribution system. **See Figure 1-7.** Service panels are wall-mounted in basements, utility rooms, and attached garages. Wherever a service panel is mounted, the panel must be easily accessible to the occupants of the residence.

RESIDENTIAL ELECTRICAL CIRCUITS

Residential electrical circuits include lighting circuits, general-purpose receptacle (small appliance) circuits, special-purpose receptacle (ranges, ovens, and GFCIs) circuits, and special hard-wired loads such as

Electricity Distribution

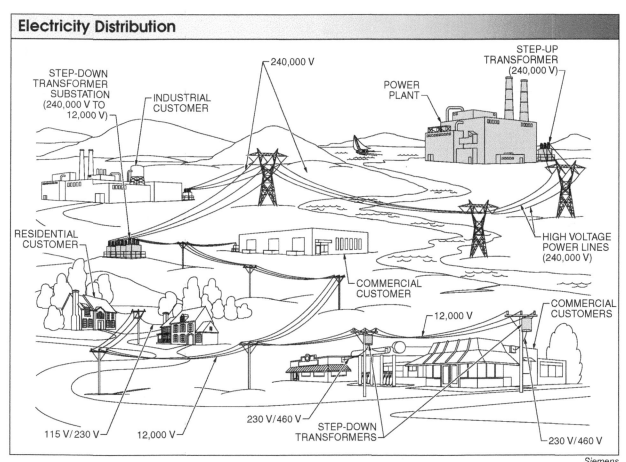

Siemens

Figure 1-5. *When electricity is distributed, electricity is controlled by step-up and step-down transformers between the generators and consumer.*

room heaters, water heaters, and air conditioner circuits. A *circuit* is a complete path (when ON) for current to take that includes electrical control devices, circuit protection, conductors, and load(s). **See Figure 1-8.** A *complete path* is a grouping of electrical devices and wires that create a path for current to take from the power source (service panel), through controls (switches), to the load (light fixtures and receptacles), and back to the power source. The service panel also provides circuit protection by fuses or circuit breakers.

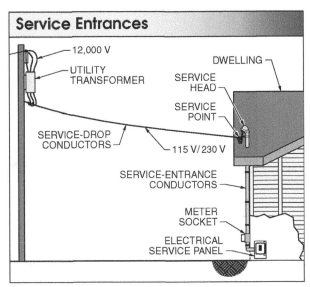

Figure 1-6. *Clearance requirements protect service-drop conductors from physical damage and protect personnel from contact with the conductors.*

GROUNDING

Grounding is an integral part of any properly operating electrical system. In residences, grounding protects the occupants by providing a safe pathway for unwanted electricity that might otherwise create a hazard. Electricity always takes the easiest flow path to ground.

Grounding is typically established at two levels: system grounding and equipment grounding. A *system ground* is a special circuit designed to protect the entire distribution system of a residence. An *equipment ground* is a circuit designed to protect individual components connected to an electrical system.

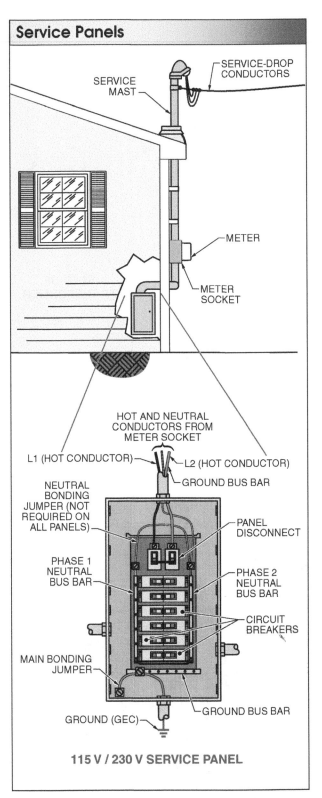

115 V / 230 V SERVICE PANEL

Figure 1-7. *Service panels provide the disconnect from the utility company, system ground, circuit breakers for hot conductors, and neutral bus bars for returning common conductors.*

Residential Electrical Systems

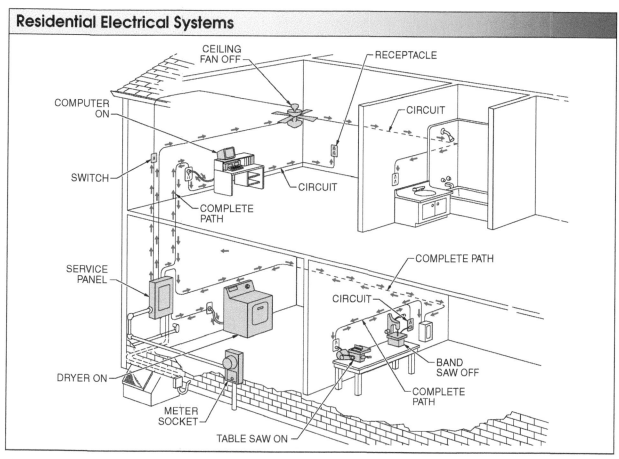

Figure 1-8. *Residential electrical systems are comprised of a number of individual circuits that, when a control (switch) is closed, create a complete path and keep the load ON.*

System Grounding

The primary function of system grounding is to protect the service entrance wiring and connections from lightning and high voltage surges. A system ground provides protection by safely routing lightning and high voltage surges away from the service entrance through the use of a highly conductive pathway to Earth. The two methods of grounding an electrical system are metal water pipe grounding and alternative electrode grounding. **See Figure 1-9.**

Water Pipe Grounding. A *water pipe ground* is a continuous underground metallic pipe that supplies a residence with water and is typically the best electrical ground for a residential electrical system. Water pipes work well as grounds because a large surface area of the pipe is in contact with the earth. The large

surface area reduces resistance and allows any unwanted electricity to easily pass through the pipe to the earth.

When a water pipe is used for grounding, the water pipe run must never be interrupted by a plastic fitting or have an open section of plumbing. Water meters are a source of open circuits when removed. To provide protection when a water meter is removed, a shunt, or meter bonding wire, must be permanently installed. A *shunt* is a permanent conductor placed across a water meter to provide a continuous flow path for ground current. **See Figure 1-10.**

A water pipe grounding electrode conductor connection must be made within 5' of a building's point of entry.

Electrical System Grounding Methods

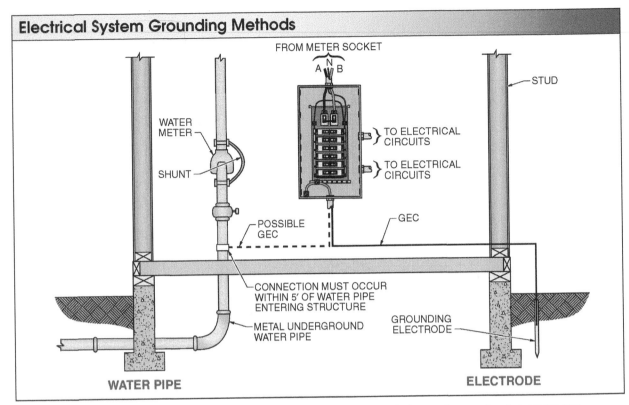

Figure 1-9. *Water pipe grounding and electrode grounding are the two main types of grounding for residences.*

Electrical Shunts

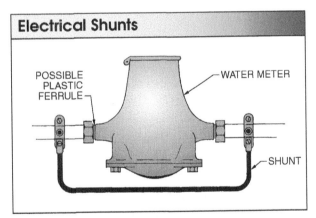

Figure 1-10. *Electrical shunts provide a continuous electrical flow path (ground) when a valve or meter is removed or when nonconducting materials are used, such as plastic ferules or rubber gaskets.*

Electrode Grounding. An *electrode* is a long metal rod used to make contact with the earth for grounding purposes. When no satisfactory grounding electrode is readily available, the common practice is to drive one or more metal rods (connected in parallel) into the ground. **See Figure 1-11.** The electrode and circuit must provide a flow path to earth with less than 25 Ω of resistance. When local soils are extremely salty, acidic, or alkaline, the local building inspector or the NEC® should be consulted to determine the correct type of electrode to use.

Equipment Grounding

The primary function of equipment grounding is to protect individual electrical devices. Equipment grounding safely grounds any devices attached to a system or plugged into receptacles inside a home. **See Figure 1-12.** When an appliance has not been properly grounded, electrical current caused by a short will seek the easiest path to earth. Unfortunately, the human body is an electrical conductor and allows current to reach earth by traveling through the body (electric shock). Properly grounded devices protect the body by harmlessly conducting unwanted electricity to ground.

Electrode Grounding

Grounding Electrode System	Minimum Size	Notes
Metal underground water pipe	10′ length	Requires supplemental electrode
Metal frame in building		Where effectively grounded
Concrete-encased electrode	2″ of concrete encasement 10′ of ½″ steel bars or 20′ of #4 Cu	
Ground ring	20′ of #2 Cu	Minimum depth 2½′

Alternative Electrodes	Minimum Size	Burial Requirements
Pipe or conduit (iron or steel shall be galvanized or metal coated)	¾″D × 8′ length	
Rods of iron and steel	⅝″D × 8′ length	
Rods of stainless steel and nonferrous rods (shall be listed if less than ⅝″D)	½″D × 8′ length	
Plate of iron or steel	¼″D × 2 sq ft	
Plate of nonferrous metal	.06″D × 2 sq ft	

Figure 1-11. *Grounding electrode systems or alternative electrodes are used for grounding residential electrical systems.*

Equipment Grounding

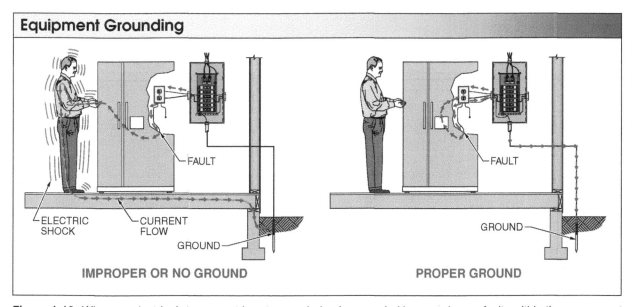

IMPROPER OR NO GROUND   **PROPER GROUND**

Figure 1-12. *When an electrical component is not grounded or is grounded improperly, any faults within the component can create a potentially hazardous situation. Touching the component can cause current to flow through the body resulting in electric shock or electrocution.*

Grounding Small Appliances. Small appliances are easily incorporated into a grounded system. Most small electrical appliances are designed with three-prong grounded plugs that match a standard three-prong grounded receptacle. **See Figure 1-13.**

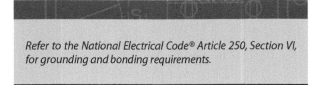

Refer to the National Electrical Code® Article 250, Section VI, for grounding and bonding requirements.

Small Appliance Grounding

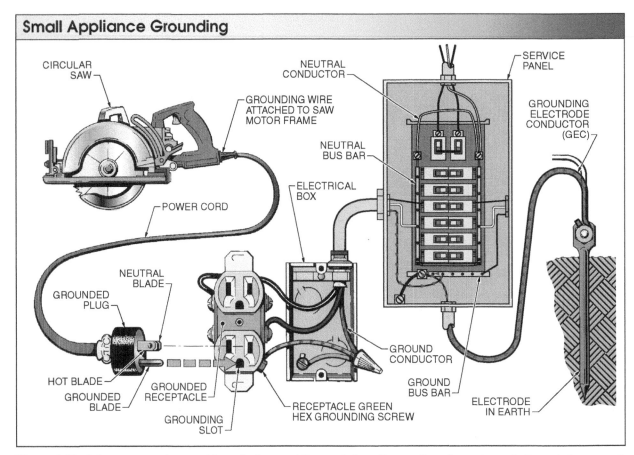

Figure 1-13. *A three-prong plug provides a fault current flow path from the small appliance to earth through the grounding system.*

The U-shaped blade of the plug and the U-shaped hole in the receptacle are the ground connections. The U-shaped blade of a plug is longer than the current-carrying blades. The added length ensures a strong ground connection while the plug is being inserted or removed from a receptacle. **See Figure 1-14.**

Manufacturers of electrical equipment such as appliances, tools, wires, and fuses manufacture products that must meet minimum required codes and standards to ensure products are built for safe operation. Exceeding minimum requirements helps establish the reputation of a company for producing reliable quality products, and also reducing product liability issues that may arise from electrical shocks, fire, and injuries.

Three-Prong Plugs

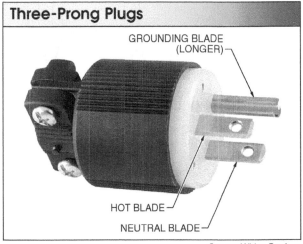

Cooper Wiring Devices

Figure 1-14. *Three-prong grounded plugs have a grounding blade that is longer than the current-carrying blades to ensure a grounded connection when the plug is being connected or disconnected.*

FUSES

A *fuse* is an electrical device used to limit the rate of current flow in a circuit. Fuses typically contain a piece of soft metal that melts, opening the circuit when the fuse is overloaded.

The three broad categories of fuses are plug, cartridge, and blade. **See Figure 1-15.** The typical homeowner generally requires 15 A or 20 A plug fuses and 30 A or 40 A cartridge fuses. Occasionally a service panel blade fuse may need to be replaced. Blade fuses typically have a 60 A or 100 A rating and are found in the main disconnect of the service panel.

For each category of fuse, several specific types of fuses can be found for specific applications. **See Figure 1-16.** Time-delay fuses are one specific type of fuse that is quite popular.

Plug fuses are no longer used, except for replacement of existing fuses in older buildings.

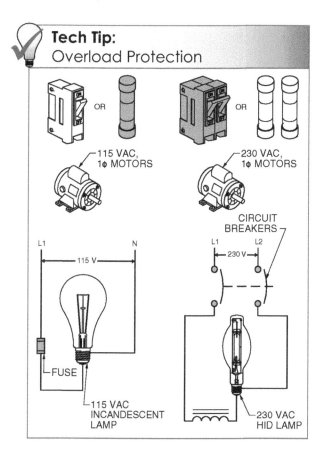

Tech Tip:
Overload Protection

Fuse Types

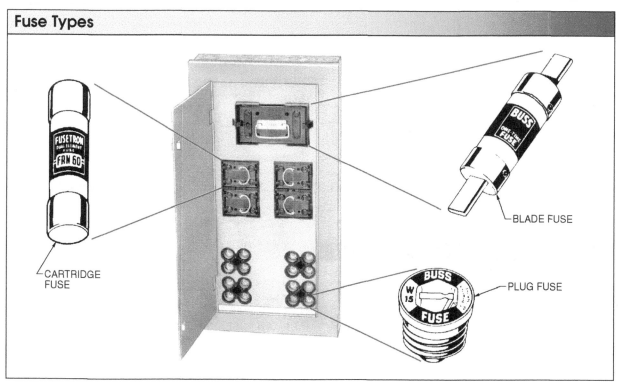

Figure 1-15. *Each type of fuse has a typical application in a residential service panel.*

Fuse Categories

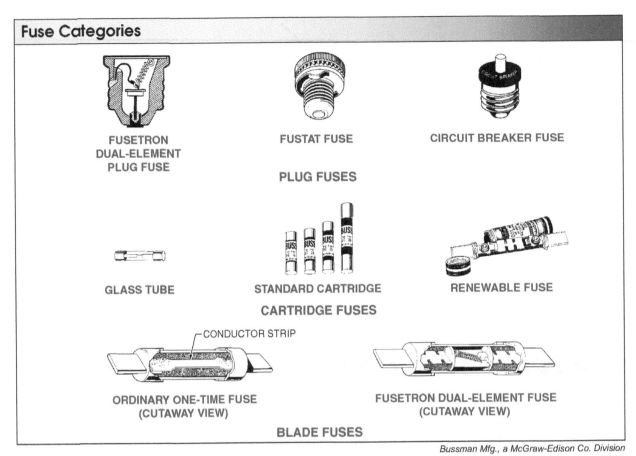

FUSETRON DUAL-ELEMENT PLUG FUSE

FUSTAT FUSE

CIRCUIT BREAKER FUSE

PLUG FUSES

GLASS TUBE

STANDARD CARTRIDGE

RENEWABLE FUSE

CARTRIDGE FUSES

CONDUCTOR STRIP

ORDINARY ONE-TIME FUSE (CUTAWAY VIEW)

FUSETRON DUAL-ELEMENT FUSE (CUTAWAY VIEW)

BLADE FUSES

Bussman Mfg., a McGraw-Edison Co. Division

Figure 1-16. *Each type of fuse can be further divided into specific applications.*

Standard Plug Fuses

A *standard plug fuse* is a screw-in type electrical safety device that contains a metal conducting element designed to melt when the current through the fuse exceeds the rated value. **See Figure 1-17.** During a serious overload condition, the melting of the conductor strip is almost instantaneous.

General specifications that apply to fuses include mountings, materials of construction, fuse types, and features. Mounting choices include solderable or surface mount, solderable with leads, and replaceable with holder or clips. Common materials of construction include glass, ceramic, and sand. Fuse types can be miniature, subminiature or micro, midget, automotive, blade type, PC board, and protective.

Standard Plug Fuses

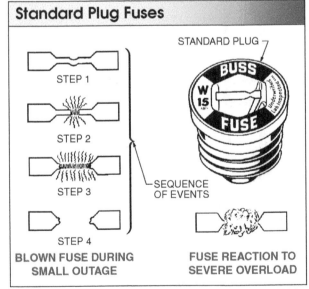

STANDARD PLUG

STEP 1

STEP 2

STEP 3

STEP 4

SEQUENCE OF EVENTS

BLOWN FUSE DURING SMALL OUTAGE

FUSE REACTION TO SEVERE OVERLOAD

Bussman Mfg., a McGraw-Edision Co. Division

Figure 1-17. *Plug fuses react differently depending on the amount of overload.*

Time-Delay Plug Fuses

A *time-delay plug fuse* is a screw-in type electrical safety device with a dual element. The first element provides the protection of a standard plug fuse for short circuits, and the second element protects against heating due to light overloads (the second element does not instantaneously melt). **See Figure 1-18.** The second element is useful in preventing nuisance tripping caused by motor-driven appliances such as refrigerators and air conditioners starting up. Manufacturers of time-delay plug fuses have created plugs with various responses to shorts and overloads.

Figure 1-18. *Time-delay plug fuses respond differently to shorts and overloads than standard plug fuses.*

Type S Plug Fuse

A *Type S plug fuse* is a screw-in type electrical safety device that has all the operating characteristics of a time-delay plug fuse plus the added advantage of

being non-tamperable. A Type S fuse is considered non-tamperable because the fuse cannot be installed into a base unless the fuse matches the size of the base. **See Figure 1-19.** Each Type S base adapter is sized for a particular size fuse. Due to the non-tamperable design, a 20 A Type S fuse will not fit into a 15 A Type S base.

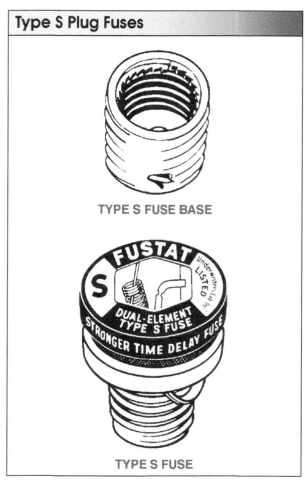

Bussman Mfg., a McGraw-Edison Co. Division

Figure 1-19. *Type S plug fuses have a distinctive base size, and once screwed into a panel, only Type S plug fuses of a specific amperage can be used.*

Fuse holders are devices for containing, protecting, and mounting fuses. Fuse holder designs include open or fully closed. Open fuse holder designs are fuse clips, fuse blocks, socket, and plug-on cap varieties. Fully enclosed designs use a fuse carrier inserted into a holder.

Cartridge Fuses

A *cartridge fuse* is a snap-in type electrical safety device that operates on the same basic heating principle as a plug fuse. However, instead of a screw base, cartridge fuses are secured by clips. Many cartridge fuses have dual elements that absorb temporary overloads (such as with motor startup) without blowing, while protecting against dangerous overloads and short circuits. **See Figure 1-20.** Cartridge fuses should be removed and installed using fuse pullers.

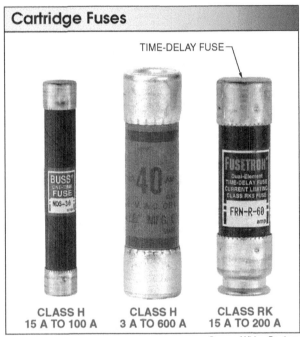

Cartridge Fuses

TIME-DELAY FUSE

CLASS H	CLASS H	CLASS RK
15 A TO 100 A	3 A TO 600 A	15 A TO 200 A

Cooper Wiring Devices

Figure 1-20. *Cartridge fuses are found with a number of different amperage ratings.*

Cartridge fuses can be purchased as one-time-use fuses or as renewable element fuses. Renewable element cartridge fuses have replaceable elements and are typically used for industrial applications.

Blade Cartridge Fuses

A *blade cartridge fuse* is a snap-in type electrical safety device that operates on the heating effect of an element. Blade cartridge fuses are typically used for high amperage rated applications found in industry. Blade fuses are secured by clips and may be one-time use or renewable. **See Figure 1-21.**

Blade Cartridge Fuses

ONE-TIME USE   TIME-DELAY

Cooper Wiring Devices

Figure 1-21. *Blade cartridge fuses, like all fuse types, are found as one-time use fuses and as time-delay fuses.*

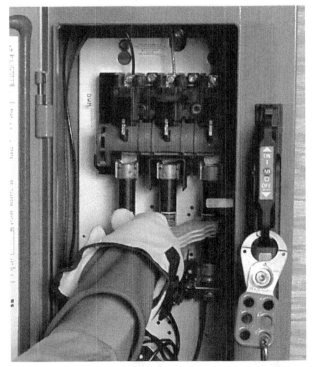

A fuse puller is used to safely remove blade fuses and cartridge fuses from electrical boxes and cabinets.

Fluke Corporation
To help prevent electrical shock, proper personal protective equipment (PPE) must be worn at all times when performing tasks on energized circuits.

Circuit Breaker Fuses

A *circuit breaker fuse* is a screw-in type electrical safety device that has the operating characteristics of a circuit breaker. The advantage of a circuit breaker fuse is that the fuse can be reset after an overload and used again and again. **See Figure 1-22.**

Circuit Breaker Fuses

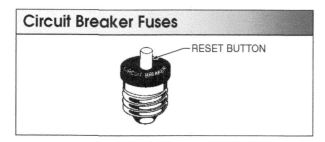

— RESET BUTTON

Figure 1-22. *Circuit breaker fuses are similar to circuit breakers in that a circuit breaker fuse can be reset after an overload and used repeatedly.*

CIRCUIT BREAKERS

Circuit breakers are similar to fuses in function. A *circuit breaker* is a fixed electrical safety device designed to protect electrical devices and individuals from overcurrents. Unlike most fuses, circuit breakers can be reset. They are the most popular overcurrent safety device. **See Figure 1-23.** The trip lever handle of a circuit breaker represents the position

of the contacts when the circuit breaker is ON and when the breaker is tripped OFF. A circuit breaker may be reset by moving the trip lever handle to the full OFF position and then returning the handle to the ON position. Individuals must ensure the source of an overload is repaired before attempting to reset a breaker. The three types of circuit breakers are thermal, magnetic, and thermal-magnetic.

Circuit Breakers

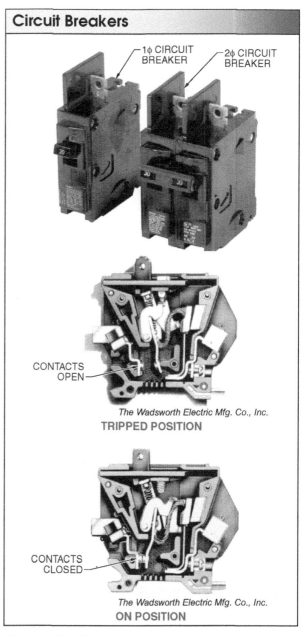

— 1φ CIRCUIT BREAKER — 2φ CIRCUIT BREAKER

CONTACTS OPEN —

The Wadsworth Electric Mfg. Co., Inc.
TRIPPED POSITION

CONTACTS CLOSED —

The Wadsworth Electric Mfg. Co., Inc.
ON POSITION

Figure 1-23. *Circuit breakers have standard voltage ratings with varying amperages, similar to fuses, but can be reset after encountering an overload.*

Thermal Circuit Breakers

A *thermal circuit breaker* is an electrical safety device that operates with a bimetallic strip that warps when overheated. The bimetallic strip is two pieces of metal made up of dissimilar materials that are permanently joined together. Because metals expand and contract at different rates, heating the bimetallic strip causes the strip to warp or curve. The warping effect of the bimetallic strip is utilized as the tripping mechanism for the thermal breaker. **See Figure 1-24.**

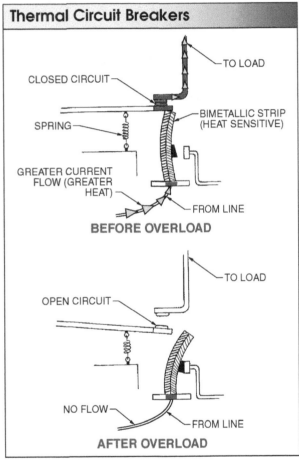

Thermal Circuit Breakers

TO LOAD

CLOSED CIRCUIT

SPRING

BIMETALLIC STRIP (HEAT SENSITIVE)

GREATER CURRENT FLOW (GREATER HEAT)

FROM LINE

BEFORE OVERLOAD

TO LOAD

OPEN CIRCUIT

NO FLOW

FROM LINE

AFTER OVERLOAD

Figure 1-24. *Thermal circuit breakers have a bimetallic strip that warps with an increase in heat caused by excessive current flow.*

Typical household wattages include a hair dryer at 1400 W, a vacuum cleaner at 1000 W, a television at 150 W, and a stereo music system at 30 W.

Once a thermal breaker has tripped, the bimetallic strip cools allowing the strip to reshape and allowing the breaker to be reset. Individuals must reset the trip lever handle to reactivate the circuit breaker and restore power to the de-energized circuit. As with all circuit breakers, the source of an overload must be removed before attempting to reset the breaker.

Magnetic Circuit Breakers

A *magnetic circuit breaker* is an electrical safety device that operates with miniature electromagnets. Electromagnets are created by passing current through a coil of wire. The greater the current, the stronger the magnetic field created by the coil. When the current through the coil exceeds the rated value of the breaker, the magnetic attraction becomes strong enough to activate the trip bar and open the circuit. **See Figure 1-25.** Once the overload is removed, the trip bar can be reset to the original position reactivating the circuit.

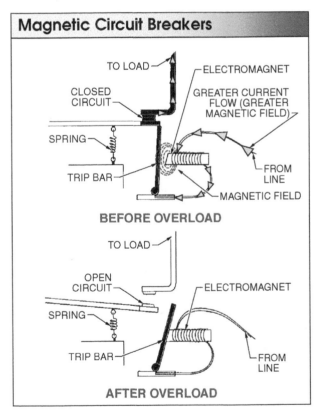

Magnetic Circuit Breakers

TO LOAD

ELECTROMAGNET

CLOSED CIRCUIT

GREATER CURRENT FLOW (GREATER MAGNETIC FIELD)

SPRING

FROM LINE

TRIP BAR

MAGNETIC FIELD

BEFORE OVERLOAD

TO LOAD

OPEN CIRCUIT

ELECTROMAGNET

SPRING

TRIP BAR

FROM LINE

AFTER OVERLOAD

Figure 1-25. *Excessive current creates a magnetic field strong enough to move the trip bar forward, causing the breaker to open.*

Thermal-Magnetic Circuit Breakers

A *thermal-magnetic circuit breaker* is an electrical safety device that combines the heating effect of a bimetallic strip with the pulling strength of a magnet to move a trip bar. **See Figure 1-26.** The magnetic portion consists of a permanent magnet or electromagnet in series with the bimetallic strip. Thermal-magnetic breakers have the fastest response times to serious overloads of any type of circuit breaker.

No matter which type of internal mechanism a circuit breaker uses, externally, most circuit breakers look the same. Circuit breakers are available in a variety of amperages, but the voltage is typically rated as 115 V for single pole residential breakers or 230 V for double pole residential breakers. To gain access to the circuit breaker connections in a service panel, the cover of the panel must be removed.

GROUND FAULT CIRCUIT INTERRUPTERS

A *ground fault circuit interrupter (GFCI)* is a fast-acting electrical device that detects low levels of leakage current to ground and opens the circuit in response to the ground fault. **See Figure 1-27.** GFCIs limit electric shock by opening circuits before an individual receives a serious injury. GFCIs do not protect against line-to-line contact.

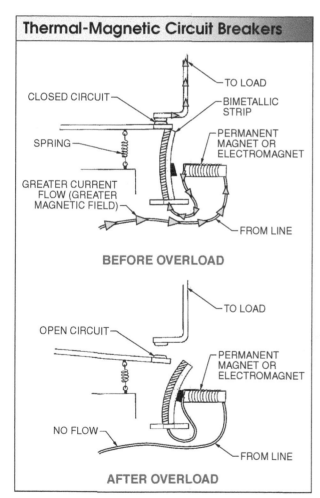

Figure 1-26. *The action of the bimetallic strip is aided by the presence of a magnet.*

Cooper Wiring Devices

Figure 1-27. *GFCIs are used for outdoor receptacle applications and anywhere where extra protection is required.*

Necessity of GFCIs

Ground fault circuit interrupters were created because of the limitations of commonly used circuit breakers. Individuals have been electrocuted by equipment in electrical systems where the fault current was not great enough to trip a standard circuit breaker. The situation exists because standard circuit breakers are designed to trip only when large currents are present (short or overload). Currents present from deteriorating insulation and minor equipment damage do not produce enough current flow to open a standard circuit breaker. As an alternative, GFCIs are specifically designed with leakage currents in mind. A GFCI can react to a current as small as $\frac{5}{1000}$ (.005) of an amp in a fraction of a second.

GFCI Operation

A sequence of events takes place when a fault current is detected by a GFCI. Typically, the load (electric shaver, drill, or garden tool) has the same amount of current flowing to the load (black or red wire) as flowing away from the load (white wire). However, in the event of a ground fault, some of the current, which normally returns to the power source through the white wire (neutral), is diverted to ground. **See Figure 1-28.**

Because a current imbalance exists between the hot and neutral wires, the sensing device detects the current difference and signals an amplifier. When the amplified signal is large enough, the amplifier activates the interrupting device. Once the interrupting device is activated, the circuit is opened and current to the load is shut off. After the ground fault has been discovered and repaired, the GFCI is reset and the process begins again. GFCIs have a test circuit built into the unit so that the GFCI can be tested on a monthly basis without a ground fault condition.

Per the NEC®, all receptacles installed in kitchen countertop areas require GFCI protection.

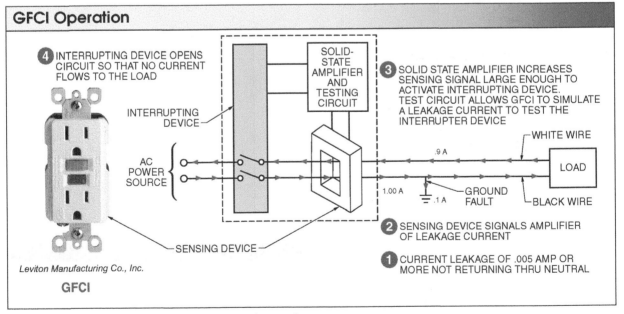

Figure 1-28. *Most GFCIs trip in 25 ms to several seconds.*

GFCI Required Locations

The National Electrical Code (NEC®) requires ground fault circuit interrupters for protection in the following areas:

1. Residential. **See Figure 1-29.**
   - Outdoor receptacles
   - Bathrooms
   - Residential garages
   - Kitchens (countertop areas)
   - Unfinished basements and crawl spaces
2. Construction Sites
3. Mobile Homes and Mobile Home Parks

   - Outdoor receptacles
   - Bathrooms
4. Swimming Pools and Fountains
   - Receptacles near pools
   - Lighting fixtures and lighting outlets near pools
   - Underwater lighting fixtures over 15 V
   - Electrical equipment used with storable pools
   - Fountain equipment operating at over 15 V
   - Cord-connected and plug-connected equipment for fountains

GFCI Installation Locations

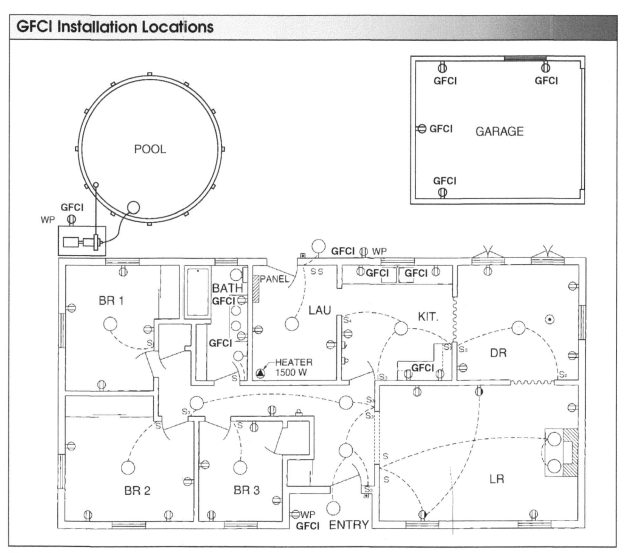

Figure 1-29. *GFCIs are installed indoors and outdoors of residences.*

Installing GFCIs

GFCIs are installed in standard receptacle boxes indoors and in weatherproof boxes outdoors. GFCI receptacles are typically installed individually. **See Figure 1-30.** GFCIs can also be installed to protect several standard receptacles in one circuit. GFCIs are considered easy to install and most manufacturers supply an installation kit to aid in installation. In addition to receptacle types, GFCIs are available as circuit breakers to protect entire circuits. Kits are also available to match GFCIs to existing circuit breaker panels.

Arc Faults

A *short circuit* is a condition that occurs when two ungrounded conductors (hot wires), or an ungrounded and a grounded conductor of a single-phase circuit, come in contact with each other. In a short circuit, current leaves the normal current-carrying path and goes around the load and back to the power source or to ground. The low-resistance path can be due to failure of circuit components or failure in the wiring of the circuit. For example, if two conductors accidentally make contact with each other, the conductors produce a dead short across the circuit, which can cause arcing. **See Figure 1-31.** The arc can create enough heat to ignite combustible material located near the conductors, causing a fire.

Short circuits can be avoided by minimizing the number of locations where electrical wire is exposed outside an enclosure, which includes reducing the number of extension cords in use.

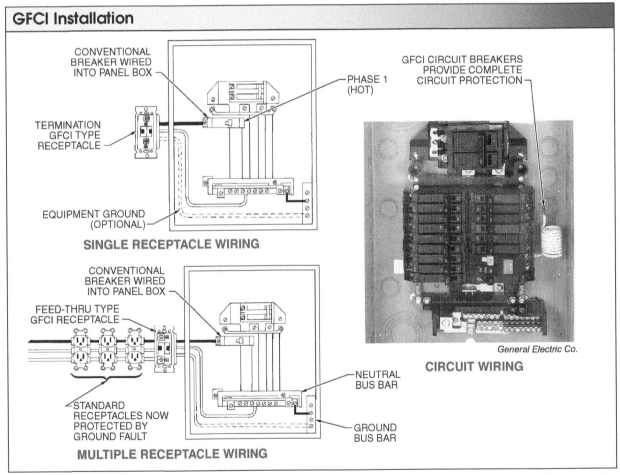

GFCI Installation

CONVENTIONAL BREAKER WIRED INTO PANEL BOX

TERMINATION GFCI TYPE RECEPTACLE

EQUIPMENT GROUND (OPTIONAL)

SINGLE RECEPTACLE WIRING

CONVENTIONAL BREAKER WIRED INTO PANEL BOX

FEED-THRU TYPE GFCI RECEPTACLE

STANDARD RECEPTACLES NOW PROTECTED BY GROUND FAULT

MULTIPLE RECEPTACLE WIRING

PHASE 1 (HOT)

NEUTRAL BUS BAR

GROUND BUS BAR

GFCI CIRCUIT BREAKERS PROVIDE COMPLETE CIRCUIT PROTECTION

General Electric Co.

CIRCUIT WIRING

Figure 1-30. *A GFCI can be wired to protect a single receptacle, multiple receptacles when wired in parallel, and entire circuits when installed as a circuit breaker in the service panel.*

Arc Faults

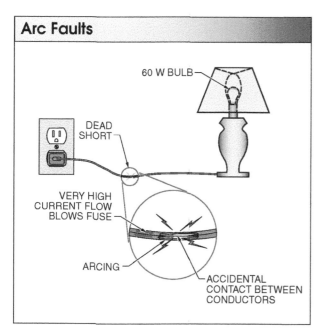

Figure 1-31. *When two conductors make accidental contact because of damaged insulation, the conductors produce a dead short across the circuit. This can result in arcing.*

Arc-Fault Circuit Interrupters. An arc-fault circuit interrupter (AFCI) is a fast-acting receptacle that detects electrical arcs and opens a circuit in response to the arc. Arc-fault circuit interrupters are installed to provide a significant level of protection against arc faults and to prevent fires. Faulty electrical wiring causes most home fires. The NEC® requires that an AFCI be installed to protect 125 V, 15 A and 20 A branch circuits in most dwellings. An AFCI is most often used in bedroom branch circuits.

Arc-fault circuit interrupters resemble ground-fault circuit interrupters (GFCIs). An AFCI is wired similar to a GFCI in that the AFCI breaker is wired to the hot and neutral of the branch circuit with a pigtail from the AFCI to connect to the neutral bus. **See Figure 1-32.** As with a GFCI, an AFCI has a test button to check its operation. Some AFCI manufacturers color code the test button, so it is a different color than the test button on a GFCI.

Arc-Fault Circuit Interrupters (AFCIs)

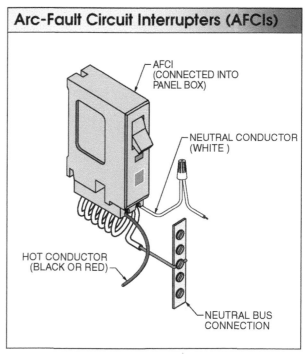

Figure 1-32. *Arc-fault circuit interrupters resemble GFCIs and are installed to provide a significant level of protection against arc faults and prevent fires.*

Name_____   Date _____

Residential Electrical Systems

_____
1. The flow of electricity through a residential electrical system is from ___ to ___.

_____
2. Using Ohm's law, current flow (A) through a circuit is ___ A when the circuit is supplied with 120 V and has 24 Ω of resistance?
   A. 0.2   **B.** 5
   C. 120   **D.** 2880

_____
3. The voltage output of commercial generators is typically ___V.
   A. 230   **B.** 1200
   C. 460   **D.** 22,000

_____
4. A(n) ___ is a grouping of electrical devices and wires that create a path for current to take from the power source (service panel), through controls (switches), to the load (light fixtures and receptacles), and back to the power source.

_____
5. ___ grounding is designed to protect the entire electrical network.

_____
6. ___ grounding is designed to protect individual electrical components.

_____
7. A(n) ___ is considered the best ground for a residential electrical system.

T   F
8. An alternative electrode must be at least 10′ in length.

T   F
9. An iron plate may be used as a made electrode.

T   F
10. A metal underground water pipe requires a supplemental electrode.

T   F
11. The minimum conduit diameter for a made electrode is ⅝″.

_____
12. A 20′ length of # ___ Cu may be used as a ground ring.

_____
13. The minimum burial depth of a ground ring is ___′.

_____
14. A(n) ___ strip trips a thermal circuit breaker.

T   F   **15.** The NEC® requires GFCI protection on construction sites.

T   F   **16.** Circuit breakers have standard voltage ratings but vary in amperage ratings.

_____   **17.** ___ fuses can be reset.

_____   **18.** ___ cartridge fuses are typically used for high-amperage-rated applications.

_____   **19.** Electromagnets are created by passing ___ through a coil of wire.

_____   **20.** A GFCI can react to current as little as ___ A in less than a fraction of a second.

_____   **21.** Standard plug fuses have a(n) ___ strip that melts when current exceeds the rated value of the fuse.

_____   **22.** A concrete-encased electrode shall be encased in ___" of concrete.

      A. 1½   **B.** 2
      C. 2½   **D.** 4

_____ ,_____   **23.** A ground ring of #2 Cu shall be a minimum of ___' in length.

      A. 10   **B.** 15
      C. 20   **D.** 40

T   F   **24.** A made electrode may be a nonferrous metal plate .06" thick × 2 sq ft.

T   F   **25.** Pipe or conduit made electrodes shall be galvanized or metal-coated.

_____   **26.** The U-shaped blade of a three-prong grounded plug is the ___ connection.

_____   **27.** Electricity always seeks the easiest ___ to ground.

_____   **28.** A meter ___ provides a continuous electrical path when a water meter is removed.

T   F   **29.** Fuses limit the rate of current flow in a circuit.

T   F   **30.** Main service panel cartridge fuses are commonly of the blade type.

T   F   **31.** A circuit breaker is reset by moving the handle to the full OFF position and then to the ON position.

T   F   **32.** GFCIs protect receptacles and plugs only.

_____   **33.** Fuse ___ should be used when removing fuses.

Grounding Electrodes

_____ **1.** The minimum length of A is ___'.

_____ **2.** The maximum angle of B is ___°.

_____ **3.** The minimum depth of C is ___'.

_____ **4.** The minimum diameter of D is ___".

_____ **5.** The minimum diameter of E is ___".

_____ **6.** The minimum length of F is ___'.

_____ **7.** The minimum size of G is ___ sq ft.

_____ **8.** The minimum thickness of G is ___" for steel.

_____ **9.** The minimum thickness of G is ___" for nonferrous metal.

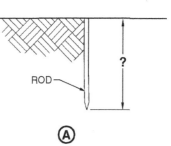

ROD

?

(A)

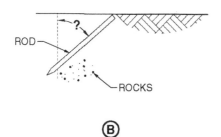

ROD

?

ROCKS

(B)

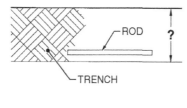

ROD

?

TRENCH

(C)

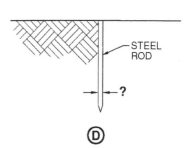

STEEL ROD

?

(D)

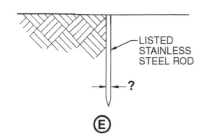

LISTED STAINLESS STEEL ROD

?

(E)

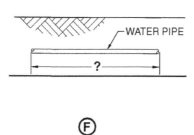

WATER PIPE

?

(F)

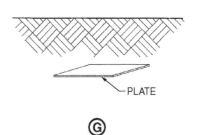

PLATE

(G)

Electrical Service

_____   **1.** Circuit breakers

_____   **2.** Meter socket

_____   **3.** Panel disconnect

_____   **4.** Ground bus bar

_____   **5.** Meter

_____   **6.** Phase 2 neutral
bus bar

_____   **7.** Service mask

_____   **8.** MBJ

_____   **9.** GEC

_____   **10.** Phase 1 neutral
bus bar

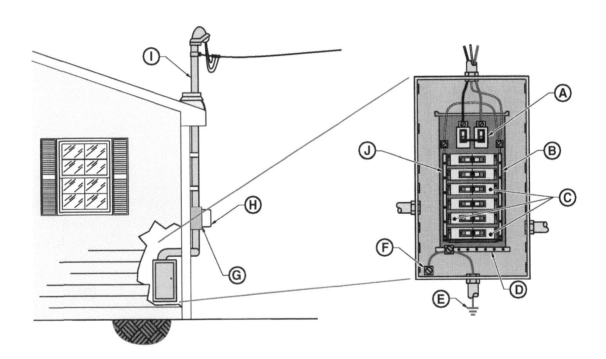

24

Grounding Systems

_____ **1.** Grounding prong

_____ **2.** Grounding rod
(electrode)

_____ **3.** Ground
conductor

_____ **4.** Load

_____ **5.** Service panel
grounding bar

_____ **6.** Receptacle

_____ **7.** GEC wire

_____ **8.** Green hex
grounding
screw

_____ **9.** Neutral
conductor

_____ **10.** Grounding slots

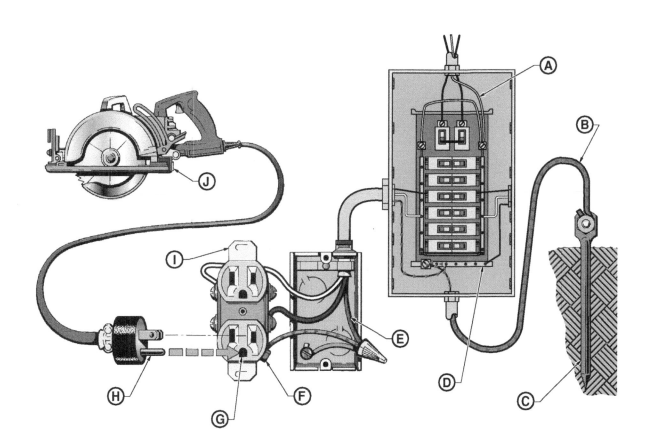

Fuses

_____

_____

_____

_____

_____

1. Glass tube fuse

2. Renewable cartridge fuse

3. Circuit breaker plug fuse

4. One-time blade fuse

5. Fusetron dual-element plug fuse

_____

_____

_____

6. Fusetron dual-element blade fuse

7. Fustat plug fuse

8. Standard cartridge fuse

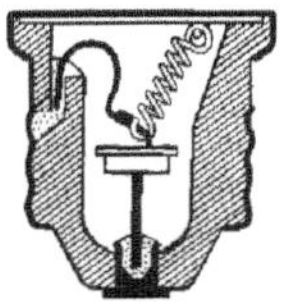

Ⓐ

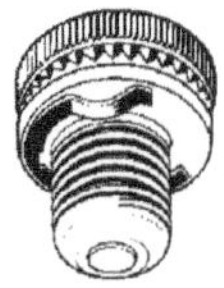

Ⓑ

Ⓒ

Ⓓ

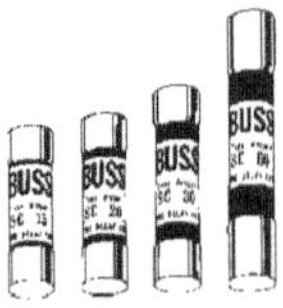

Ⓔ

Ⓕ

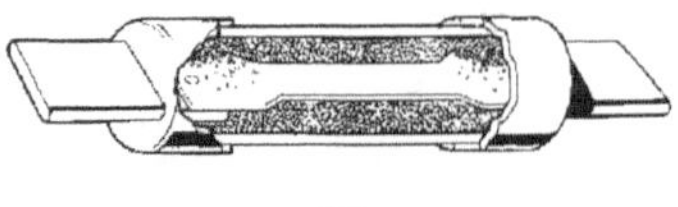

Ⓖ

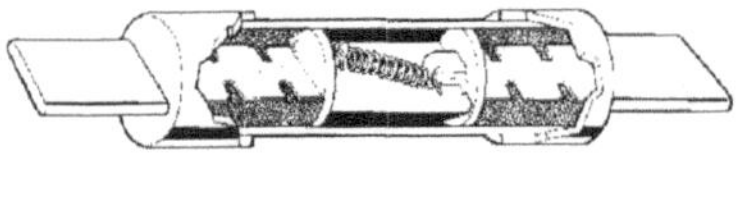

Ⓗ

Electrical Tools and Safety 2

Chapter 2 discusses the quality of tools and how to organize and use tools safely. Chapter 2 explains general electrical safety rules and safety on the construction site. Individuals must follow all OSHA and NFPA 70E safety rules and conform to all codes and rules when performing electrical work.

useful in the work the electrician performs. Taking time to create a tool inventory allows individuals to pick and choose the best types and manufacturers of tools for the type of work performed.

QUALITY TOOLS

"Get the right tool for the job" has been said over and over again. It was discovered long ago that, to do a job correctly, quality tools and instruments are required. Electrical work especially requires quality tools. Experience has shown that individuals are more productive when they use quality tools and instruments. **See Figure 2-1.** Quality tools last longer and more consistently provide quality results. By comparison, inferior tools result in lost time and money due to breakage and unnecessary mistakes.

Electricians are not required to buy every electrical tool immediately on entering an apprenticeship, because quality tools are expensive. Typically, an apprentice or journeyman electrician purchases a few basic quality tools at first and then adds other tools as needed. By acquiring quality tools over time, electricians create a selection of quality tools that are

Ideal Industries, Inc.

Quality tools make a job easier by being reliable, whenever and wherever needed.

Electrical Tools and Instruments

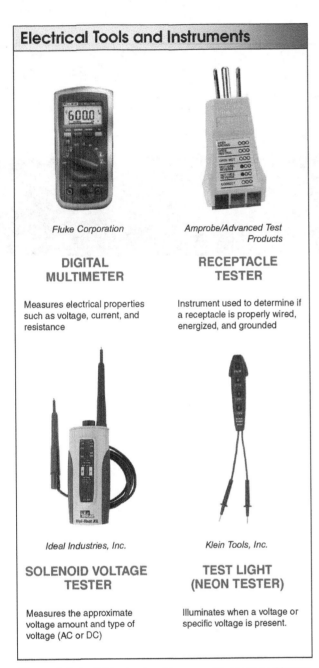

Fluke Corporation

DIGITAL MULTIMETER

Measures electrical properties such as voltage, current, and resistance

Amprobe/Advanced Test Products

RECEPTACLE TESTER

Instrument used to determine if a receptacle is properly wired, energized, and grounded

Ideal Industries, Inc.

SOLENOID VOLTAGE TESTER

Measures the approximate voltage amount and type of voltage (AC or DC)

Klein Tools, Inc.

TEST LIGHT (NEON TESTER)

Illuminates when a voltage or specific voltage is present.

Figure 2-1. *Individuals should use quality tools and instruments because inferior tools result in lost time and money due to breakage and unnecessary mistakes.*

ORGANIZED TOOL SYSTEMS

Because electrical tools are expensive to replace, individuals must protect their investment. To be effective, electrical tools must be available when needed, and protected from damage caused by the environment and daily use. An organized tool system must provide both a central location for retrieval of electrical tools and a means of protecting the tools during storage.

Electrical tools are organized in several ways depending upon where and how frequently the tools are used. When tools are used at a repair bench or at home for hobbies, a pegboard may be appropriate. When electrical tools are used at a construction site, a leather pouch is typically used to carry the most commonly used tools. When tools are to be transported to another job site or home, a toolbox is typically used.

Pegboards

Pegboard is a thin constructed board typically available in 4′ x 8′ sheets that is perforated with equally spaced holes for accepting hooks. **See Figure 2-2.** Pegboard is constructed of heavy-duty tempered material that is best suited for the weight of tools. When pegboard is mounted, outlines of tools can be made at the hooks for inventory purposes. Tool outlines can be painted onto the pegboard or cut out of self-adhesive shelf paper.

Electrical Tool Pouches

An *electrical tool pouch* is a small, open tool container (pouch) for storing a few commonly used electrical tools. An electrical tool pouch is typically made of heavy-duty leather, and is designed to be used with a belt that holds the pouch in place. **See Figure 2-3.**

Tool pouches can be simple in design and hold only a few tools, or they can be relatively large and designed to hold a wide selection of tools. The type of pouch chosen depends on the type of work an individual is planning to perform.

A typical electrician's tool pouch includes assorted cutting pliers, adjustable pliers, an adjustable wrench, nutdrivers, a torpedo level, regular and Phillips screwdrivers, a measuring tape, and a wire stripper/crimper.

Organizing Tool Storage

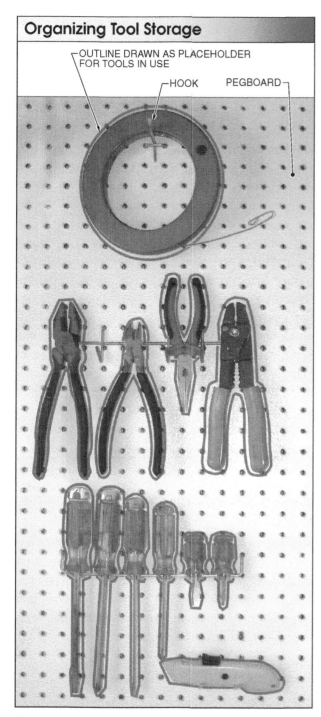

Figure 2-2. *When pegboards are used to store tools, outlines of the tools painted on the pegboard make it easy to check that all tools are put away.*

Toolboxes made from composite material are the strongest. Steel boxes are 55% as strong and aluminum boxes, 40%.

Electrical Tool Pouches

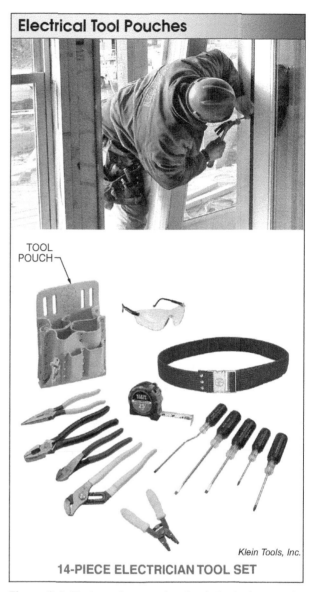

Klein Tools, Inc.

14-PIECE ELECTRICIAN TOOL SET

Figure 2-3. *Tool pouches can be simple in design or relatively large, depending on the type of work an individual is planning to perform.*

Toolboxes

Most individuals prefer to store tools in a quality toolbox. **See Figure 2-4.** A well-designed toolbox can be locked and will keep tools clean and dry. In addition, toolboxes, whether fixed or portable, provide a place where all tools can be collected. To ensure a complete inventory of tools at the end of the day or after a job, a list of all tools owned can be kept in the toolbox as a checklist.

Toolboxes

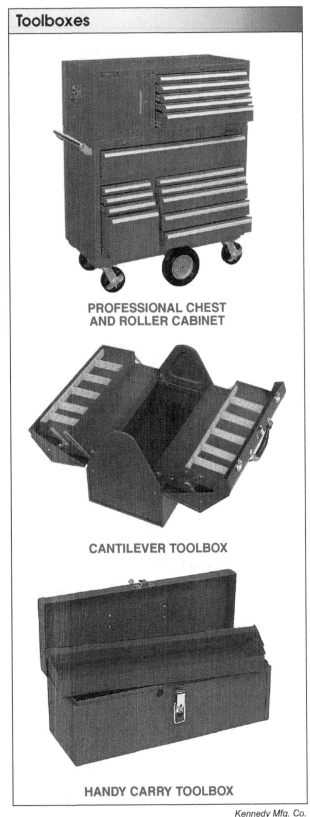

PROFESSIONAL CHEST AND ROLLER CABINET

CANTILEVER TOOLBOX

HANDY CARRY TOOLBOX

Kennedy Mfg. Co.

Figure 2-4. *Tools can be collected in a toolbox. Toolboxes keep tools clean and dry, and can be locked for security.*

Whichever system an individual chooses, organization is a must. An organized tool system ensures that tools are kept clean, dry, and at the fingertips of the individual when needed.

TOOL SAFETY

All tools can be dangerous when left in the wrong place, such as on a ladder. Many accidents are caused by tools falling off ladders, shelves, and scaffolds that are being used or moved.

Every tool should have a designated place in a toolbox. Do not carry tools in clothes pockets unless the pocket is designed for the tool. Keep pencils in a pocket designed for pencils; do not place pencils behind an ear or under a hardhat or cap.

Keep tools with sharp points away from the edges of benches or work areas, because individuals brushing against a tool can cause the tool to fall and injure a leg or foot. When carrying edged and sharply pointed tools, carry the tool with the cutting edge or the point down and outward from the body. An individual setting a tool down should place the tool back in a toolbox, or at least in a safe location. Electrical tools are divided into three categories: hand, power, and testing.

Hand Tools

Tool safety requires tool knowledge. Tool safety not only requires that the proper type of hand tool be chosen, but also that the correct size of tool be chosen. Individuals should use quality hand tools, and use the tools only in the manner in which they were designed to be used. **See Figure 2-5.**

Proper Hand Tool Use. Individuals must understand how to correctly use each tool owned. Tools must not be forced into a use beyond normal function. Questions must be asked when the proper and safe operation of a tool is not understood. Individuals are often tempted to use a screwdriver for a chisel or pliers as a wrench, but the correct tool will perform the task faster and safer. **See Figure 2-6.** The cost of a tool and the time taken to purchase a tool will prove far less costly than a serious accident caused by using a tool incorrectly.

Hand Tools . . .

Ideal Industries, Inc.

FUSE PULLER

Safely removes cartridge-type fuses

Klein Tools, Inc.

FISH TAPE

Used to pull wire through conduit and wires around obstructions in walls

Klein Tools, Inc.

ELECTRICIAN'S KNIFE

Removes insulation from nonmetallic cable and service conductors

Greenlee Textron Inc.

WIRE STRIPPER

Removes insulation from small-diameter wire

Klein Tools, Inc.

CRIMPING PLIERS

Used to crimp die marked insulated and noninsulated solderless terminals

The Stanley Works

DIAGONAL CUTTING PLIERS

Useful for cutting wire and cables that are difficult to reach

The Stanley Works

HACKSAW

Cuts heavy cable, pipe, and conduit

Klein Tools, Inc.

LONG-NOSE PLIERS

Useful for bending wire and positioning small components

The Stanley Works

HAMMER

Used to mount electrical boxes with nails, and determine height of receptacle boxes

Klein Tools, Inc.

ADJUSTABLE WRENCH

Tightens thick items such as hex head bolts and nuts, conduit couplings, and lag bolts

The Stanley Works

SIDE-CUTTING PLIERS

Used to cut cable, remove knockouts, twist wire, and debur conduit

Klein Tools, Inc.

TORPEDO LEVEL

Useful in leveling thermostats and conduit bends

Klein Tools, Inc.

UTILITY KNIFE

Cuts and scores drywall for drywall operations

Klein Tools, Inc.

TONGUE-AND-GROOVE PLIERS

Used to tighten box connectors, locknuts, and conduit couplings

The Stanley Works

FLATHEAD SCREWDRIVER

Used to install and remove threaded fasteners

Figure 2-5. *(continued . . .)*

. . . Hand Tools

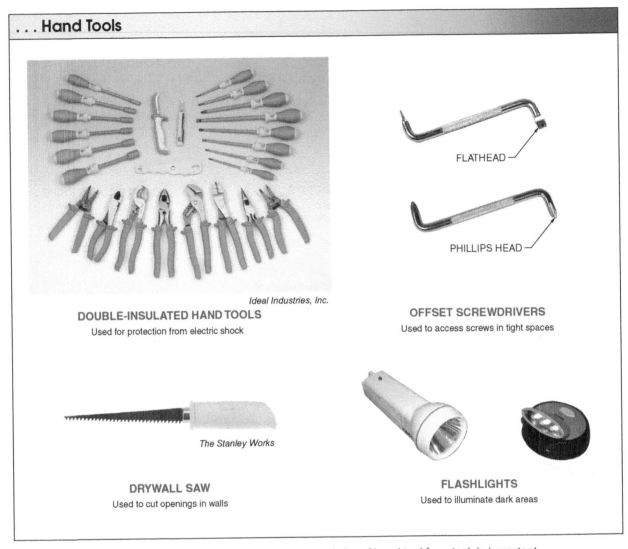

Ideal Industries, Inc.

DOUBLE-INSULATED HAND TOOLS
Used for protection from electric shock

FLATHEAD

PHILLIPS HEAD

OFFSET SCREWDRIVERS
Used to access screws in tight spaces

The Stanley Works

DRYWALL SAW
Used to cut openings in walls

FLASHLIGHTS
Used to illuminate dark areas

Figure 2-5. *Understanding how to choose the correct type and size of hand tool for a task is important.*

Maintaining Hand Tools in Proper Operation. Periodic checks of hand tools will aid in keeping tools in good condition. Always inspect a tool before using the tool. Do not use a tool that is in poor or faulty condition. Tool handles must be free of cracks and splinters and be fastened securely to the working part of the tool. Damaged tools are not only dangerous to the user but are also less productive than tools in good working condition. When inspection indicates a dangerous condition, repair or replace the tool immediately.

Cutting tools must be sharp and clean, because dull tools are dangerous. The extra force exerted when using dull tools often results in losing control of the tool. Dirt or oil on a tool can cause the tool to slip on the workpiece and cause injury.

Power Tools

Do not use any power tools without knowing the operation, methods of use, and safety precautions of the power tool. **See Figure 2-7.** Authorization from a supervisor can be required before using a power tool.

Safe Use of Hand Tools

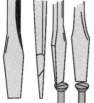

SELECT A SCREWDRIVER OF LENGTH AND TIP FITTED TO THE WORK. SCREWDRIVERS ARE SPECIFIED BY THE LENGTH OF THE BLADE. THE TIP SHOULD BE STRAIGHT AND NEARLY PARALLEL SIDED. IT SHOULD ALSO FIT THE SCREW SLOT AND BE NO WIDER THAN THE SCREW HEAD.

IF THE TIP IS TOO WIDE IT WILL SCAR THE MATERIAL AROUND THE SCREW HEAD. IF THE SCREWDRIVER IS NOT HELD IN LINE WITH THE SCREW IT WILL SLIP OUT OF THE SLOT AND MAR BOTH THE SCREW AND THE WORK.

IF THE TIP IS ROUNDED OR BEVELED IT WILL RAISE OUT OF THE SLOT, SPOILING THE SCREW HEAD. REGRIND OR FILL THE TIP TO MAKE IT AS SHOWN AT TOP. THE TWO COMMON SCREWDRIVER HEADS ARE PHILLIPS AND FLAT HEAD.

TO CUT HORIZONTALLY ACROSS THE GRAIN WITH THE WORK HELD IN THE VISE, PRESS THE FOREFINGER AND THUMB TOGETHER ON THE CHISEL TO ACT AS A BRAKE. TO AVOID SPLINTERING THE CORNERS, CUT HALFWAY FROM EACH EDGE TOWARD THE CENTER. REMOVE THE CENTER STOCK LAST.

WOOD CHISEL

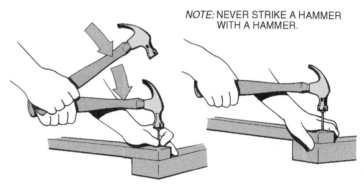

NOTE: NEVER STRIKE A HAMMER WITH A HAMMER.

WOOD HAMMER

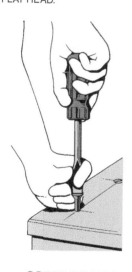
SCREWDRIVER

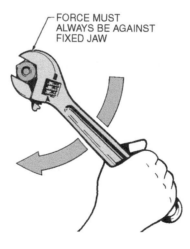
FORCE MUST ALWAYS BE AGAINST FIXED JAW

ADJUSTABLE WRENCH

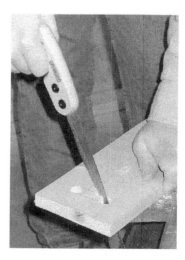
COMPASS SAW

Figure 2-6. *A screwdriver sometimes is used as a chisel, or pliers as a wrench, but the correct tool will perform the task faster and safer.*

Power Tools

Milwaukee Electric Tool Corp.

ELECTRIC POWER DRILL

Used with drill bits to drill medium-size and large holes

Milwaukee Electric Tool Corp.

BATTERY POWERED DRILL

Used with drill bits to drill holes and with assorted bits for working fasteners

Milwaukee Electric Tool Corp.

ELECTRIC DIE GRINDER

Used to cut wood, drywall, and metal

Milwaukee Electric Tool Corp.

CARBIDE TIPPED BIT

Used to drill into masonry

Milwaukee Electric Tool Corp.

TWIST DRILL BIT

Used for drilling small holes

Milwaukee Electric Tool Corp.

STEP DRILL BIT

Used to drill multiple size holes in wood, metal, and plastic

SPADE BIT

Used to drill medium-size holes

Advantage Drills Inc.

SHIP AUGER BIT

Used to rough bore holes

Milwaukee Electric Tool Corp.

HOLESAW

Used to drill large holes

Klein Tools, Inc.

IMPACT DRIVER AND SOCKET BITS

Used to install and remove fasteners quickly

Advantage Drills Inc.

GRINDING WHEELS AND BITS

Used to grind wood, plastic, and metal

DeWALT Industrial Tool Co.

CUTOUT SAW

Used to cut openings in walls

LASER LEVEL

Used to provide a level plane over a specific distance

Figure 2-7. *Power tools must only be used with a proper understanding of tool operation and the safety precautions associated with the tool.*

Grounding. All power tools are grounded except for power tools that are approved double-insulated. Non-double insulated power tools must have a three-conductor cord. A three-prong plug connects to a grounded receptacle (outlet). **See Figure 2-8.** OSHA and local codes must be consulted for the proper grounding specifications of power tools. A proper ground ensures that a faulty power tool causing an electrical short will trip a circuit breaker or blow a fuse in the circuit.

Non-current-carrying metal parts that are connected to ground include all metal boxes, raceways, enclosures, and equipment. Unwanted current exists because of insulation failure or because a current-carrying conductor makes contact with a non-current-carrying part of the system. In a properly grounded system, the unwanted current flow trips fuses or circuit breakers. Once the fuse or circuit breaker is tripped, the circuit is open and no additional current flows.

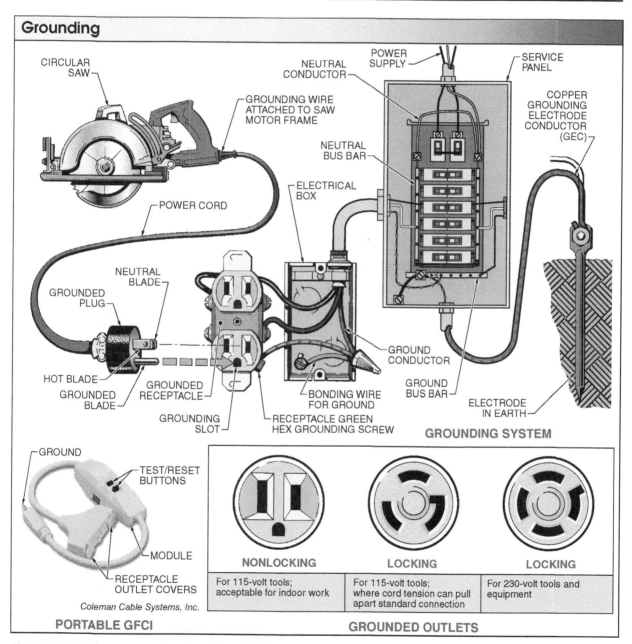

Figure 2-8. *A proper ground ensures that a faulty power tool causing an electrical short will trip a circuit breaker or blow a fuse in the circuit.*

Double-insulated tools have two prongs and will have a notation on the specification plate that the tool is double-insulated. Double-insulated tools are relatively safe but grounded tools are typically used on job sites. Electrical parts in the motor of a double-insulated tool are surrounded by extra insulation to help prevent shock; therefore, the tool does not have to be grounded. Both the interior and exterior of double-insulated tools must be kept clean of grease, dirt, and water that might conduct electricity.

Safety Rules for Power Tools

When handling power tools, the following safety rules must be followed:

- Power tools must be inspected and serviced by qualified repair personnel at regular intervals as specified by the manufacturer or OSHA.
- Know and understand all the safety recommendations of the manufacturer.
- Inspect electrical cords to verify that the cords are in good condition.
- Verify that all safety guards are in place and in proper working order. Do not remove, displace, or jam guards or any other safety devices. **See Figure 2-9.**
- Make all adjustments, blade changes, and inspections with the power OFF (cord disconnected).
- Before connecting a tool to a power source, ensure that the on/off switch is in the OFF position.
- Wear safety goggles and a dust mask when using power tools. **See Figure 2-10.**
- Ensure that the material to be worked is free of obstructions and securely clamped.
- Be attentive (focused on the work) at all times.
- A change in the sound of a power tool during operation normally indicates trouble. Investigate any change in sound immediately.
- When work is completed, shut the power OFF. Wait until a portable tool stops before leaving or laying down the tool.
- When a power tool is defective, remove the tool from service. Alert other personnel to the situation.
- Avoid operating power tools in locations where sparks could ignite flammable materials.
- Unplug power tools by pulling on the plugs; never pull on the cord.

Tool Guards

Figure 2-9. *Do not remove, displace, or jam guards or any other safety devices on a power tool.*

Power Tool Dust

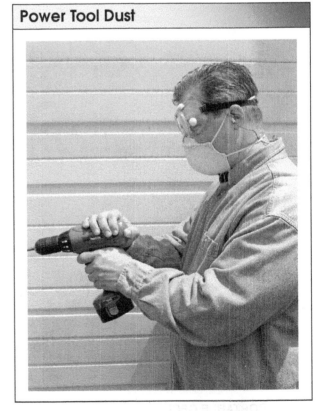

Figure 2-10. *When using power tools for tasks that may result in potentially hazardous dust or other air contaminants, wear safety goggles and a dust mask.*

TEST INSTRUMENTS

Proper equipment and procedures are required when taking measurements with test instruments. To take a measurement, the test instrument must be set to the correct measuring position and properly connected into the circuit to be tested; the displayed measurement is then read and interpreted. There is always the possibility that the test instrument will be set to the incorrect function or incorrect range, and/or will be misread. Test instruments not properly connected to a circuit increase the likelihood of an improper measurement and may create an unsafe condition. When taking measurements with test instruments, the following precautions should be observed:

- Follow all electrical safety practices and procedures.
- Check and wear personal protective equipment (PPE) for the procedure being performed.
- Perform only authorized procedures.
- Follow all manufacturer recommendations and procedures.

Voltage Indicators

A *voltage indicator* is a test instrument that indicates the presence of voltage when the test tip touches, or is near, an energized hot conductor or energized metal part. The tip glows and/or the device creates a sound when voltage is present at the test point. Voltage indicators are used to test receptacles, fuses, circuit breakers, cables, and other devices in which the presence of voltage must be detected. **See Figure 2-11.** The hot side (short slot or black wire) of a standard receptacle is typically used to test a voltage indicator for proper operation.

Voltage indicators are available in various voltage ranges (a few volts to hundreds of volts) and in the different voltage types (AC, DC, AC/DC) for testing various types of circuits.

Voltage Indicator Advantages/Disadvantages. The advantages of voltage indicators are that voltage indicators are inexpensive, are small enough to carry in a pocket, are easy to use, are nonconductive, and indicate a voltage without touching any live parts of a circuit, even through conductor insulation. One disadvantage of voltage indicators is that voltage indicators indicate only that voltage is present, not the actual amount of voltage. Another disadvantage of voltage indicators is that voltage indicators may not provide an indication that voltage is present, even when voltage is present. Voltage indicators can be fooled by wires or cables that are shielded.

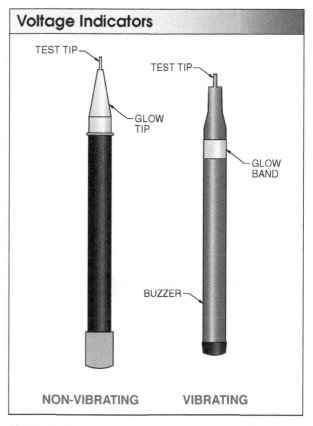

Voltage Indicators

TEST TIP

TEST TIP

GLOW TIP

GLOW BAND

BUZZER

NON-VIBRATING   **VIBRATING**

Figure 2-11. *A voltage indicator is a test instrument that indicates the presence of voltage when the test tip touches, or is near, an energized hot conductor or energized metal part.*

If a voltage indicator does not illuminate, check the indicator on a known live circuit to determine if the instrument is faulty.

Voltage Indicator Applications. Voltage indicators are used for many applications such as finding a break in an extension cord or wire, or determining when a receptacle is hot (energized). One important application of a voltage indicator is in making a preliminary test to determine if any metal parts or wires are hot before beginning service on a circuit or component. A fault occurs when any hot (energized) conductor touches a metal part that is not grounded. Faults occur because of a nick in conductor insulation, or when an exposed metal terminal screw contacts another metal part when a switch or receptacle is loose or not connected properly.

Voltage Indicator Measurement Procedures

Before taking any measurements using a voltage indicator, ensure the indicator is designed to take measurements on the circuit being tested. Refer to the operating manual of the test instrument for all measuring precautions, limitations, and procedures. Always wear required personal protective equipment and follow all safety rules when taking a measurement. **See Figure 2-12.** To test for voltage using a voltage indicator, apply the following procedures:

SAFETY PROCEDURES
• Follow all electrical safety practices and procedures.
• Check and wear personal protective equipment (PPE) for the procedure being performed.
• Perform only authorized procedures.
• Follow all manufacturer recommendations and procedures.

1. Verify that the voltage indicator has a voltage rating higher than the highest potential voltage in the circuit being tested. Residential circuits are typically 115 V. When circuit voltage is unknown, slowly bring the voltage indicator near the conductor or slot being tested. The voltage indicator will glow and/or sound when voltage is present. The brighter a voltage indicator glows, the higher the voltage or the closer the voltage indicator is to the voltage source.

2. Place the tip of the voltage indicator on or near the wire or device being tested. When testing an extension cord for a break, test several points along the wire. Expect the voltage tester to turn on and off

when moved along a cord that has twisted wire conductors, because the hot wire will change position along the cord.

3. Remove the voltage indicator from the test area.

4. When the voltage indicator does not indicate the presence of voltage by glowing or making a sound, do not assume that there is no voltage and start working on exposed components of a circuit. Always take a second test instrument (voltmeter or multimeter) and measure for the presence of voltage before working around or on exposed wires and electrical components.

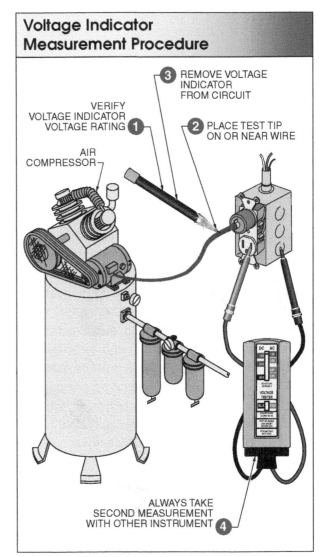

Figure 2-12. *Always verify that a voltage indicator or any test instrument has a voltage rating higher than the highest potential voltage in the circuit being tested.*

Test Lights

Electrical circuits can be tested safely and inexpensively with a test light. **See Figure 2-13.** Most hardware stores have test lights at a low cost. A *test light (neon tester)* is a test instrument that is designed to illuminate in the presence of 115 V and 230 V circuits. A soft glow indicates a 115 V circuit; a brighter glow indicates a 230 V circuit. Many residential electrical problems can be solved by determining if voltage is present at a location in the circuit.

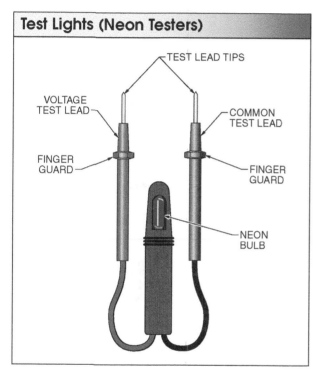

Test Lights (Neon Testers)

TEST LEAD TIPS

VOLTAGE TEST LEAD

COMMON TEST LEAD

FINGER GUARD

FINGER GUARD

NEON BULB

Figure 2-13. *A test light (neon tester) is a test instrument that is designed to illuminate in the presence of 115 V and 230 V circuits.*

Test Light Advantages/Disadvantages. The advantages of using test lights are that test lights are inexpensive, small enough to carry in a pocket, and easy to use. The disadvantages of test lights are that test lights have a limited voltage range and cannot determine the actual voltage of a circuit, only that voltage is present in a circuit. Test lights that have a wider voltage range are better than test lights that have only one voltage rating. For example, a neon test light rated for 90 VAC to 600 VAC is better than a test light rated for only 120 VAC. Neon test lights

have a very high resistance so the neon bulb draws very little current when taking measurements. The small current draw allows neon bulbs to have a longer life expectancy than other test lights. Another disadvantage of neon test lights is that neon test lights must not be used to test ground fault circuit interrupters (GFCIs).

Test Light Applications

Test lights are primarily used to determine when voltage is present in a circuit (the circuit is energized), such as when testing receptacles. When testing a receptacle, the test light bulb illuminates when the receptacle is properly wired and energized.

When a receptacle is properly wired, a test light bulb will illuminate when the test light leads are connected between the neutral slot and the hot slot. A test light bulb will also illuminate when the leads are connected between the ground slot and the hot slot. If the test light illuminates when the leads are connected from the neutral slot to the ground slot, the hot (black) and neutral (white) wires are reverse wired. The situation of having the hot and neutral wires reversed (nonpolarized) is a safety hazard and must be corrected.

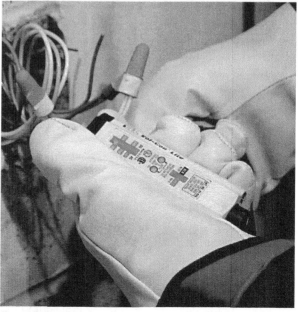

For additional protection from electric shock when testing energized circuits with a test instrument, rubber insulating gloves and leather protective gloves should be worn.

If a test light illuminates when the test leads are connected to the neutral slot and hot slot but does not light when connected to the ground slot and hot slot, the receptacle is not grounded. When a test light illuminates, but is dimmer than when connected between the neutral slot and hot slot, the receptacle has an improper ground (having high resistance). Improper grounds are also a safety hazard and must be corrected.

Test Light Measurement Procedures

Before using a test light, always check the test light on a known energized circuit that is within the test light's rating to ensure that the test light is operating correctly.

Before taking any measurements using a test light, ensure the test light is designed to take measurements on the circuit being tested. Refer to the operating manual of the test instrument for all measuring procedures, limitations, and precautions. **See Figure 2-14.** To test for voltage using a test light, apply the following procedures:

SAFETY PROCEDURES
• Follow all electrical safety practices and procedures.
• Check and wear personal protective equipment (PPE) for the procedure being performed.
• Perform only authorized procedures.
• Follow all manufacturer recommendations and procedures.

1. Verify that the test light has a voltage rating higher than the highest potential voltage in the circuit. Care must be taken to guarantee that the exposed metal tips of the test light leads do not touch fingers or any metal parts not being tested.
2. Connect one test lead of the test light to the neutral side of the circuit or ground. When testing a circuit that has a neutral or ground, connect to the neutral or ground side of the circuit first.
3. Connect the other test lead of the test light to the other side (hot side) of the circuit. Voltage is present when the test light bulb illuminates. Voltage

is less than the rating of the test light when the test light is dimly lit, and is higher than the rating of the test light when the test light glows brighter than normal. Voltage is not present in a circuit or present at a very low level when a test light does not illuminate.

4. Remove the test light from the circuit.

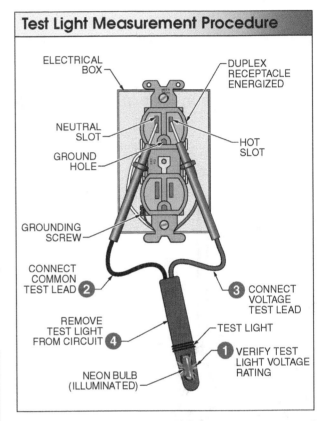

Figure 2-14. *Care must be taken to ensure that the exposed metal tips of the test light leads or any test instrument do not touch fingers or any metal parts not being tested.*

Continuity Testers

A *continuity tester* is a test instrument that is used to test a circuit for a complete path for current to flow. A closed switch that is operating correctly has continuity while an open switch does not have continuity. **See Figure 2-15.** A continuity tester or test instrument with continuity test mode indicates when an electrical device or circuit has a complete path by emitting an audible sound (beep). Indication of a complete path is used to determine the condition of a device, such as a switch, as open or closed.

Continuity Testers

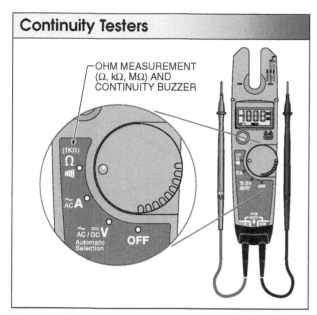

OHM MEASUREMENT (Ω, kΩ, MΩ) AND CONTINUITY BUZZER

Figure 2-15. *A continuity tester is a test instrument that is used to test a circuit for a complete path for current flow.*

Continuity Tester Advantages and Disadvantages. The main advantage of using a continuity tester or continuity test mode of a test instrument is the audible response (beep). An audible response allows an individual to concentrate on the testing procedures without looking at the test instrument.

The main disadvantage of using a continuity tester is that a continuity tester only emits a sound when continuity is detected. Continuity testers only operate on circuits that have very low resistance (40 Ω or less) and will not indicate the actual resistance of a circuit or device.

Continuity Tester Applications

Continuity testers are used to test for a complete flow path (open, broken, or closed) in any de-energized low-resistance device. A continuity tester is a common test tool to use when testing single-pole, 3-way, and 4-way switches. For example, a 3-way switch has one terminal that is the "common" terminal. Depending upon the manufacturer, the common terminal can be located in any position on a 3-way switch. The common terminal is often identified on a 3-way switch by a different color. On specialty switches, such as automobile switches, the common terminal is not

identified. Continuity testers are used to determine the common terminal and proper switch operation.

Continuity Tester Measurement Procedures

Continuity is tested with a test instrument set to the continuity test mode. Before taking any continuity measurements using a continuity tester, ensure the meter is designed to take measurements on the circuit being tested. Refer to the operating manual of the test instrument for all measuring procedures, limitations, and precautions. Always wear required personal protective equipment and follow all safety rules when taking the measurement. **See Figure 2-16.** To take continuity measurements with a continuity tester, apply the following procedures:

SAFETY PROCEDURES

- Follow all electrical safety practices and procedures.
- Check and wear personal protective equipment (PPE) for the procedure being performed.
- Perform only authorized procedures.
- Follow all manufacturer recommendations and procedures.

1. Set the continuity tester function switch to continuity test mode as required. Most test instruments have the continuity test mode and resistance mode sharing the same function switch position.
2. With the circuit de-energized, connect the test leads across the component being tested. The position of the test leads is arbitrary.
3. When there is a complete path (continuity), the continuity tester beeps. When there is no continuity (open circuit), the continuity tester does not beep.
4. After completing all continuity tests, remove the continuity tester from the circuit or component being tested and turn the instrument OFF to prevent battery drain.

A continuity tester must only be used on de-energized circuits or components. Any voltage applied to a continuity tester causes damage to the test instrument and/or harm to the individual. Always test a circuit for voltage before taking a continuity test.

Continuity Tester Measurement Procedure

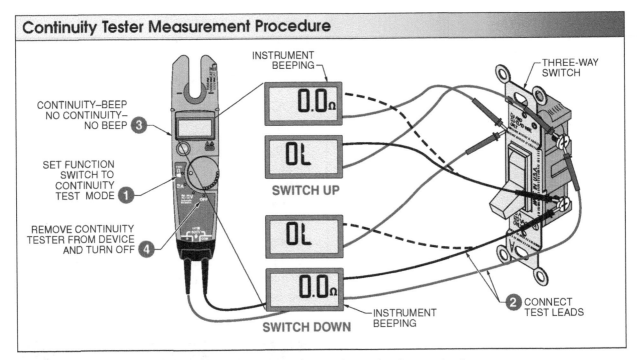

Figure 2-16. *When checking the continuity of a circuit, the circuit must be de-energized.*

Voltage Testers

A *voltage tester* is an electrical test instrument that indicates the approximate amount of voltage and the type of voltage (AC or DC) in a circuit by the movement of a pointer (and vibration, on some models). When a voltage tester includes a solenoid, the solenoid vibrates when the tester is connected to AC voltage. Some voltage testers include a colored plunger or other indicator such as a light that indicates the polarity of the test leads as positive or negative when measuring a DC circuit. **See Figure 2-17.**

Voltage Tester Advantages/Disadvantages. The main advantage of voltage testers is that individuals can concentrate on placing the test leads of the instrument instead of reading the instrument. A disadvantage of solenoid voltage testers is that solenoid voltage testers affect electronic equipment measurements. Solenoid voltage testers and multimeters cannot be used to test GFCIs because the solenoid causes the tester to have high resistance to current flowing. Nonsolenoid voltage testers are low impedance (low resistance) and can be used to test GFCIs. GFCIs are designed to trip at approximately 6 mA (0.006 A).

Voltage Testers

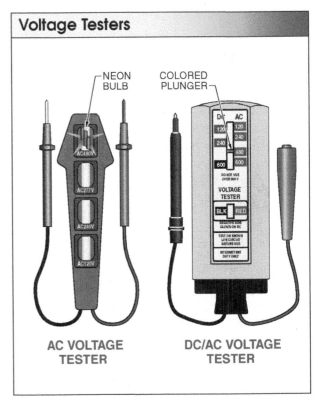

Figure 2-17. *A voltage tester is an electrical test instrument that indicates the approximate amount of voltage and the type of voltage (AC or DC) in a circuit by the movement of a pointer (and vibration on some models).*

Voltage Tester Applications. Voltage testers are used to take measurements any time the voltage of the circuit being tested is within the rating of the tester and an exact voltage measurement is not required. Exact voltage measurements such as 118.7 V are not required to determine when a receptacle is hot, a system is grounded, fuses or circuit breakers are good or bad, or when a circuit is 115 V, 230 V, or 460 V.

Voltage Tester Measurement Procedures

Before using a voltage tester or any voltage measuring instrument, always check the voltage tester on a known energized circuit that is within the voltage rating of the voltage tester to verify proper operation.

Before taking any voltage measurements using a voltage tester, ensure the tester is designed to take measurements on the circuit being tested. Refer to the operating manual of the test instrument for all measuring procedures, limitations, and precautions. Always wear required personal protective equipment and follow all safety rules when taking the measurement. **See Figure 2-18.** To take a voltage measurement with a voltage tester, apply the following procedures:

SAFETY PROCEDURES
• Follow all electrical safety practices and procedures.
• Check and wear personal protective equipment (PPE) for the procedure being performed.
• Perform only authorized procedures.
• Follow all manufacturer recommendations and procedures.

1. Verify that the voltage tester has a voltage rating higher than the highest potential voltage in the circuit being tested.

2. Connect the common test lead to the point of testing (neutral or ground).

3. Connect the voltage test lead to the point of testing (ungrounded conductor). The pointer of the voltage tester indicates a voltage reading and vibrates when the current in the circuit is AC. The indicator shows a voltage reading and does not vibrate when the current in the circuit is DC.

4. Observe the voltage measurement displayed.

5. Remove the voltage tester from the circuit.

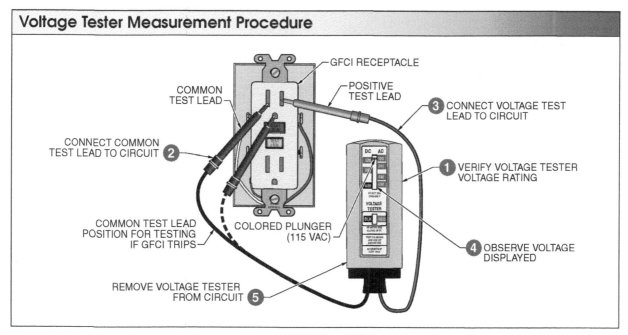

Voltage Tester Measurement Procedure

Figure 2-18. *The pointer of a voltage tester indicates the voltage reading and vibrates when the current in a circuit is AC.*

Branch Circuit Identifiers

A *branch circuit identifier* is a two-piece test instrument consisting of a transmitter that is plugged into a receptacle and a receiver that provides an audible indication when located near the circuit to which the transmitter is connected. **See Figure 2-19.** A branch circuit identifier is used to identify a particular circuit breaker. Before working on any electrical circuit, the circuit must be de-energized. Normally, branch circuits are de-energized by turning OFF the circuit breakers of the circuit and applying a lockout/tagout device. Often, the circuit breaker in the circuit to be de-energized is not clearly marked or identifiable. Turning OFF an incorrect circuit breaker may unnecessarily require loads to be reset.

Branch Circuit Identifiers

AEMC® instruments

Figure 2-19. *A branch circuit identifier provides an audible indication to identify the circuit breaker or fuse for a particular receptacle within an electrical circuit.*

Branch Circuit Identifier Test Procedures. Before performing any test using a branch circuit identifier, it must be ensured that the branch circuit identifier is designed for the circuit to be tested. The test instrument manufacturer's operating manual should be consulted for any measuring precautions, limitations, and procedures. The required PPE should be worn,

and all safety rules should be followed when taking the measurement. **See Figure 2-20.** To identify a circuit (circuit breaker) using a branch circuit identifier, apply the following procedures:

SAFETY PROCEDURES

- Follow all electrical safety practices and procedures.
- Check and wear personal protective equipment (PPE) for the procedure being performed.
- Perform only authorized procedures.
- Follow all manufacturer recommendations and procedures.

1. Turn ON the branch circuit transmitter and receiver of the branch circuit identifier.
2. Plug the transmitter of the branch circuit identifier into the receptacle (outlet) that is to be identified.
3. To verify that the receiver is operational, test the receiver of the branch circuit identifier at the same receptacle to which the transmitter is connected.
4. Use the receiver to identify any part of the circuit or the circuit breaker that controls the circuit being tested.
5. When the circuit being tested requires power to be turned OFF, open the circuit breaker and lockout/tagout the breaker of the circuit being tested.
6. Use a voltmeter or DMM to verify that the power is OFF before working on the identified circuit.
7. After work has been performed on the circuit, reset the breaker to the ON position.
8. Verify with the voltmeter or DMM that power is restored to the branch circuit.

Digital Multimeters

A *digital multimeter* is an electrical test instrument that can measure two or more electrical properties and display the measured properties as numerical values. Basic digital multimeters measure resistance (continuity) and AC or DC voltage and current. Advanced digital multimeters include special functions such as measuring frequency, capacitance, and temperature. **See Figure 2-21.**

Branch Circuit Identifier Test Procedure

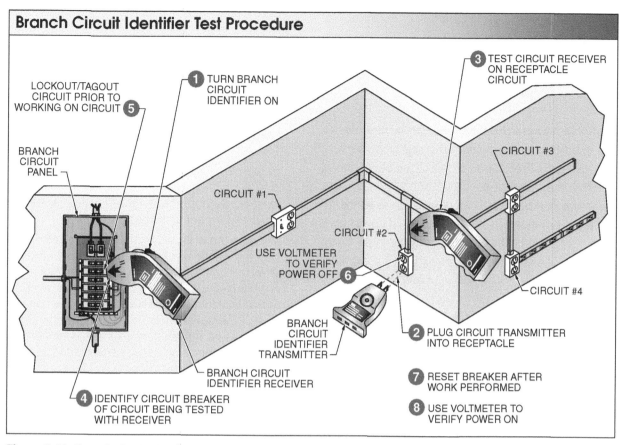

LOCKOUT/TAGOUT CIRCUIT PRIOR TO WORKING ON CIRCUIT **5**

1 TURN BRANCH CIRCUIT IDENTIFIER ON

3 TEST CIRCUIT RECEIVER ON RECEPTACLE CIRCUIT

BRANCH CIRCUIT PANEL

CIRCUIT #3

CIRCUIT #1

CIRCUIT #2

USE VOLTMETER TO VERIFY POWER OFF **6**

CIRCUIT #4

BRANCH CIRCUIT IDENTIFIER TRANSMITTER

BRANCH CIRCUIT IDENTIFIER RECEIVER

2 PLUG CIRCUIT TRANSMITTER INTO RECEPTACLE

4 IDENTIFY CIRCUIT BREAKER OF CIRCUIT BEING TESTED WITH RECEIVER

7 RESET BREAKER AFTER WORK PERFORMED

8 USE VOLTMETER TO VERIFY POWER ON

Figure 2-20. *Branch circuit identifiers reduce the time required to identify circuits in an unmarked panel.*

Digital Multimeters

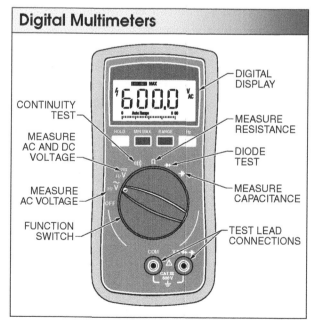

DIGITAL DISPLAY

CONTINUITY TEST

MEASURE RESISTANCE

MEASURE AC AND DC VOLTAGE

DIODE TEST

MEASURE AC VOLTAGE

MEASURE CAPACITANCE

FUNCTION SWITCH

TEST LEAD CONNECTIONS

Figure 2-21. *A digital multimeter is an electrical test instrument that can measure two or more electrical properties and display the measured properties as numerical values.*

Tech Tip: Correct Test Lead Connections

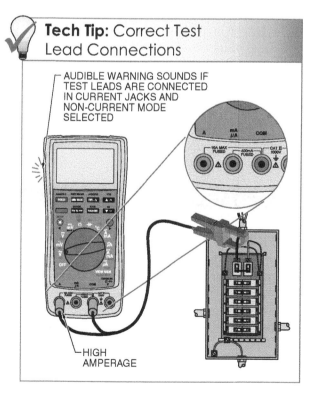

AUDIBLE WARNING SOUNDS IF TEST LEADS ARE CONNECTED IN CURRENT JACKS AND NON-CURRENT MODE SELECTED

HIGH AMPERAGE

Digital Multimeter Advantages/Disadvantages. The main advantage of digital multimeters compared to analog multimeters is the ability of a digital multimeter to record measurements in addition to making it easy to read measured displayed values.

Digital Multimeter Applications. Digital multimeters are used to take measurements any time the voltage, current, or resistance of the circuit being tested is within the rating of the meter and an exact measurement is expected. Exact voltage measurements are required to determine the amount of power consumed by a circuit, if a brownout is occurring, if fuses or circuit breakers are good or bad, or if a circuit is 12 V, 24 V, 115 V, 230 V, 460 V, or any voltage.

Digital Multimeter Measurement Procedures

Before using a digital multimeter or any measuring instrument, always check the digital multimeter on a known energized circuit that is within the voltage rating of the meter to verify proper operation.

Before taking any voltage measurements using a digital multimeter, ensure the meter is designed to take measurements on the circuit being tested. Refer to the operating manual of the test instrument for all measuring procedures, limitations, and precautions. Always wear required personal protective equipment and follow all safety rules when taking the measurement. **See Figure 2-22.** To take a voltage measurement with a digital multimeter, apply the following procedures:

SAFETY PROCEDURES

- Follow all electrical safety practices and procedures.
- Check and wear personal protective equipment (PPE) for the procedure being performed.
- Perform only authorized procedures.
- Follow all manufacturer recommendations and procedures.

---

1. Set the function switch of the voltmeter to AC voltage.

2. Plug the black test lead into the common jack (marked COM).
3. Plug the red test lead into the voltage jack (marked V).
4. Measure the voltage into the transformer. Connect the voltmeter between the common and hot terminals on the primary side of the transformer. The problem is located upstream from the transformer when there is no voltage present or the voltage is not at the correct level. The problem may be a blown fuse or open circuit. Voltage must be reestablished to the transformer before the transformer can be tested.
5. Measure the voltage out of the transformer. Connect the voltmeter between the common and hot terminals on the secondary side of the transformer. Voltage must be present at the secondary side terminals when voltage is supplied to the primary side of the transformer. The transformer is open and must be replaced when there is no voltage reading on the secondary side terminals.
6. Remove the voltmeter from the circuit.

Safety Rules for Test Instruments

Proper equipment and procedures are required when taking measurements with test instruments. To take a measurement, the test instrument must be set to the correct measuring position and properly connected into the circuit to be tested; the displayed measurement is then read and interpreted. There is always the possibility that the test instrument will be set to the incorrect function or incorrect range, and/or will be misread. Test instruments not properly connected to a circuit increase the likelihood of an improper measurement and may create an unsafe condition.

When handling test instruments, the following safety rules must be followed:

- Choose the correct test instrument for an application. Any test instrument is designed for taking measurements on some type of application, but not for all applications.
- Test instruments that are damaged (cracked or with missing parts or worn insulation) must not be used.
- Inspect the test leads of a test instrument on a regular basis (each time used).

Digital Multimeter Measurement Procedure

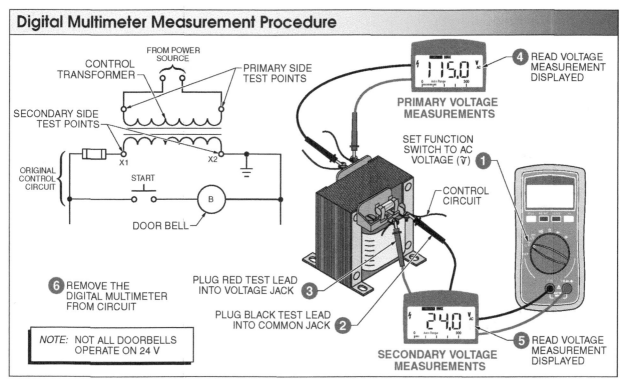

Figure 2-22. *Depending on the electrical property being tested, the test leads of a multimeter may have to be connected to a circuit in a specific manner.*

- The battery door of test instruments must be kept closed during instrument operation.
- Test the fuses of a test instrument on a regular basis. **See Figure 2-23.**
- When taking a measurement, keep fingers behind the finger guards on the test leads.
- Test instruments must not be used in hazardous environments (explosive gas, vapor, or dust) unless specifically designed for hazardous environments.
- Before use, verify that the test leads of a test instrument are in the proper jacks—the jacks that correspond to the setting of the function switch—before taking any measurements. **See Figure 2-24.**
- Before use, verify the operation of the test instrument by measuring a known voltage.
- Before taking a measurement, verify that the function switch of a test instrument matches the desired measurement and the connections of the test leads to the circuit.
- To take a measurement, connect the common (negative) test lead of a test instrument before connecting the voltage (positive) test lead. To disconnect, disconnect the positive test lead first.

Test Instrument Fuses

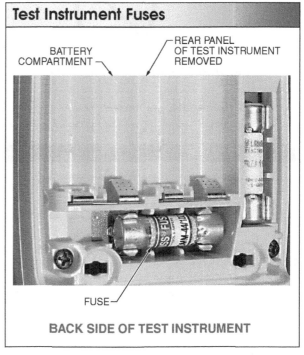

BACK SIDE OF TEST INSTRUMENT

Figure 2-23. *The fuses of a test instrument must be checked on a regular basis to ensure they are in good condition.*

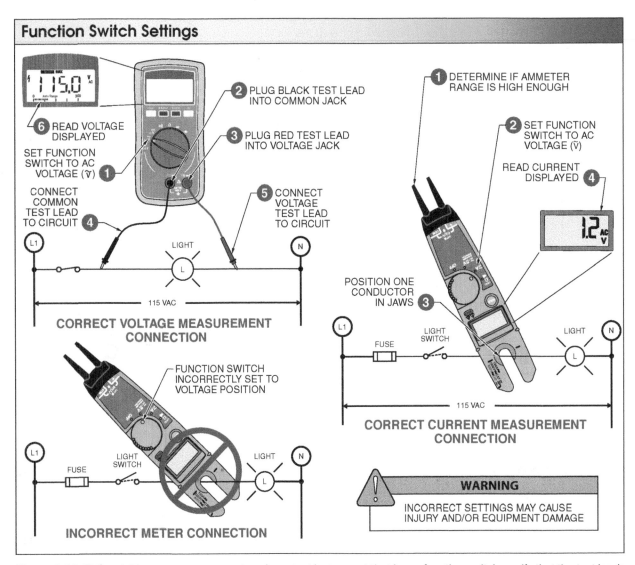

Function Switch Settings

6 READ VOLTAGE DISPLAYED

1 SET FUNCTION SWITCH TO AC VOLTAGE ($\bar{v}$)

2 PLUG BLACK TEST LEAD INTO COMMON JACK

3 PLUG RED TEST LEAD INTO VOLTAGE JACK

4 CONNECT COMMON TEST LEAD TO CIRCUIT

5 CONNECT VOLTAGE TEST LEAD TO CIRCUIT

115 VAC

CORRECT VOLTAGE MEASUREMENT CONNECTION

1 DETERMINE IF AMMETER RANGE IS HIGH ENOUGH

2 SET FUNCTION SWITCH TO AC VOLTAGE ($\bar{v}$)

4 READ CURRENT DISPLAYED

3 POSITION ONE CONDUCTOR IN JAWS

FUSE   LIGHT SWITCH   LIGHT

115 VAC

CORRECT CURRENT MEASUREMENT CONNECTION

FUNCTION SWITCH INCORRECTLY SET TO VOLTAGE POSITION

FUSE   LIGHT SWITCH   LIGHT

INCORRECT METER CONNECTION

WARNING

INCORRECT SETTINGS MAY CAUSE INJURY AND/OR EQUIPMENT DAMAGE

Figure 2-24. *Before taking any measurements using a test instrument that has a function switch, verify that the test leads of the test instrument are in the proper jacks that correspond to the setting of the function switch.*

To clean a test instrument, periodically wipe the case of the instrument with a damp cloth and mild detergent. Do not use abrasives or solvents.

ELECTRICAL SAFETY

The safety of an individual working with electricity is far more advanced today than 20 years ago. State and local safety laws, NFPA 70E, and OSHA safety rules are in place to create a safe working environment. **See Figure 2-25.** Individuals using common sense, safeguards, proper equipment, and safety practices are safe when working on electrical equipment.

An individual must understand electricity and be able to apply the principles of safety to all tasks performed. When an individual performs tasks without regard to safety, the individual is disregarding the safety of fellow workers. Electricity must be respected, but respect for electricity comes from understanding electricity. Whenever an individual has a doubt about electricity, the individual must ask for assistance from an electrician. All unsafe conditions, equipment, tools, or work practices must be reported to appropriate persons as soon as possible.

Electrical Safety Organizations

NFPA 70E	29 CFR 1926 OSHA	
Standard for Electrical Safety in the Workplace	**U.S. Department of Labor**	-1926.400 - Introduction
- 90 - Electrical Safety in the Workplace	Occupational Safety & Health Administration	-1926.402 - Applicability
-100 - Safety-Related Work Practices		-1926.403 - General Requirements
-200 - Safety-Related Maintenance Requirements	**Construction Industry**	-1926.404 - Wiring Design & Protection
-300 - Safety Requirements for Special Equipment	Safety and Health Regulations	-1926.405 - Wiring Methods & Equipment for general use
-400 - Installation Safety Requirements	for Construction	-1926.406 - Specific Purpose Installations
	1926 Subpart K - Electrical	-1926.407 - Hazardous
		-1926.416 - General Requirements
		-1926.417 - Lockout and Tagging of Circuits

Figure 2-25. *State and local safety laws, NFPA 70E, and OSHA safety rules are in place to create a safe working environment.*

Electrical Shock

Electrical shock is the condition that occurs when an individual comes in contact with two conductors of a circuit or when the body of an individual becomes part of an electrical circuit. In either case, a severe shock can cause the heart and the lungs to stop functioning. Also, due to the heat produced by current flow, severe burns can occur where the electricity (current) enters and exits the body. **See Figure 2-26.**

Individuals using portable power tools must ensure that the power tool is safe and in proper operating condition. Portable power tools that are not double insulated must have a third wire on the plug for grounding in case a short occurs. Theoretically, when electric power tools are grounded, fault current flows through the third wire to ground instead of through the body to ground.

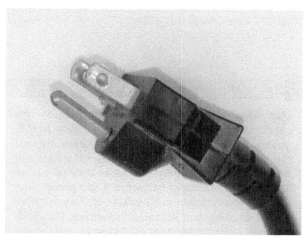

For protection against electrical shock, power tools must be used with a cord that has a grounded (three-prong) plug.

Electrical Shock Effects

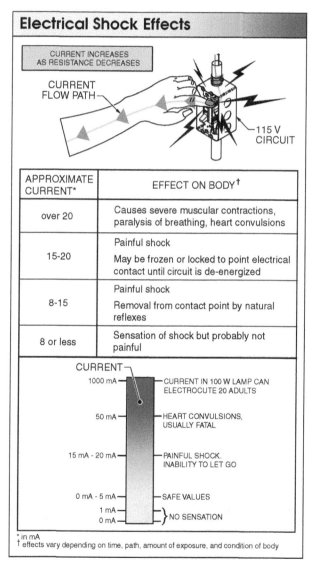

APPROXIMATE CURRENT*	EFFECT ON BODY†
over 20	Causes severe muscular contractions, paralysis of breathing, heart convulsions
15-20	Painful shock May be frozen or locked to point electrical contact until circuit is de-energized
8-15	Painful shock Removal from contact point by natural reflexes
8 or less	Sensation of shock but probably not painful

* in mA
† effects vary depending on time, path, amount of exposure, and condition of body

Figure 2-26. *Possible effects of electrical shock include the heart and lungs ceasing to function, and/or severe burns where the electricity (current) enters and exits the body.*

Out of Service Protection

Before any repair is performed on a piece of electrical equipment, an individual must be absolutely certain the source of electricity is locked out and tagged out (out of service) by a lockout/tagout device. **See Figure 2-27.** Whenever an individual leaves a job for any reason, upon returning to the job the individual must ensure that the source of electricity is still locked out and tagged out. **See Figure 2-28.**

Lockout/Tagout Kits

Ideal Industries, Inc.

LOCKOUT/TAGOUT KIT

Figure 2-27. *Lockout/tagout devices are available for locking out and tagging out any type of electrical equipment.*

Always read the owner's manual before using an electrical appliance or tool with an extension cord. Many electrical components have specifications for the length and diameter of extension cords. These extension cords can be used without overheating and causing a fire.

Lockout/Tagout

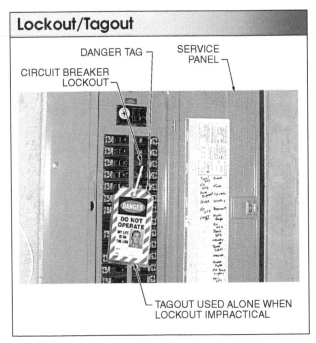

Figure 2-28. *Whenever an individual leaves a job for any reason, upon returning to the job the individual must ensure that the source of electricity is still locked out and tagged out.*

Safety Color Codes and Safety Labels

Federal law (OSHA) has established specific colors to designate cautions and dangers that are encountered when performing electrical tasks. **See Figure 2-29.** Individuals must understand the significance of colors, signal words, and symbols as related to electrical hazards.

CAUTION

Use extreme caution when working with electrical conductors and power lines. According to U.S. government statistics, electrical injuries are not the most common type of accidental injury but have the highest fatality rate when they do occur. The biggest electrical hazard is a live power line, and the injury rates are highest in construction, manufacturing, and services industries. These industries account for only 7% of the workforce but 44% of electrical injuries.

Safety Labels and Signal Words

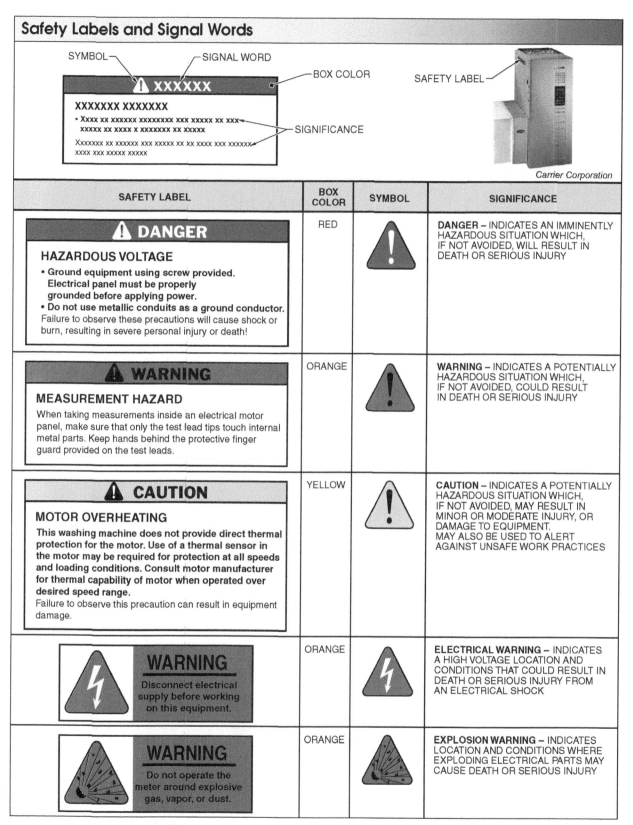

Figure 2-29. *Federal law (OSHA) has established specific colors to designate cautions and dangers that individuals must understand.*

ELECTRICAL CODES

The *National Electrical Code® (NEC®)* is a book of electrical standards that indicate how electrical systems must be installed and how work must be performed. The NEC® is a 775-page book that states the standards for all residential, commercial, and industrial electrical work. The NEC® is amended every three years to stay current with new safety issues, electrical products, and procedures.

In most areas the NEC® and local code requirements are the same, but the NEC® sets only minimum requirements. Local codes, particularly in large cities, may be stricter than those of the NEC®. In the case of stricter local codes, the local code must be followed.

Notes explaining some of the NEC® requirements for general lighting and appliance circuits can be shown on electrical plans. Numbers and notes on electrical plans refer to specific articles and tables in the NEC® code. **See Figure 2-30.** When more specific requirements (codes) are established by local city councils, the local codes are obtained from the local building inspector.

NEC® Code Book		
Chapter	**Articles**	**Description**
	1	Introduction
1	2	Definitions
2	10	Wiring and Protection
3	43	Wiring Methods and Materials
4	21	Equipment for General Use
5	27	Special Occupancies
6	23	Special Equipment
7	10	Special Conditions
8	4	Communication Systems
9	-	Tables

Figure 2-30. *The National Electrical Code® (NEC®) is a book of electrical standards that indicate how electrical systems must be installed and how work must be performed.*

PERSONAL PROTECTIVE EQUIPMENT

Personal protective equipment (PPE) is clothing and/or equipment worn by individuals to reduce the possibility of injury in the work area. The use of personal protective equipment is required whenever work occurs on or near energized exposed electrical circuits. The National Fire Protection Association standard NFPA 70E, *Standard for Electrical Safety in the Workplace*, addresses electrical safety requirements for employee workplaces that are necessary for safeguarding employees in pursuit of gainful employment.

NFPA 70E

For maximum safety, personal protective equipment and safety requirements for test instrument procedures must be followed as specified in NFPA 70E, OSHA Standard Part 1910 Subpart 1–*Personal Equipment* (1910.132 through 1910.138), and other applicable safety mandates. Personal protective equipment includes protective clothing, head protection, eye protection, ear protection, hand and foot protection, back protection, knee protection, and rubber insulated matting. **See Figure 2-31.**

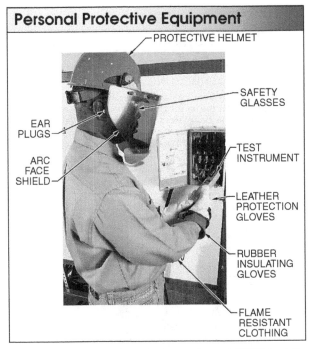

Personal Protective Equipment

PROTECTIVE HELMET
SAFETY GLASSES
EAR PLUGS
ARC FACE SHIELD
TEST INSTRUMENT
LEATHER PROTECTION GLOVES
RUBBER INSULATING GLOVES
FLAME RESISTANT CLOTHING

Figure 2-31. *Personal protective equipment (PPE) is clothing and/or equipment worn by individuals to reduce the possibility of injury in the work area and is specified by the National Fire Protection Association Standard NFPA 70E, Standard for Electrical Safety in the Workplace.*

Personal Protective Equipment Safety Rules

The following personal protective equipment safety rules must be followed when performing tasks on electrical systems:

- Wear protective clothing for the task being performed.
- Wear safety glasses, goggles, or an arc face shield, depending on the task being performed. **See Figure 2-32.**

Eye Protection

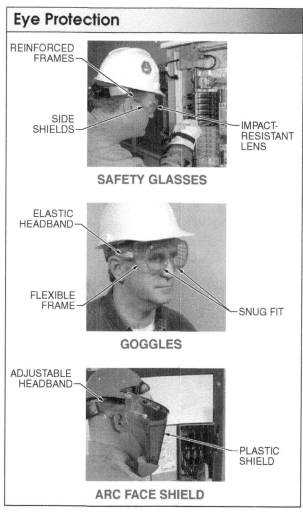

Figure 2-32. *Always wear safety glasses, goggles, or an arc face shield, depending on the electrical task being performed.*

- Wear thick-soled rubber work shoes for protection against sharp objects and insulation against electrocution. Wear work shoes with safety toes when required for the task. **See Figure 2-33.**

Electrical Work Shoes

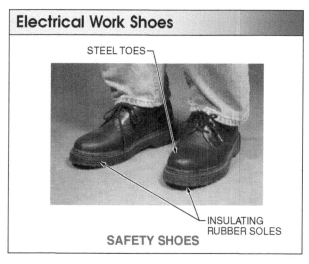

Figure 2-33. *Wear thick-soled rubber work shoes for protection against sharp objects and insulation against electrocution. Certain tasks require work shoes with steel toes.*

- Wear rubber boots in damp locations.
- Wear an approved safety helmet (hard hat). Confine long hair and be careful to avoid having long hair near powered tools.
- Wear electrical gloves and cover gloves when taking measurements on energized circuits. Electrical gloves provide protection against electrocution and cover gloves protect electrical gloves from damage. **See Figure 2-34.**

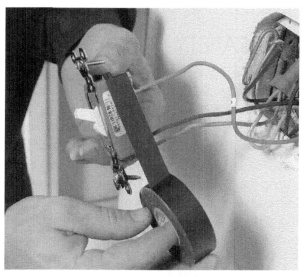

To ensure safe wiring of residential circuits, electric tape should be used to cover the terminals on switches and receptacles.

Hand Protection from Electrocution

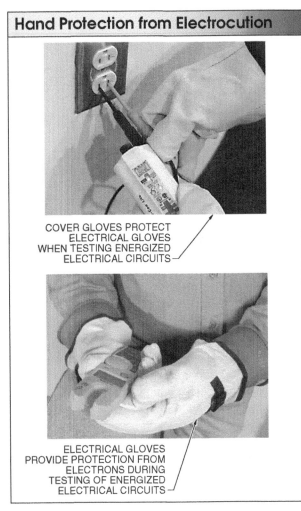

COVER GLOVES PROTECT
ELECTRICAL GLOVES
WHEN TESTING ENERGIZED
ELECTRICAL CIRCUITS

ELECTRICAL GLOVES
PROVIDE PROTECTION FROM
ELECTRONS DURING
TESTING OF ENERGIZED
ELECTRICAL CIRCUITS

Figure 2-34. *Always wear electrical gloves and cover gloves when taking measurements on energized circuits.*

FIRE SAFETY

The chance of fire is greatly reduced by good house-keeping. Keep debris in designated containers and away from buildings.

If a fire should occur, the first thing an individual must do is provide an alarm. The fire department must be called and all personnel on the job must be alerted. During the time before the fire department arrives, a reasonable effort should be made to contain the fire. In the case of a small fire, portable fire extinguishers (available around the site) are used to extinguish the fire. The stream from fire extinguishers must be directed to the base of fires. Each type of fire is designated by a class letter. **See Figure 2-35.**

Classes of Fire

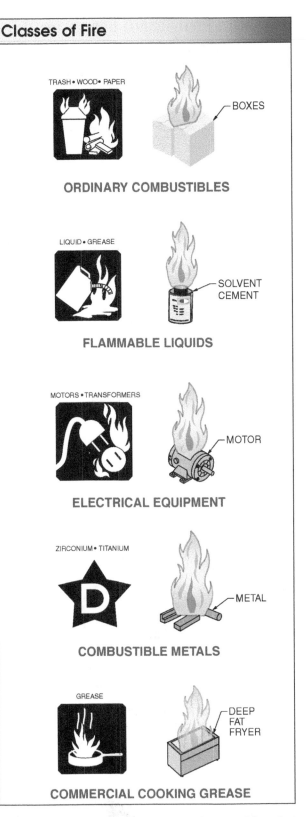

TRASH • WOOD • PAPER

BOXES

ORDINARY COMBUSTIBLES

LIQUID • GREASE

SOLVENT
CEMENT

FLAMMABLE LIQUIDS

MOTORS • TRANSFORMERS

MOTOR

ELECTRICAL EQUIPMENT

ZIRCONIUM • TITANIUM

D

METAL

COMBUSTIBLE METALS

GREASE

DEEP
FAT
FRYER

COMMERCIAL COOKING GREASE

Figure 2-35. *There are five common classes of fires that fire extinguishers are typically purchased to fight: Class A, Class B, Class C, Class D, and Class K.*

There are five common classes of fires that fire extinguishers are typically purchased to fight. Fire extinguishers are labeled with operating instructions for extinguishing fires. **See Figure 2-36.**

Fire Extinguishers

FIRE EXTINGUISHERS ARE LABELED WITH OPERATING INSTRUCTIONS

FIRE EXTINGUISHER CAN BE USED ON CLASS A, B, AND C FIRES

Kidde

Figure 2-36. *Most fires are fought with fire extinguishers containing water, CO_2, dry chemicals, halon gas, or wet chemicals.*

Class A Fires

Class A fires occur in wood, clothing, paper, rubbish, and other such items. Class A fires are typically controlled with water. (Symbol: green triangle.)

Class B Fires

Class B fires occur in flammable liquids such as gasoline, fuel oil, lube oil, grease, thinner, and paints. The agents required for extinguishing Class B fires dilute or eliminate the air to the fire by blanketing the surface of the fire. Chemicals such as foam, CO_2, dry chemical, and Halon are used on Class B fires. (Symbol: red square.)

Class C Fires

Class C fires occur in facilities near or in electrical equipment. The extinguishing agent for Class C fires must be a nonconductor of electricity and provide a smothering effect. CO_2, dry chemical, and halon gas extinguishers are typically used. (Symbol: blue circle.)

Class D Fires

Class D fires occur in combustible metals such as magnesium, potassium, powdered aluminum, zinc, sodium, titanium, zirconium, and lithium. The extinguishing agent for Class D fires is a dry powdered compound. The powdered compound creates a smothering effect that is somewhat effective. (Symbol: yellow star.)

Class K Fires

Class K fires occur in kitchens with grease. The extinguishing agent for Class K fires is a wet chemical (potassium acetate).

Fire class symbols appear on each fire extinguisher. A red sign typically points to the location of a fire extinguisher. Because not all fire extinguishers can be used on all types of fires, individuals must be aware of how to identify which fire extinguisher to use on what type of fire.

When firefighters are called, be prepared to direct the firefighters to the location of the fire. Also, inform the firefighters of any special problems or conditions that exist, such as downed electrical wires or chemical leaks from electrical equipment.

ənⁱⁱ.

Electrical Tools and Safety

TEST

Name_____ Date _____

Electrical Tools and Safety

_____ 1. ___ are appropriate for bending wire and positioning small components.

_____ 2. ___ are for removing insulation from small-diameter wire.

_____ 3. ___ are appropriate for cutting cable, removing knockouts, twisting wire, and deburring conduit.

T   F   4. Neon testers are used to determine the presence of 115 V–230 V in circuits.

T   F   5. All power tools must be grounded unless the power tool is double-insulated.

_____ 6. Appropriate electrical protective clothing includes items such as safety glasses, insulating gloves, a hard hat, and work shoes with ___ toes.

T   F   7. A continuity tester indicates the condition of a circuit with lights.

_____ 8. A fish tape may be used to ___ .
     A. pull wire through conduit   **C.** both A and B
     B. route wire through wall cavities   **D.** none of the above

_____ 9. ___ are used to tighten box connectors, locknuts, and small conduit couplings.

_____ 10. A(n) ___ is used to safely remove cartridge fuses.

_____ 11. Because of their length, most ___ can be used to determine vertical wall placement of receptacle boxes.

T   F   12. A keyhole saw can be used in drywall to cut openings for outlet boxes.

T   F   13. Cable strippers are often used to strip nonmetallic sheathed cable.

_____ 14. An electrician's pouch is usually made of heavy-duty ___.

_____ 15. Screwdrivers are specified by ___ width.

_____ **16.** When using a(n) ___ wrench, force should always be applied against the fixed jaw.

_____ **17.** A continuity tester ___.
- **A.** has a self-contained battery
- **C.** both A and B
- **B.** tests for unbroken circuits
- **D.** none of the above

_____ **18.** ___ are bits are commonly used with electric drills for rough boring holes.

_____ **19.** ___ are used to tighten box connectors, lockouts, and conduit couplings.

_____ **20.** OSHA has designated the color ___ to indicate portable containers that are dangerous because they contain flammable liquids.

T   F   **21.** Electrical shock occurs when an individual comes in contact with two conductors of a circuit or when the body of an individual becomes part of a circuit.

T   F   **22.** Unsafe conditions must be reported to a supervisor.

T   F   **23.** A utility knife is often used to cut small pipe and conduit.

_____ **24.** All adjustments to power tools must be made with the tool ___.

_____ **25.** Many accidents are caused by ___ falling off ladders, shelves, and scaffolding that are being moved.

_____ **26.** Double-insulated tools have ___ prong(s).
- **A.** one
- **C.** three
- **B.** two
- **D.** four

_____ **27.** The first thing to do when a fire occurs is to give a(n) ___.

_____ **28.** The two most common screwdriver heads are Phillips and ___.

T   F   **29.** Generally, double-insulated tools must not be used in wet working conditions.

T   F   **30.** Power tools must be inspected by an OSHA representative every six months.

Fire Classification

_____ **1.** Class A fire

_____ **2.** Class B fire

_____ **3.** Class C fire

_____ **4.** Class D fire

WOOD

MOTOR

PAINT

METAL

Ⓐ   Ⓑ   Ⓒ   Ⓓ

Color Codes

_____ **1.** Warning labels

_____ **2.** Caution labels

_____ **3.** Electrical warning labels

_____ **4.** Danger labels

_____ **5.** Explosion warning labels

Fire Safety

_____ **1.** A water extinguisher can be used with a Class ___ fire.

T   F   **2.** Multipurpose dry chemical extinguishers can be used with all classes of fires.

T   F   **3.** Pump tank extinguishers with plain water should be used for Class B fires.

T   F   **4.** Carbon dioxide extinguishers should be used for Class A fires.

_____ **5.** The stream from extinguishers should be directed at the ___ of flames.

_____ **6.** Carbon dioxide extinguishers may be used for Class B and Class ___ fires.

_____ **7.** Dry powdered compound extinguishers are used on Class ___ fires.

_____ **8.** When a fire occurs, the first thing an individual must do is ___.

T   F   **9.** The extinguishing agent for a Class D fire must be a dry powdered compound.

T   F   **10.** Water should not be used on Class C fires.

Tool Safety

T   F   **1.** Cutting tools should be sharp and clean.

T   F   **2.** Tools should not be carried in a pocket unless the pocket is designed for that tool.

T   F   **3.** Pencils should be carried behind an ear.

T   F   **4.** When carrying sharply pointed tools, carry the point away from the body.

T   F   **5.** Tools should be used only for the tasks for which they are designed.

T   F   **6.** Power tools should be inspected and serviced regularly.

T   F   **7.** Power tool guards may be used to facilitate work.

T   F   **8.** A hammer should never be struck against another hammer.

T   F   **9.** Always shut off power to an electrical tool when the work is completed.

T   F   **10.** All adjustments, blade changes, etc., to power tools should be made when the power is OFF and the cord is disconnected.

T   F   **11.** A change in sound during tool operation can indicate trouble with a power tool.

T   F   **12.** Material on which work is to be done should always be checked to see that it is free of obstructions.

T   F   **13.** Safety goggles must always be worn when performing work.

T   F   **14.** Before using a power tool, check the electrical cord to verify condition.

T   F   **15.** The safety recommendations of the manufacturer must always be read before using a new power tool.

Electrical Plans 3

Chapter 3 provides an understanding of electrical symbols, abbreviations, and diagrams that are used when working in the residential electrical industry. The ability to identify commonly used electrical symbols, recognize electrical abbreviations, and read electrical circuit diagrams is required for efficient understanding of residential electrical circuits, as well as wiring and troubleshooting circuits.

LANGUAGE OF ELECTRICAL PRINTS

All trades have a specific language that must be understood in order to transfer information efficiently within the trade. Trade languages include symbols, drawings or diagrams, schematics, words, phrases, and abbreviations. To perform electrical work, individuals must understand the electrical language, have an understanding of how electrical components function, and have an understanding of the relationships among components and devices in a circuit. With an understanding of the electrical language, an individual is able to read drawings and diagrams, understand circuit operation, and troubleshoot various problems. Electrical prints that are used to convey electrical information include pictorial drawings, electrical layouts, schematic diagrams, and line diagrams.

Pictorial Drawings

A *pictorial drawing* is a drawing that shows the length, height, and depth of an object in one view. Pictorial drawings indicate physical details such as holes and shoulders of an object as seen by the eye. **See Figure 3-1.**

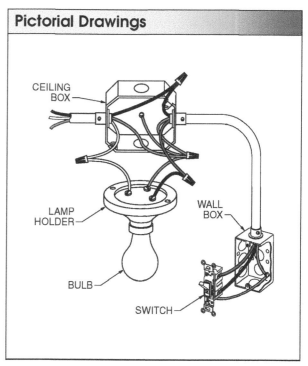

Pictorial Drawings

CEILING BOX

LAMP HOLDER

WALL BOX

BULB

SWITCH

Figure 3-1. *Pictorial drawings show the physical details of electrical components as seen by the eye.*

Electrical Abbreviations and Symbols

A *symbol* is a graphic element that represents a component, device, or quantity. Symbols are used to represent electrical components and devices on electrical and electronic diagrams. An *abbreviation* is a letter or combination of letters that represents a word. **See Figure 3-2.**

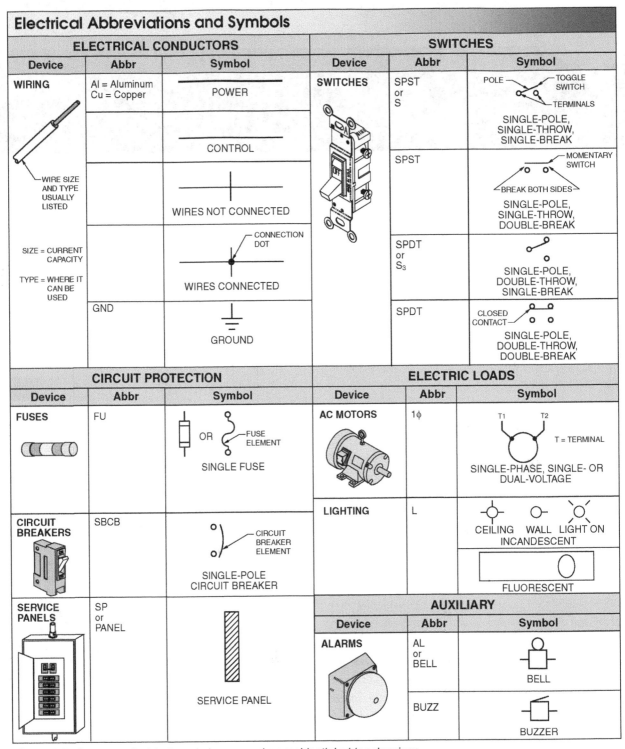

Figure 3-2. *Common electrical symbols are used on residential wiring drawings.*

Electrical Layouts

An *electrical layout* is a drawing that indicates the connections of all devices and components in a residential electrical system. Electrical layouts represent as closely as possible the actual location of each part of an electrical circuit. Electrical layouts often include details of the type of wire and the kind of hardware used to fasten wires and cables to terminals and boxes. **See Figure 3-3.**

An electrical layout typically indicates the devices and components of a system with rectangles or circles. The location or layout of all system devices and components is accurate on the electrical diagram for the particular dwelling. All connecting wires are shown connected from one part to another. Individuals should use electrical layouts when constructing a dwelling and when performing maintenance.

Schematic Diagrams

A *schematic diagram* is a drawing that indicates the electrical connections and functions of a specific circuit arrangement using graphic symbols. Schematic diagrams do not indicate the physical size or appearance of any component or device found in a circuit. Typically, schematic diagrams are associated with electronic circuits and not with residential electrical systems.

Schematic diagrams are used to show the wiring required for the proper operation of a component (light or motor). **See Figure 3-4.** Schematic diagrams are essential when troubleshooting because schematic diagrams enable an individual to trace a circuit and all functions without regard to the actual size, shape, or location of the component, device, or part.

The CSI MasterFormat™ is a master list of numbers and titles for organizing information about construction requirements, products, and activities into a standard sequence. The CSI MasterFormat™ is broken up into divisions that define the broad area of construction. An estimator must be able to read and interpret architectural and shop drawings such as wiring diagrams and schematic diagrams. Familiarity with various lines, drawing scales, symbols, and abbreviations is required to develop an accurate bid.

Electrical Layouts

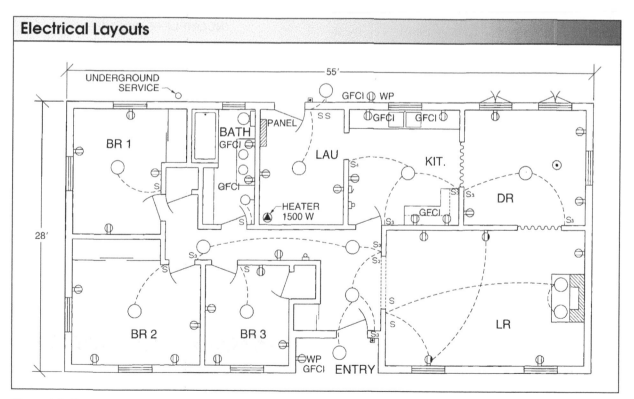

Figure 3-3. *Electrical layouts show as accurately as possible the actual location of each component of an electrical circuit.*

Schematic Diagrams

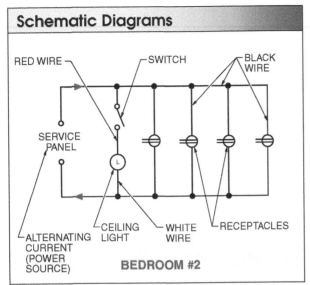

Figure 3-4. *Schematic diagrams show the components of an electrical system laid out so that the circuit is easy to read instead of showing the actual position of the components.*

ELECTRICAL CIRCUITS

An *electrical circuit* is an assembly of conductors (wires), electrical devices (switches and receptacles), and components (lights and motors) through which current flows. An electrical circuit is complete (a closed circuit) when current flows from the power source (service panel) to the load (lights and motors), and back to the power source. A circuit is not complete (an open circuit) when current does not flow. A broken wire, a loose connection, or a switch in the OFF position stops current from flowing in an electrical circuit. **See Figure 3-5.**

Residential Electrical Circuit Devices and Components

All industrial and residential electrical circuits include five basic parts. The five parts include a component (load) that converts electrical energy into some other usable form of energy such as light, heat, or motion; a source of electricity; conductors to connect the individual devices and components; a method of controlling the flow of electricity (switch); and protection devices (fuses or circuit breakers) to ensure that the circuit operates safely within electrical limits.

Electrical Circuits

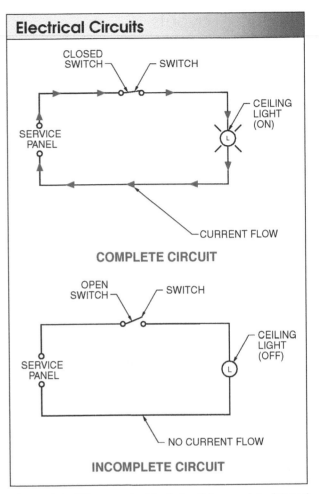

Figure 3-5. *When an electrical circuit is complete (closed circuit), current flows to the load (load is ON) and back to the power source (service panel). When a circuit is not complete (open circuit), current does not flow through the load (load is OFF).*

Electrical circuit devices and components can be shown using line diagrams, wiring diagrams, and/or schematic diagrams. For example, an automobile interior lighting circuit includes the five components of a typical electrical circuit. **See Figure 3-6.** The source of electricity is the battery, conductors can be either chassis wires or the car frame, the control device is a plunger-type door switch, the load is the interior light, and a fuse is the protection device.

Standard practices for drawing electrical and electronic diagrams are detailed in Electrical and Electronic Diagrams, ANSI Y14.15.

Electrical Circuit Components

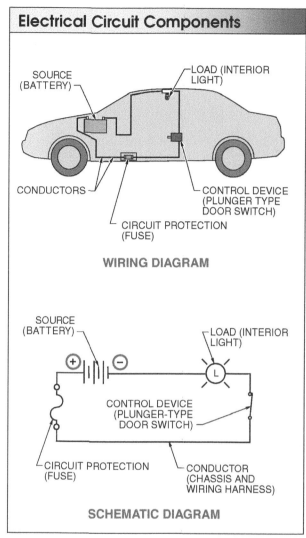

Figure 3-6. *All electrical circuits include a power source, conductors, a load, a control device, and overload protection.*

Tech Tip:
Circuit Application

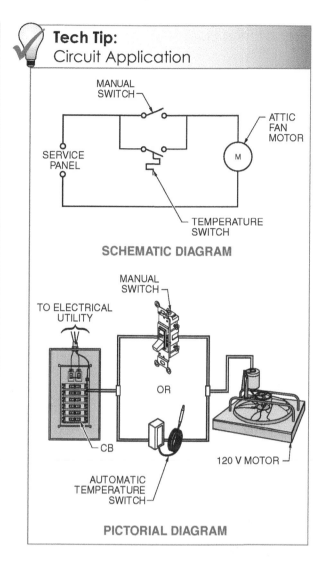

Residential electrical circuit wiring diagrams can become quite complicated. **See Figure 3-7.** All residential circuits use the service panel as the power source for the dwelling. Conductors are the wires pulled through conduit or metallic and nonmetallic cables. The most common control device for a residence is a single-pole switch. Control devices can be single-pole switches, three-way switches, motion detectors, timers, or thermostats. Loads found in residential circuits include lamps, motors, televisions, radios, and appliances. The typical overcurrent device used to protect a residential electrical circuit is a circuit breaker.

Ruud Lighting, Inc.

Photoelectric and timer switches increase the security and convenience of landscape lighting.

Residential Electrical Circuit Components

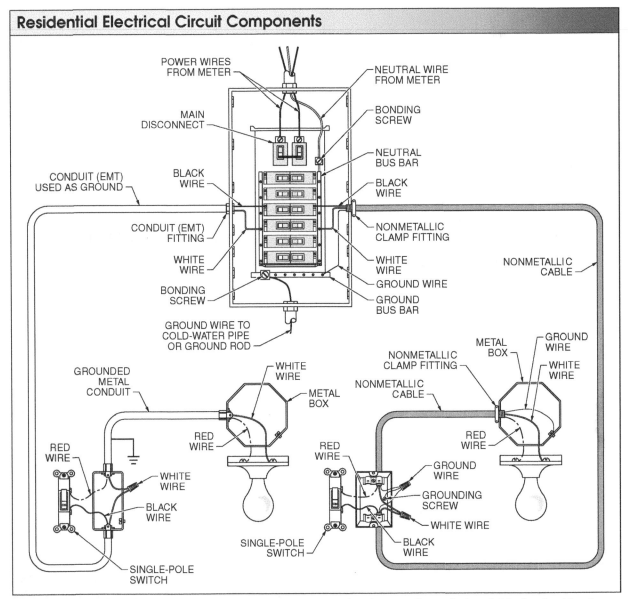

Figure 3-7. *Residential electrical circuits include a power source (service panel), conductors (wiring), a load (such as a light), control devices (switches), and overload protection (fuses or circuit breakers).*

Manually and Automatically Controlled Circuits

A *manually controlled circuit* is any circuit that requires a person to initiate an action for the circuit to operate. **See Figure 3-8.** Line diagrams can be used to illustrate manually controlled circuits that have a single-pole switch (S) controlling a light bulb in a fixture. The voltage level of the circuit is typically indicated at the top of a line diagram. Circuit voltages in a residence are typically 115 VAC, but can be 12 VAC, 18 VAC, 24 VAC, or 230 VAC.

Dark nodes are used on line diagrams or schematics to indicate an electrical connection. When a node is not present, the wires only cross each other and are not electrically connected. Line diagrams are read from left (L1) to right (L2).

Flipping a single-pole switch (S1) to the ON position allows current to pass from the hot conductor (L1), through the closed contacts of the wall switch (S1), through the lamp (L), and on through the common conductor (L2), forming a complete path (circuit) for current to flow and allow the light to illuminate.

Manually Controlled Circuits

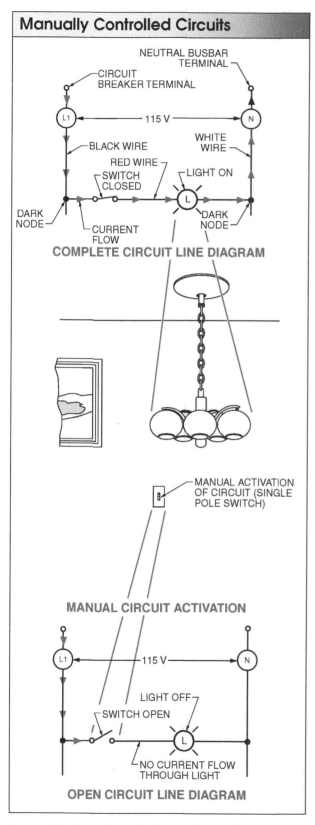

Figure 3-8. *Manually controlled circuits require a person to initiate an action for the circuit to operate.*

Flipping the switch to the OFF position opens the contacts of the switch, stopping the flow of current to the lamp and turning the light OFF.

Automatically controlled devices have replaced many functions that were once performed manually. As a part of modern technology, automated control circuits are designed to replace manual devices. Any manually controlled circuit can be converted to automatic operation. For example, an electric motor on a sump pump can be turned ON and OFF automatically by adding an automatic control device such as a float or pressure switch. **See Figure 3-9.** Float and pressure switch control circuits are used to turn sump pumps OFF and ON to prevent flooding automatically. An automatic control circuit turns on a pump when water reaches a predetermined level, which removes the water. The float or pressure switch senses the change in water level and automatically stops the pump.

Automatically Controlled Circuits

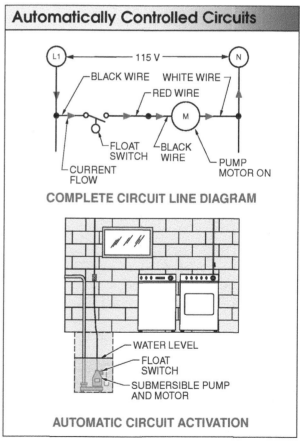

Figure 3-9. *Automatically controlled circuits do not require the interaction of a person to control a function, such as turning a pump ON.*

When a sump pump uses an automatic float switch circuit, float switch contacts FS1 determine when current passes through the circuit to start the motor. Current passes through float switch contacts FS1 and the motor to L2 when float switch contacts FS1 are closed. To start the pump motor, current must flow from the power source, through the load, and back to the power source, making a complete circuit. The pump moves water until the water level drops enough to open the float switch contacts FS1 and shut the pump motor OFF. A power failure or the manual opening of the contacts prevents the pump from automatically moving water even after the water reaches the predetermined level.

Another example of automatic control is a photocell controller. **See Figure 3-10.** Photocell controllers automatically detect the absence of light and respond by sending commands to loads (lights). A photocell controller responds to the absence of light by turning lights inside or outside a dwelling ON. Photocell controllers are typically installed on motion lights near doorways, garages, and windows.

Photocell controllers can also transmit OFF commands in response to approaching daylight. A photocell controller responds to light by turning connected lighting loads inside or outside the dwelling OFF. Photocell controllers are mounted to standard octagon electrical boxes or directly to controlled lighting.

ELECTRICAL PLANS

An *electrical plan* is a drawing and list that indicates what devices are to be used, where the electrical devices are to be placed, and how the devices are to be wired. **See Figure 3-11.** Electrical plan drawings indicate the use of switches, lights, and receptacles with symbols. Symbols always represent real-world devices. Because electrical symbols are easy to draw, symbols are used in place of actual pictorial drawings of electrical devices. Electrical plans are easier to draw and understand when the plans are drawn using symbols. Electrical plans drawn with detailed pictorial drawings of each device and component are difficult to understand.

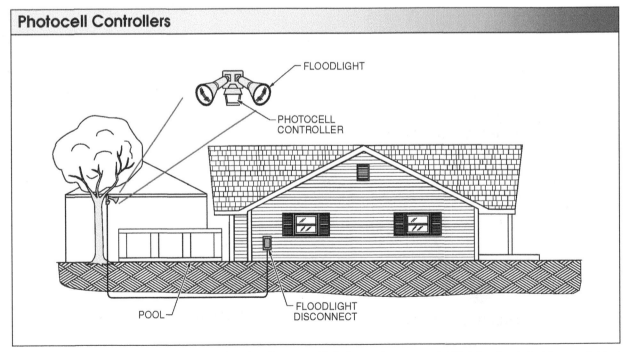

Figure 3-10. *Photocell controllers respond to the absence of light by turning connected lighting loads inside or outside a dwelling ON, and respond to light by turning lights OFF.*

Electrical Plans

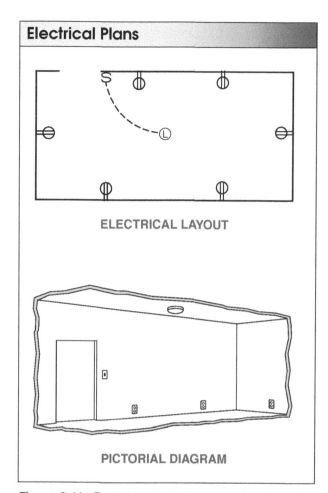

ELECTRICAL LAYOUT

PICTORIAL DIAGRAM

Figure 3-11. *Basic electrical plans provide information about what type of device is used where in a room. Pictorial diagrams can be used to show the room after wiring and drywalling.*

Understanding electrical device symbols is the key to understanding electrical plans. **See Figure 3-12.** Architects always use symbols when creating electrical plans. Electrical device symbols and the operation of the devices must be understood to properly wire circuits from electrical plans. When interpreting electrical plans, unfamiliar symbols must be identified before any work is started. **See Figure 3-13.**

The more time spent analyzing electrical plans, the easier it becomes to understand electrical symbols and the functions of a specific circuit. **See Figure 3-14.** Most electrical plans used in constructing new homes today are drawn to provide the same types of information from one home to another.

Electrical Plan Information

Electrical plans are important because electrical plans indicate what tasks an individual must perform. Electrical plans provide information similar to a good road map. Road maps indicate the location of roads, towns, and points of interest. A good electrical plan indicates the location of all major electrical devices and connections. Electrical plans are also used to determine the number of openings in walls and ceilings and to estimate the amount of wire and conduit that is required for the job. Electrical plans must provide enough information to indicate the best and shortest way to complete a job.

Division 26 of the CSI MasterFormat™ specification system includes descriptions of electrical site work, conduit, panelboards, lighting, telecommunication systems, and HVAC systems.

Trus Joist, A Weyerhaeuser Business
When wood I-joists are installed, conductors are routed through prefabricated knockouts.

Electrical Devices and Symbols

Figure 3-12. *Understanding electrical device symbols is the key to understanding electrical plans.*

Residential Electrical Symbols

GENERAL OUTLETS

Symbol	Description
◯ or Ⓛ	Lighting (wall)
or ⌖	Lighting (ceiling)
[◯]	Ceiling lighting outlet for recessed fixture (Outline shows shape of fixture.)
⊂◯▭	Continuous wireway for fluorescent lighting on ceiling, in coves, cornices, etc. (Extend rectangle to show length of installation.)
Ⓛ	Lighting outlet with lamp holder
Ⓛ$_{PS}$	Lighting outlet with lamp holder and pull switch
Ⓕ	Fan outlet
Ⓙ	Junction box
Ⓓ	Drop-cord equipped outlet
–Ⓒ	Clock outlet

To indicate wall installation of above outlets, place circle near wall and connect with line as shown for clock outlet.

AUXILIARY SYSTEMS

Symbol	Description
▣	Pushbutton
⌐	Buzzer
⊶▢	Bell
⊶▢	Combination bell-buzzer
CH	Chime
◇	Annunciator
D	Electric door opener
M	Maid's signal plug
▢	Interconnection box
Ⓣ	Thermostat
▶	Outside telephone
▷	Telephone
R	Radio outlet
TV	Television outlet

CONVENIENCE OUTLETS

Symbol	Description
⊖	Duplex convenience receptacle
⊖$_3$	Triplex convenience outlet (Substitue other numbers for other variations in number of plug positions.)
⊖	Duplex convenience outlet — split wired
⊖$_{GR}$	Duplex convenince outlet for grounding-type plugs
⊖$_{WP}$	Weatherproof convenience outlet
⊖ X″	Multioutlet assembly (Extend arrows to limits of installation. Use appropriate symbol to indicate type of outlet. Also indicate spacing of outlets as X inches.)
⊖$_S$	Combination switch and convenience outlet
⊖R	Combination radio and convenience outlet
⊙	Floor outlet
⊖$_R$	Range outlet
⬤$_{DW}$	Special-purpose outlet. (Use subscript letters to indicate function. DW–dishwasher, CD–clothes dryer, etc.)

MISCELLANEOUS

Symbol	Description
◣	Heating panel
▨	Service panel
▬	Distribution panel
– – – –	Switch leg indication. (Connects outlets with control points.)
Ⓜ	Motor
⏚	Ground connection
┤├	2-conductor cable
┤╫├	3-conductor cable
┤╫╫├	4-conductor cable
←	Cable returning to service panel
∿	Fuse
⌐o͞o	Circuit breaker
⊖$_{a, b}$	Special outlets. (Any standard symbol given above may be used with the addition of subscript letters to designate some special variation of standard equipment for a particular architectural plan. When so used, the variation should be explained in the key to symbols and, if necessary, in the specifications.)
⊖$_{a, b}$	
⬤$_{a, b}$	
▢$_{a, b}$	
⊖ GFCI	Ground-fault circuit interrupter
⊖ WP	Weatherproof

SWITCH OUTLETS

Symbol	Description	Symbol	Description
S–S$_1$	Single-pole switch	S$_D$	Dimmer switch
S$_2$	Double-pole switch	S$_P$	Switch and pilot light
S$_3$	Three-way switch	S$_{WP}$	Weatherproof switch
S$_4$	Four-way switch		

Figure 3-13. *When interpreting electrical plans, unfamiliar symbols must be identified before any work is started.*

Residential Electrical Plans

ELECTRICAL PLAN — NEC©

A 210.7(A). All receptacles installed on 15 A and 20 A circuits shall be located in accordance with type of room and use.
B 210.8(A). All 125 V, 1ɸ receptacles shall have GFCI protection when installed in: (1) bathrooms, (2) garages, (3) outdoors where there is direct grade access, (4) crawl spaces below grade level and unfinished basements, (5) kitchen countertops, (6) boathouses.
C 210.23. Individual branch circuits can supply any load for which they are rated.
D 210.52(A). General provicions for installing receptacle outlets in dwelling units.
210.52(B). For small appliance load, including refrigeration equipment, in kitchen, pantry, and breakfast room, two or more 20 A appliance circuits shall be provided. Such circuits shall have no other outlets. Countertop outlets shall be supplied at each wall counter that is 12″ or wider.
E 210.52(E). All receptacles installed outdoors must be of GFCI type.
F 210.52(H). Hallways 10′ in length or more shall have a receptacle provided.
G 220.12. Unit lighting load for dwelling occupancies shall not be less than 3 VA per sq ft. See Table 220.12 in determining load on "VA per sq ft" basis, outside dimensions of the building shall be used, excluding open porches, garages, unused spaces, unless adaptable for future use.
H 210.11(C)(2). At least one 20 A circuit shall be provided for a laundry receptacle.
I 220.4(D). For general illumination one 15 A branch circuit is required for every 600 sq ft area.

$$(3 \text{ VA} \times 600 \text{ sq ft} = 1800 \text{ VA } \tfrac{1800}{120} = 15 \text{ A})$$

J 220.52(A). Feeder load for two small appliances circuits to be 3000 VA (1500 VA for each two-wire circuit).
220.52(B). Each laundry circuit feeder load calculation to be at least 1500 VA.

K 230.42(A). Service entrance conductors per 220 and 310.15.
L 410.8. Fixtures in Clothes Closets.
M 422.11(E). Branch circuit to a single non-motor appliance not to exceed OCPD rating. If an appliance is rated at 13.3 A or less, the next standard OCPD is required, not to exceed 150% of the appliance rating.
N 725. Remote control, signaling, and Power Limited circuits.

ELECTRICAL KEY PLAN

Letter	NEC©	Subject
A	210.7(A)	Receptacle Location
B	210.8(B)	GFCI Location
C	210.23	Receptacles
D	210.52(A)(B)	Circuits (Load Rating)
E	210.52(E)	GFCI Location (Outdoors)
F	210.52(H)	Receptacle Location (Hallways)
G	220.12	Lighting Load
H	210.11(C)(2)	Laundry Circuit
I	220.4(D)	Circuits (General Purpose)
J	220.16	Circuits (Small Appliance)
K	230.42	Service Entrance
L	410.8	Fixtures (Clothes Closet)
M	422.28	Circuits (Branch)
N	725	Signaling and Power Limited Circuits

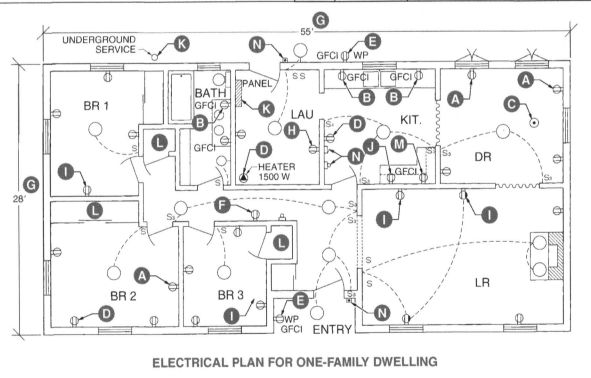

ELECTRICAL PLAN FOR ONE-FAMILY DWELLING

Figure 3-14. *Residential electrical plans must follow NEC® standards and local electrical codes.*

DEVELOPING ELECTRICAL PLANS

A complete electrical plan or set of plans is developed in two main parts: component plan(s) and wiring plan(s).

Component Plans

A *component plan* is a group of schedules that state the required locations for receptacles, lights, and switches according to the NEC® and local codes. To be effective, a component plan must include devices that satisfy all the electrical requirements for light, heat, and power to a given area.

To aid in the determination of which electrical devices are required for a system, electrical component lists are created with the component plan for each area of a residence such as the living room, kitchen, bedrooms, and bathrooms. A *component list* is a list of electrical equipment indicating manufacturer, specifications, and how many of each electrical component are required for a room or area. A partial component list is a component schedule. **See Figure 3-15. See Appendix.** Careful study of an electrical component list or schedule will aid in developing a checklist for work to be performed. Additional ideas can be obtained from component plans that were not initially recognized as being required from the electrical plan. Sample component plans and component lists are available in home planning books and magazines. Public and school libraries and the internet are sources for finding home component plans, lists, and schedules.

Component Plan Schedules

Area	Convenience Receptacles	Special-Purpose Outlets	General Lighting	Major Appliances	General Switching for All Areas
Bedrooms	No space along a wall should be more than 6′ from a receptacle outlet. Any wall space 2′ or larger should have a minimum of 1 receptacle. NEC® 210.52 (A).	TV outlet, intercom, speakers (music), and telephone jack	Ceiling, wall or valence light, lamp switched at receptacle	Room air conditioner, electric baseboard heat	Switches are typically placed opposite the hinged side of door. When there are two or more entrances to a room, multiple switching should be used. Door switch can be used for closet light so that the switch turns the light ON and OFF as the door opens and closes. Switches with pilot lights should be used on lights, fans, and other electrical devices which are in locations not readily observable.
Living Room	No space along a wall should be more than 6′ from a receptacle outlet. Any wall space 2′ or larger should have a minimum of 1 receptacle. NEC® 210.52 (A).	TV outlet, intercom, speakers (music)	Ceiling fixture, recessed lighting, valence light, lamp switched at receptacle, possible dimmer	Room air conditioner, built-in stereo system, electric baseboard heating	
Family Room	No space along a wall should be more than 6′ from a receptacle outlet. Any wall space 2′ or larger should have a minimum of 1 receptacle. NEC® 210.52 (A).	TV outlet, intercom, speakers (music), bar area (ice maker, blenders, small refrigerator, hot plate), telephone jack, thermostat	Ceiling fixture, recessed lighting, valence lights, studio spot lights, lamp switched at receptacle, fluorescent light, possible dimmer	Room air conditioner, built-in stereo system, electric baseboard heating	
Dining Room	No space along a wall should be more than 6′ from a receptacle outlet. Any wall space 2′ or larger should have a minimum of 1 receptacle. NEC® 210.52 (A).	Elevated receptacles (48″) for buffet tables, speakers (music)	Ceiling chandelier, recessed lighting, valence lighting for china hutch	Room air conditioner, electric baseboard heating	

Figure 3-15. *Component lists and schedules are created for each area of a residence such as the living room, kitchen, bedrooms, and bathrooms.*

To develop a component plan, apply the following procedures:

• Use as many resources as possible for ideas.
• Keep the electrical plan as simple and straightforward as possible, so the plan can be easily understood.

Wiring Plans

Once the locations of receptacles, lights, and switches have been determined and indicated on a component plan, a wiring plan can be drawn. **See Figure 3-16.** A wiring plan is a floor plan drawing that indicates the placement of all electrical devices and components and the wiring required to connect all the equipment into circuits. Wiring plans are the basis for all electrical work performed at a residence. The purpose of a wiring plan is to group individual electrical devices into specific circuits. A wiring plan is essential when determining the number of components per circuit (load) and the best route for wiring to take. Time spent studying a wiring plan aids in anticipating problems and reducing the amount of materials required. Slash marks indicate the number of conducting wires in cables or conduit and an arrow symbolizes the circuit connection to the service panel.

Pictorial drawings (construction layouts) illustrate how wiring will look after all circuit wiring is installed. **See Figure 3-17.**

The NEC® is updated every three years. Individuals must consult the latest edition.

Manufacturer Specifications

When an individual installs an electrical appliance such as washing machine, dryer, or electric oven, the individual must follow the directions or specifications provided by the manufacturer. Manufacturer specifications indicate the size of wire that must be used and the amount of overload protection that must be provided. The specifications set by manufacturers must agree with or surpass the standards of the NEC®.

Electrical Plans

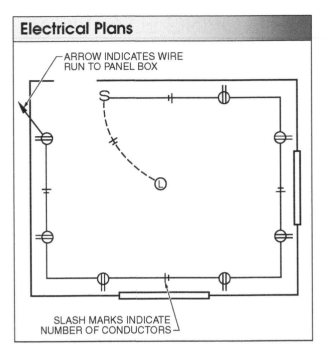

Figure 3-16. *A wiring plan provides a detailed layout of cable or wire (in conduit), routes, and number of conductors.*

Tech Tip:
Prints

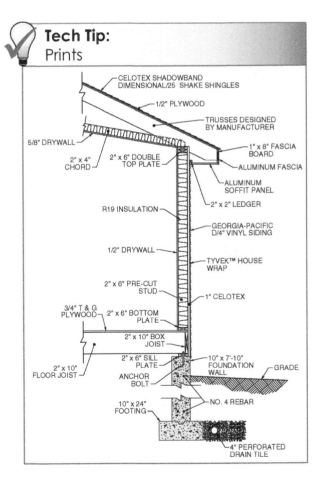

Pictorial Drawings (Construction Layouts)

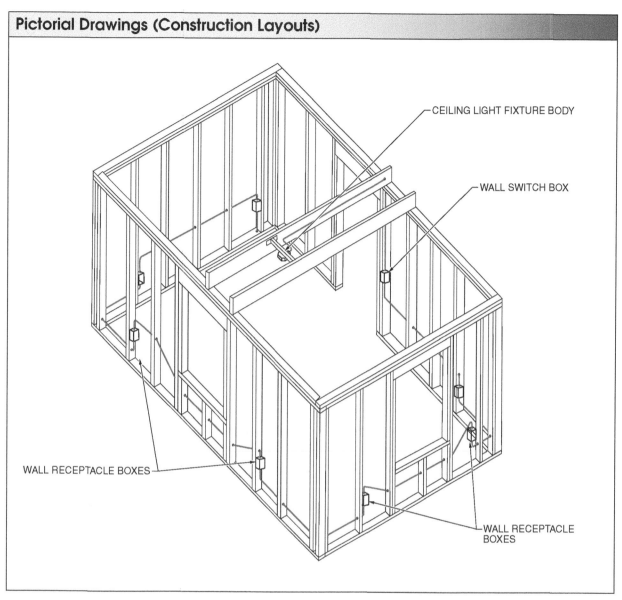

CEILING LIGHT FIXTURE BODY

WALL SWITCH BOX

WALL RECEPTACLE BOXES

WALL RECEPTACLE BOXES

Figure 3-17. *Pictorial drawings (construction layouts) indicate how a circuit will appear when wiring is completed.*

After connecting each circuit breaker to the load center (breaker box), it is a good practice to identify each breaker for the circuit it controls in order to aid in future troubleshooting. **See Figure 3-18.**

Electrical plans can be used to retrieve information on the existing layout of circuits and devices in areas such as kitchens.

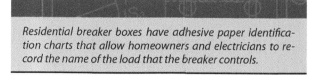

Residential breaker boxes have adhesive paper identification charts that allow homeowners and electricians to record the name of the load that the breaker controls.

Circuit Breaker Identification

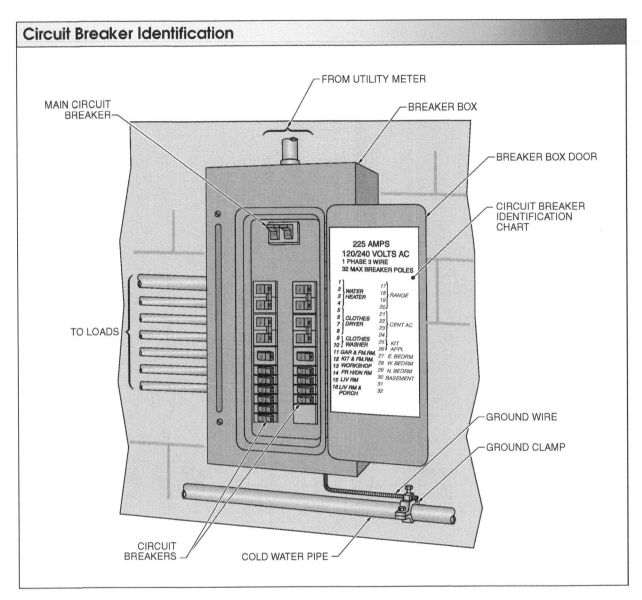

Figure 3-18. *Circuit breaker identification charts for residential loads are typically located on the inside of the breaker box door.*

Name_____ Date _____

Electrical Plans

_____ **1.** An electrical ___ is a drawing that shows where electrical devices are to be placed.

_____ **2.** The ___ plan provides a detailed layout of the cable routes and the number of conductors.

_____ **3.** Residential electrical system installations must comply with ___.
- **A.** the NEC®
- **B.** local building codes
- **C.** electrician's union requirements
- **D.** both A and B

_____ **4.** ___ are used to represent switches, lights, and receptacles on electrical plans.
- **A.** Pictorial drawings
- **B.** Orthographic drawings
- **C.** Symbols
- **D.** none of the above

_____ **5.** The NEC® sets ___ requirements for residential electrical installations.
- **A.** minimum
- **B.** maximum
- **C.** cost
- **D.** none of the above

T   F **6.** The overload protection required for an electrical appliance is provided by manufacturer specifications.

T   F **7.** A component plan may be used to determine the best required location for switches, lights, and receptacles.

T   F **8.** Electrical plans show all necessary framing details of the structure.

_____ **9.** Switches are typically placed opposite the ___ side of a door.

_____ **10.** All 115 V, 1φ receptacles installed in bathrooms shall have ___ protection.

_____ **11.** At least one ___ A circuit shall be provided for a laundry circuit.

T   F **12.** In general, service entrance conductors shall have an ampacity of not less than the sum of non continuous plus 125% of continuous loads.

T   F   **13.** Ruled lines extending from a receptacle symbol indicate a wall receptacle.

T   F   **14.** Subscript letters can be added to a standard symbol to designate a variation of a standard piece of equipment.

15. Hallways over ___ shall have a receptacle provided.
- **A.** 4′ wide
- **B.** 8′ high
- **C.** 10′ long
- **D.** none of the above

16. Switch outlets are represented on plans by ___.
- **A.** circles with letters
- **B.** circles without letters
- **C.** letters with or without subscripts
- **D.** none of the above

T   F   **17.** Individual branch circuits can supply any load for which the branch circuit is rated.

T   F   **18.** A cable returning to a service panel is represented by a dashed line on plans.

19. The branch circuit load used during calculations for laundry circuits is to be ___.

20. The symbol for a thermostat is a T inside a(n) ___.

Electrical Symbols

1. Lighting outlet

2. Fuse

3. Clock outlet

4. Switch leg

5. Fan outlet

6. Service panel

7. Interconnecting telephone

8. Range outlet

9. Motor

10. Ground connection

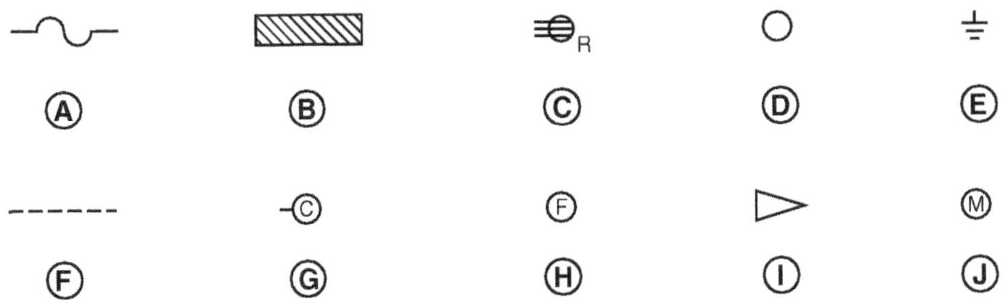

(A)   (B)   (C)   (D)   (E)

(F)   (G)   (H)   (I)   (J)

Porch Entrance

Draw the appropriate electrical symbols as indicated.

1. GFCI weatherproof duplex convenience outlet

2. Recessed ceiling outlet

3. Single-pole switch controlling the recessed ceiling outlet

4. Chime

5. Closed-circuit television camera

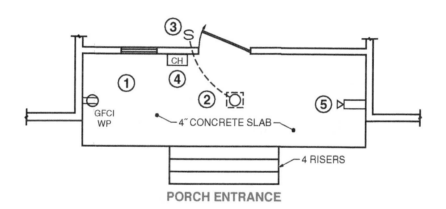

PORCH ENTRANCE

One-Family Dwelling

_____ **1.** Symbols at ___ represent duplex convenience outlets.

_____ **2.** ___ duplex convenience outlets are located on either side of the kitchen sink.

_____ **3.** BR 2 contains one switched ___ in the ceiling.

_____ **4.** A(n) ___ W heater is located in the laundry.

_____ **5.** Lighting outlets in the hallway ceiling are controlled by ___ switches.

_____ 6. The one-family dwelling contains approximately ___ sq ft (disregarding the entry).

_____ 7. The service panel is located in the ___.

_____ 8. A switch to the right of the bathroom door controls a ___-mounted lighting outlet.

_____ 9. A(n) ___ outlet is shown at H in the diagram.

_____ 10. The electrical service to the dwelling is brought in ___.

_____ 11. The two duplex convenience outlets on the switch leg in the living room are __-wired.

_____ 12. Pushbuttons at the front and back doors control a(n) ___ and a bell in the kitchen.

_____ 13. Each bedroom has a total of ___ duplex convenience outlets.

_____ 14. ___ GFCI convenience outlets are located at E in the diagram.

_____ 15. The hallway has a total of ___ duplex convenience outlet(s).

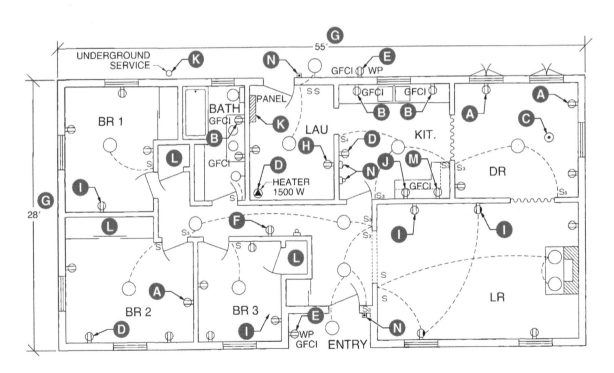

80

Partial Floor Plan

Draw the appropriate electrical symbols as indicated.

1. Six duplex convenience outlets

2. Two three-way switches controlling an overhead light

3. Ceiling lighting outlet with lampholder

4. Lighting outlet with lampholder and pull switch

5. Television outlet

6. Thermostat

7. Interconnecting telephone

8. Ceiling fan

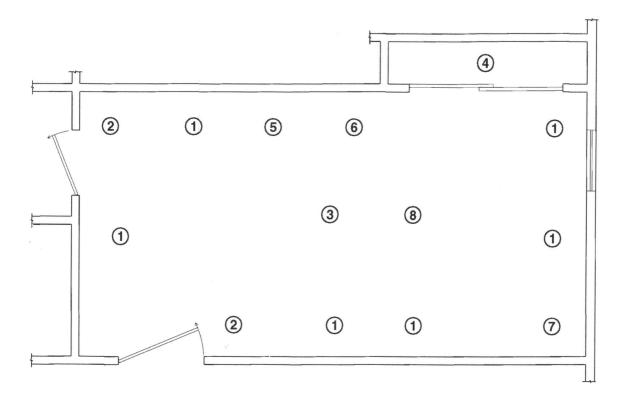

Symbols

_____	1.	◯
_____	2.	▣
_____	3.	⌒
_____	4.	▶⊢
_____	5.	☐

_____	6.	⊙
_____	7.	o☐
_____	8.	D
_____	9.	■
_____	10.	Ⓜ

Making Electrical Connections 4

Chapter 4 provides an overview of how to make electrical connections that are mechanically and electrically secure. Splices and connections are identified and procedures are provided on how to avoid the troublesome points of splices and connections. Common splices and connections (soldered and solderless) are shown with procedures and proper tool usage. With the use of so many electronic devices in the home, an occasion may arise where loose connections need to be reattached by soldering. Most soldering is performed on copper wire and splices. It is important to know the difference among the types of soldering processes and tools used to solder wires.

ELECTRICAL SUPPLIES

Typical electrical supplies required to wire a residence are hand tools, solder, soldering tools, tape, solderless connectors, crimping tools, and wire markers.

Tape

Although many types of tape can be used as insulation on electrical wiring, plastic tape, which insulates against voltages up to 600 V per wrap, is typically used for residential wiring. **See Figure 4-1.**

Electrical Tape

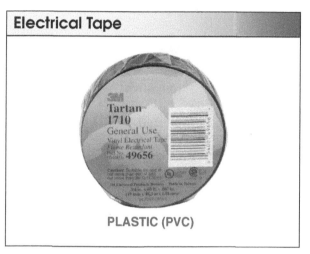

PLASTIC (PVC)

Figure 4-1. *Plastic electrical tape is used almost universally on residential wiring.*

Wall plates for switches and receptacles should be tested for levelness after installation.

Wire Nuts

Wire nuts are designed to hold several electrical wires firmly together and to provide an insulating cover for the connection. Wire nuts are available in several sizes. **See Figure 4-2.** The size of a wire nut is determined by the number and size of wires (conductors) to be connected. Manufacturers of wire nuts use color coding to indicate the maximum number of conductors allowed per connection, but the color coding scheme can vary by manufacturer.

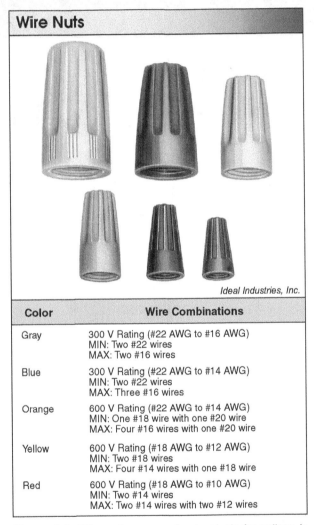

Wire Nuts

Ideal Industries, Inc.

Color	Wire Combinations
Gray	300 V Rating (#22 AWG to #16 AWG) MIN: Two #22 wires MAX: Two #16 wires
Blue	300 V Rating (#22 AWG to #14 AWG) MIN: Two #22 wires MAX: Three #16 wires
Orange	600 V Rating (#22 AWG to #14 AWG) MIN: One #18 wire with one #20 wire MAX: Four #16 wires with one #20 wire
Yellow	600 V Rating (#18 AWG to #12 AWG) MIN: Two #18 wires MAX: Four #14 wires with one #18 wire
Red	600 V Rating (#18 AWG to #10 AWG) MIN: Two #14 wires MAX: Two #14 wires with two #12 wires

Figure 4-2. *Wire nuts are used extensively for splices in outlet and junction boxes.*

A wire nut is a solderless plastic connector that uses a tapered metal coil spring to twist wires together. Wire nuts are typically used to connect wire sizes AWG No. 22 through No. 8.

Wire Markers

A *wire marker* is a preprinted peel-off sticker designed to adhere to insulation when wrapped around a conductor. Wire markers resist moisture, dirt, and oil and are used to identify conductors that have the same color but different uses. For example, the two hot black conductors (L1 and L2) of a home can each be marked with a different lettered and numbered wire marker. **See Figure 4-3.** Wire markers are still used even when different-colored conductors are used. Using wire markers in addition to color coding further clarifies the uses of all conductors.

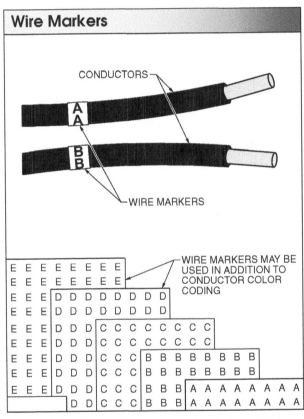

Wire Markers

Figure 4-3. *Wire markers are used to further clarify the meaning of a conductor in addition to color coding.*

Solder and the Soldering Process

Solder is an alloy consisting of specific percentages of two or more metals. In electrical work, an alloy usually consists of tin (Sn) and lead (Pb). Most lead-based solders are being replaced with lead-free (Pb-free) solders. *Soldering* is the process of making a sound electrical and mechanical joint between certain metals by joining them with solder. A good solder joint provides a metallurgical bond to metals.

A soldering iron tip transfers thermal energy from a solder station through its tip to the solder connection. The conductivity of a soldering iron tip determines how fast the thermal energy can be transferred from the heater to the solder joint. The shape and size of the soldering iron tip also affect its performance. Temperature is important, but it is not the major factor in determining the supply of heat. The length and size of the soldering iron tip determine heat flow capability. The actual shape of the soldering iron tip establishes how well heat is transferred from it to the connection.

Most metals react with oxygen when exposed to air, especially if the metal is heated. Both oxygen exposure and heat are present during soldering. Solder flux is formulated to remove a thin film of oxide from the metal and keep further oxidation from taking place. Solder flux is also used to make the solder and metal surfaces bond more easily with each other. During the soldering process, heat from the moving soldering iron tip liquefies the solder and simultaneously vaporizes the flux to remove oxide film. **See Figure 4-4.**

Soldering Iron Tips

Good soldering skills are essential for repairing PC boards. Careless work creates unnecessary damage. When soldering, the soldering iron tip must heat the metal to solder-melting temperature before actual soldering can take place. The tip should be held directly against one side of the component lead. **See Figure 4-5.** Solder should be applied from the opposite side. The solder-melting temperature is reached in a matter of 1 sec to 5 sec. Therefore, the soldering iron tip and the solder should be applied simultaneously.

CAUTION: Be sure to apply solder to the opposite side of the component lead, not directly to the soldering iron tip. If the soldering procedure is performed correctly, the solder will flow across the joint, firmly bonding the component lead to the PC board.

Selecting Soldering Iron Tips. Selecting the proper tip will improve heat transfer to the soldering application. The correct tip should have a similar shape and dimension to the object being soldered. Flat tips produce a larger contact area with a connection than do conical tips. Flat tips also tend to transfer heat more efficiently. The tip width must match the width of the object being soldered. A wide range of tip sizes and shapes are available for use with multiple applications. **See Figure 4-6.**

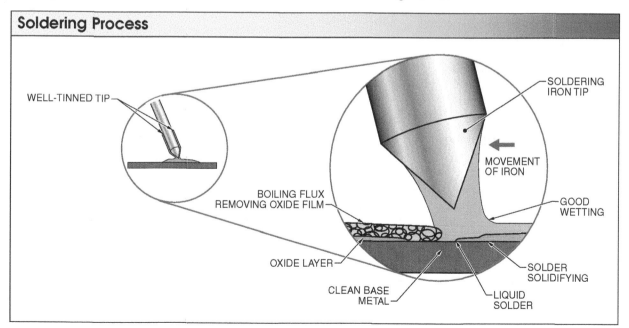

Soldering Process

WELL-TINNED TIP

SOLDERING IRON TIP

MOVEMENT OF IRON

BOILING FLUX REMOVING OXIDE FILM

GOOD WETTING

OXIDE LAYER

CLEAN BASE METAL

LIQUID SOLDER

SOLDER SOLIDIFYING

Figure 4-4. *During the soldering process, heat from the moving soldering iron tip liquefies solder and simultaneously vaporizes flux to remove oxide film.*

Soldering Application

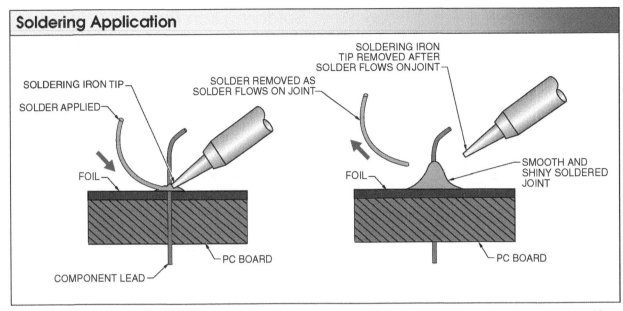

Figure 4-5. *A soldering iron tip should be held to one side of a component lead while solder is applied to the opposite side.*

Soldering Tips

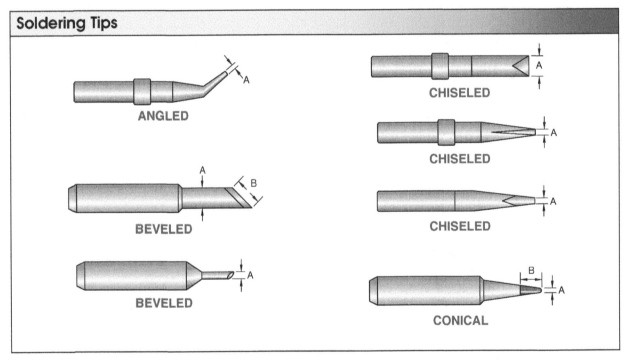

Figure 4-6. *A wide range of tip sizes and shapes are available for use with multiple applications. The tip of the soldering iron transfers heat from the soldering station to the work. The shape and size of the tip are mainly determined by the type of work to be performed.*

Care of Soldering Iron Tips. One of the most common causes of tip failure is the loss of the protective layer of solder, which causes the tip working surface to become oxidized. This is commonly referred to as a de-tinned tip. A de-tinned tip will minimize the ability of the tip to accept solder. Without a properly tinned tip, it is virtually impossible to transfer sufficient heat to metals to be joined. **See Figure 4-7.**

Tinned Soldering Tips

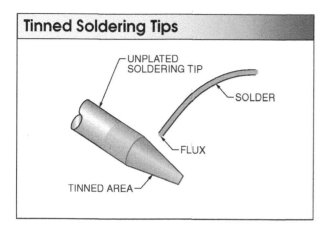

Figure 4-7. *Without a properly tinned soldering tip, it is virtually impossible to transfer sufficient heat to metals to be joined.*

An oxide tip will also lose its wettability and cannot supply heat to some of the part being soldered. Wetting occurs when solder covers the tip and acts as a heat transfer medium. Wetting means the molten solder leaves a continuous permanent film on the metal surface. Wetting can only be done properly on the clean surface of a tip. If the tip does not have good wettability, it will only contact the part over a tiny area and will not be able to transfer heat efficiently.

All dirt and grease must be removed from the metal surface of the tip. Using light abrasives and/or fluxes to remove these contaminants produces highly solderable and wettable surfaces. Soldering tips may be cleaned with a sponge, tip tinner, or polishing bar. **See Figure 4-8.**

Cleaning with Sponges. Sponges with holes can provide more surface area and allow contaminants to fall to the bottom of a sponge tray. The best method of dampening a sponge is to expose the entire sponge to the water source and then squeezing to remove excess water. It should be damp but not dripping wet. Synthetic sponges are not recommended for soldering tips because they may include elements or contaminants that could reduce the life of the tip.

Cleaning with Tip Tinners. A tip tinner is used to remove light oxidation from soldering tips. Cleaning with a tip tinner is necessary when a sponge does not work to remove oxidation. A tip is rotated in the tip tinner until a bright tinning appears on the

surface of the tip. Solder should be applied immediately to re-tin the surface of the tip. Tip tinner should not be overused because it will reduce the useful life of the tip.

Cleaning Soldering Tips

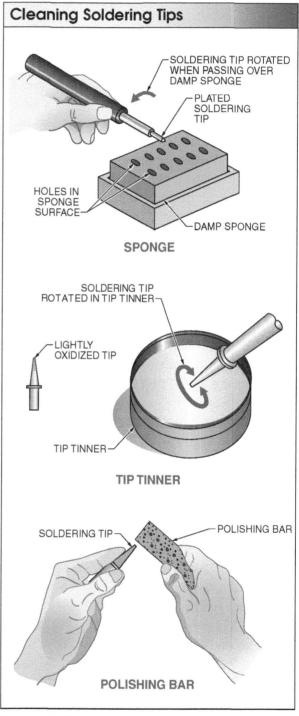

Figure 4-8. *Soldering tips may be cleaned with a sponge, tip tinner, or polishing bar.*

Cleaning with Polishing Bars. When heavy oxidation cannot be removed by using a tip tinner, cleaning the tip may require the use of an abrasive polishing bar. The polishing bar is applied to the soldering iron when cool, not hot. Care must be taken not to remove the iron plating. Once cleaned, the tip should be re-tinned immediately.

Lead Solders

The two most popular lead solder alloys used for electronic connections are Sn63/Pb37 and Sn60/Pb40. The Sn63/Pb37 alloy is considered to be a eutectic alloy. A *eutectic alloy* is an alloy that has one specific melting temperature with no intermediate stage. In other words, solder goes directly from a solid to a liquid with no "plastic" or "elastic" phase in between. For example, lead has a melting point of 621°F (327°C), and tin melts at 450°F (232°C). Alloying 63% tin with 37% lead forms a eutectic alloy that melts at 374°F (190°C). *Note:* The melting point of this particular alloy is lower than either of the parent metals.

The Sn60/Pb40 alloy is a relatively inexpensive general-purpose alloy that has a very small plastic regional range of approximately 8°F. It also has a low melting point of 374°F (190°C).

These alloys allow rapid solder time that can prevent excessive temperature from being applied to a component. They also help prevent unreliable solder joints. Soldering must be focused on the initial solder connection.

A solder connection is an alloy that combines metallurgically with the surface of another metal to form an intermetallic layer. The intermetallic layer forms the electrical and mechanical connections. Any joint that exhibits a high resistance will adversely affect the operation of a circuit.

Solder joints that are overheated generally have a dull, bumpy exterior. A good lead solder joint is smooth and shiny. **See Figure 4-9.** Overheated joints are often mistakenly called "cold solder joints." A *cold solder joint* is a defective solder joint that results when the parts being joined do not exceed the liquid temperature of the solder. A cold solder joint can be identified by jagged shapes on the surface of

a joint. Overheated and cold solder joints indicate a lack of proper wetting techniques when solder was applied. Jagged, irregular shapes of a solder joint are an indication that the solder did not flow properly within the joint.

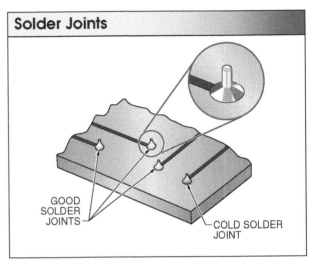

Solder Joints

GOOD SOLDER JOINTS

COLD SOLDER JOINT

Figure 4-9. A good solder joint is smooth and shiny.

Eutectic tin-lead solders Sn63/Pb37 and Sn60/Pb40 have been in use for a long time. Tin-lead solder has many advantages including a melting point of 374°F (190°C). This temperature is high enough to remain solid in normal applications but still low enough to not seriously damage a PC board and its components. Tin-lead solder also has a fairly long shelf life when properly stored. Even though there are advantages to using lead-based solder, modern concerns over lead exposure in the environment have forced the industry to seek lead-free alternatives.

Avoiding Bad Solder Joints

Bad solder joints result in improper connections of components and can lead to damaged components and circuits. Therefore, efforts should be made to prevent bad solder joints. To achieve a proper solder joint, the following conditions should be avoided:

- too little solder on a joint, which produces a weak joint and causes a component to work loose over time
- too much solder on a connection, which could bridge component leads and cause unwanted conducting paths

- too little heat while soldering, which produces a cold solder joint and results in a poor connection
- too much heat while soldering, which may damage a component
- no flux or too little flux while soldering, which does not allow all contaminants to be removed and can result in a poor connection

Soldering Safety

Tests have shown that soldering with lead and lead-free alloys creates a level of breathable airborne pollutants that exceeds the safety threshold set by OSHA. In particular, lead soldering may generate airborne lead oxide and fumes from colophony.
Lead oxide is produced from lead-based solder, and colophony is produced from solder flux.

Because exposure to these fumes may be a health hazard, suitable measures to protect health must be carried out. When reasonable, exposure should be avoided. If that is not practical, then the process should be adequately controlled using fume extractors and proper ventilation techniques.

It is necessary to follow instructions for safe working practices provided by the manufacturer, including the correct use and adjustment of control measures such as local extraction ventilation. For safe soldering, it is important to observe the following rules:

- A soldering iron should always be returned to its stand when not in use and should never be set down on a workbench.
- Soldering should be done in a well-ventilated area. The smoke formed while soldering is mostly created from the flux and can be toxic or irritating. Keeping the head to the side of the work will prevent smoke inhalation.
- Hands should be washed after handling solder. Solder may contain lead, which is a poisonous metal.
- Safety glasses should be worn for protection from splatter and lead clippings.

While newly built residential structures have the most modern types of wire connectors installed, older structures may have terminations that occasionally require the use of soldering tools to repair soldered connections.

SPLICES

A *splice* is the joining of two or more electrical conductors by mechanically twisting the conductors together or by using a special splicing device. Individuals must be careful when making splices because splices can cause electrical problems. Splices must be able to withstand any reasonable mechanical strain that might be placed on the connection. Electrical splices must also allow electricity to pass through the connection as if the connection were one wire.

Wire amperage is a measurement of how much current a wire can carry safely. Ampacity varies according to the size of the wire. When wiring a circuit, wiring that has amperage ratings that exceed circuit demands must be installed. For dedicated appliance circuits, the wattage rating of the appliance should be checked to ensure the appliance does not exceed the maximum wattage rating of the circuit. **See Figure 4-10.**

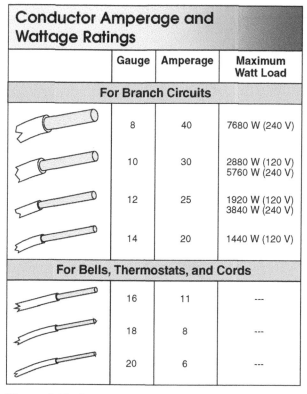

Conductor Amperage and Wattage Ratings			
	Gauge	Amperage	Maximum Watt Load
For Branch Circuits			
	8	40	7680 W (240 V)
	10	30	2880 W (120 V) / 5760 W (240 V)
	12	25	1920 W (120 V) / 3840 W (240 V)
	14	20	1440 W (120 V)
For Bells, Thermostats, and Cords			
	16	11	---
	18	8	---
	20	6	---

Figure 4-10. *The amperage and wattage ratings of conductors vary with the AWG size of the wire.*

Types of Splices

Many types of wire splicing can be found in a residence. Typical residential wire splices are pigtail, Western Union, T-tap, portable cord, and cable.

Pigtail Splices. Because a pigtail splice is simple to make, it is the most commonly used electrical splice. **See Figure 4-11.** *Note:* There are two ways to end pigtail splices: bent over and cut off.

When a pigtail splice is taped, the ends must be bent over so the sharp wire points do not penetrate the tape. When wire nuts are used instead of tape, the ends of the conductors are cut off.

When more than two wires are joined by a pigtail splice, all wires must be twisted together securely before the wire nut is installed. **See Figure 4-12.** Twisting the wires together insures that all the wires are properly fastened before installing a wire nut.

Pigtail Splices

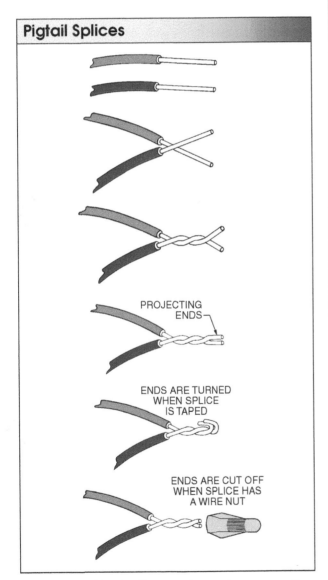

PROJECTING ENDS

ENDS ARE TURNED WHEN SPLICE IS TAPED

ENDS ARE CUT OFF WHEN SPLICE HAS A WIRE NUT

Figure 4-11. *Pigtail splices are the most common splices used in residential wiring.*

Multiple-Wire Pigtail Splices

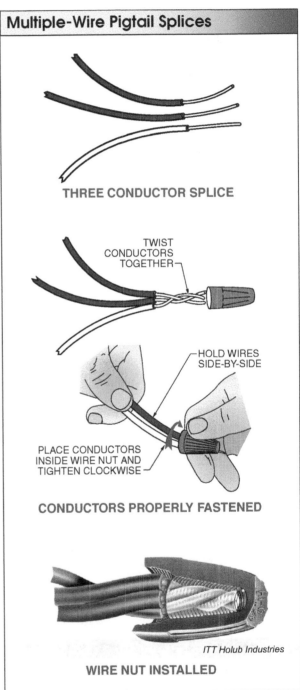

THREE CONDUCTOR SPLICE

TWIST CONDUCTORS TOGETHER

HOLD WIRES SIDE-BY-SIDE

PLACE CONDUCTORS INSIDE WIRE NUT AND TIGHTEN CLOCKWISE

CONDUCTORS PROPERLY FASTENED

ITT Holub Industries

WIRE NUT INSTALLED

Figure 4-12. *Multiple-wire pigtail splices require that the wires be twisted together prior to the wire nut being installed.*

Western Union Splices. A *Western Union splice* is a type of splice that is used when the connection must be strong enough to support long lengths of heavy wire. **See Figure 4-13.** When a splice is to be taped, care must be taken to eliminate any sharp edges from the wire ends. In the 1800s, Western Union splices were used to repair telegraph wires.

Western Union Splices

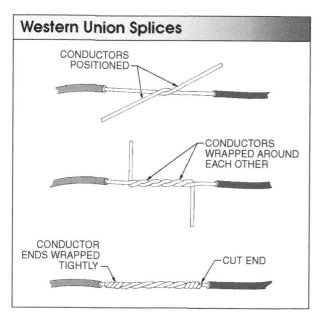

Figure 4-13. *Western Union splices are used where considerable strain is placed on the connection.*

T-Tap Splices. A T-tap splice is a type of splice that allows a connection to be made without cutting the main wire. **See Figure 4-14.** A T-tap splice is one of the most difficult splices to perform correctly. A certain amount of practice is required to make a T-tap connection look neat. Good technique must be used to ensure proper T-tap splicing.

Prior to splicing or connecting conductors together, refer to National Electrical Code® (NEC®), Requirements for Electrical Installations. Several different methods are listed, including brazing, soldering, welding with a fusible metal alloy, and connecting with a device identified for electrical use such as a wire nut. The NEC® also covers terminal types, temperature limitations, equipment provisions, and separate connector provisions for electrical splices.

Portable Cord Splices. Portable cord splices are a weak type of splice because there is no connector to hold the conductors together. **See Figure 4-15.**

CAUTION

Portable cord splices should be used for emergency purposes only.

Portable cords with stranded wires or solid wires can be spliced if the conductors are 14 AWG or larger. Electrified tape cannot be used so the splice must retain the insulation and outer covering properties of the portable cord.

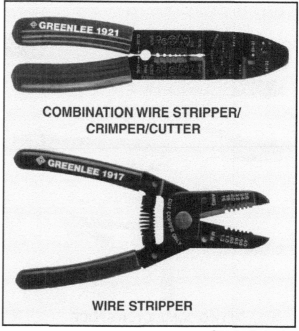

COMBINATION WIRE STRIPPER/ CRIMPER/CUTTER

WIRE STRIPPER

Greenlee Textron Inc.

Wire strippers and stripper/crimper/cutter tools are used to remove insulation from electrical wires, crimp terminals, and shear small-diameter bolts.

Cable Splices. Large stranded cables are not often used in residential wiring. However, cable splices are used in other applications such as for battery jumper cables and welding cables. **See Figure 4-16.** When jumper cables or welding cables are broken, the cables can be temporarily repaired by using cable splicing.

T-Tap Splices

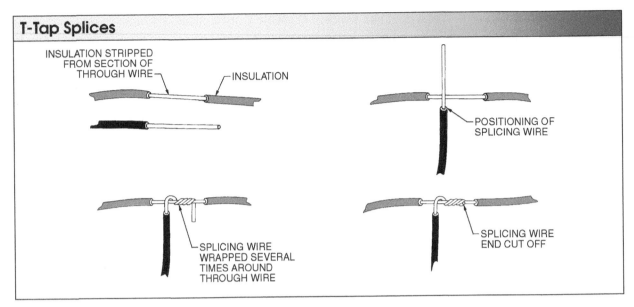

Figure 4-14. *T-tap splices are used to splice a conductor to a through wire.*

Portable Cord Splices

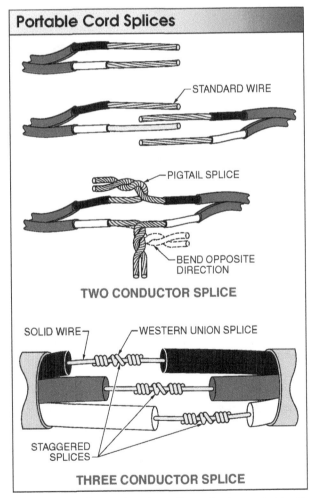

Figure 4-15. *Staggering the splices of a large sized portable cord reduces the bump created.*

Taping Splices

Taping is required to protect splices from oxidation (formation of rust) and to insulate individuals against electrical shock. Taping must provide at least as much insulative and mechanical protection for the splice as the original insulation. Although one wrap of tape (plastic or vinyl) provides insulative protection up to 600 V, several wraps are required to provide a strong mechanical protection. **See Figure 4-17.** When plastic tape is used, the tape is stretched as it is applied. Stretching secures the tape more firmly.

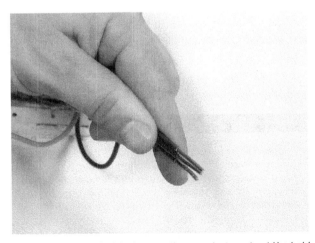

To make a proper electrical connection, conductors should be held as close together as possible.

Cable Splices

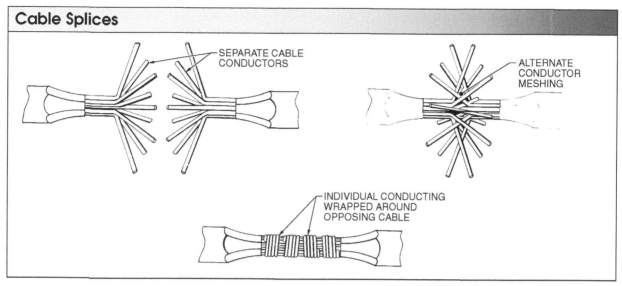

Figure 4-16. *Large stranded cables must be carefully spliced and firmly secured using cable splicing procedures.*

Taping Splices

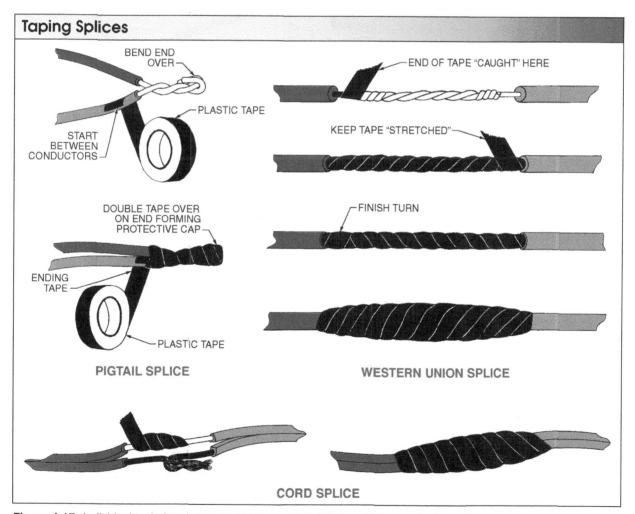

Figure 4-17. *Individual techniques are used to tape pigtail, Western Union, and cable splices.*

Soldering Splices

Once the decision to solder an electrical splice has been made and the insulation has been stripped off the wire, the splice should be soldered as soon as possible. The longer a metal conductor is exposed to dirt and air, the greater the oxidation present on the wire, and the less chance an individual has of achieving a properly soldered connection. Clean metal surfaces are required to allow molten solder to flow freely around the connection.

The metal surfaces of the conductors are cleaned with light sandpaper, emery cloth, or by applying flux to the surfaces. Solder typically is in wire form and is melted with heat from a soldering device such as a soldering iron, soldering gun, or propane torch. **See Figure 4-18.**

is used, individuals must apply solder to the splice on the side opposite the point where heat is being applied. **See Figure 4-19.** Melting solder flows toward the source of heat. Thus, if the top of the wire is hot enough to melt the solder, the bottom of the wire closest to the heat source will draw the solder down through all the wires. The splice must be allowed to cool naturally without movement. Once cooled, the splice is cleaned of any excess flux with a damp rag and taped.

> ## CAUTION
>
> *Do not blow on or dip the soldered connection in water to cool. Rapid cooling reduces the strength of solder connections.*

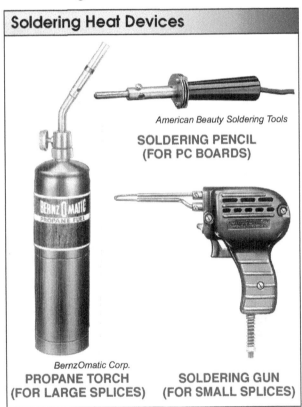

Soldering Heat Devices

American Beauty Soldering Tools

SOLDERING PENCIL (FOR PC BOARDS)

BernzOmatic Corp.

PROPANE TORCH (FOR LARGE SPLICES)   **SOLDERING GUN (FOR SMALL SPLICES)**

Figure 4-18. *Heat for soldering splices is provided by soldering irons, soldering guns, or propane torches.*

Electric soldering irons and soldering guns are used when electricity is available. A propane torch is used to solder large wires or when there is no electricity at the job site. Whatever heating method

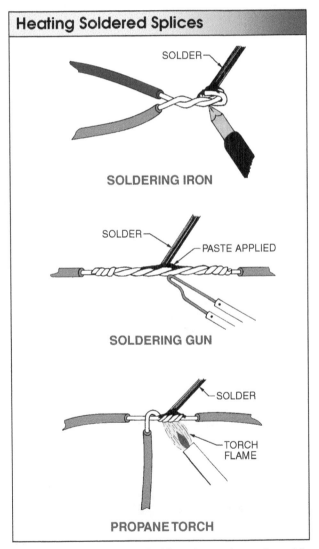

Heating Soldered Splices

SOLDER

SOLDERING IRON

SOLDER — PASTE APPLIED

SOLDERING GUN

SOLDER

TORCH FLAME

PROPANE TORCH

Figure 4-19. *Heat is applied from beneath a splice while solder is applied from above.*

Because solderless connectors save time and are easy to use, an individual is no longer required to solder each and every splice. Solderless connections not only take less time to make connections, but also require less skill. However, individuals should understand how to solder. No matter what type of splice is being used, all splices must be accessible and made in an approved junction box.

SOLDERLESS CONNECTORS

A *solderless connector* is a device used to join wires firmly without the help of solder. Because solderless connectors are convenient and save time, several types have been developed. Some of the most commonly used solderless connectors approved by Underwriters Laboratories® are split-bolt connections, screw terminals, back-wired connectors, wire nuts, and crimps.

Split-Bolt Connectors

A *split-bolt connector* is a type of solderless mechanical connection used for joining large cables such as in service entrances. A split bolt is slipped over the wires to be connected so that a nut can be attached. Split-bolt connectors must be made of the same material as conductors to prevent corrosion. **See Figure 4-20.**

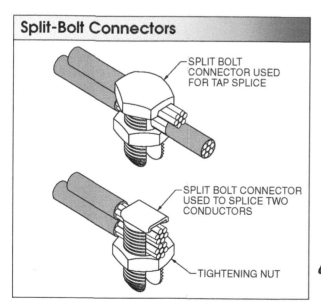

Figure 4-20. *Split-bolt connectors are used to splice large cables.*

Screw Terminals. Screw terminals provide a mechanically and electrically secure connection. Since wiring is always attached to electrical equipment with right-hand screws, the wire should be bent around a screw in a clockwise direction. **See Figure 4-21.** When the screw method of wire connections is used, the screw draws the wire tight without pushing the wire away from the terminal.

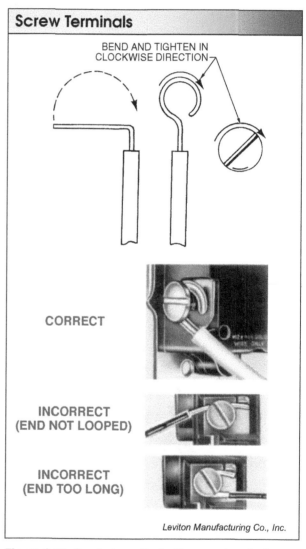

Leviton Manufacturing Co., Inc.

Figure 4-21. *Conductors attached to screw terminals must be bent so that the wire can be tightened in a clockwise direction.*

⚠	DANGER

To prevent electrical shock, never work on energized equipment when connecting conductors to screw terminals.

Back-Wired (Quick) Connectors

A *back-wired connector,* also known as a quick connector, is a mechanical connection method used to secure wires to the backs of switches and receptacles. The wires are held in place by either spring or screw tension, with screw tension being the most secure. **See Figure 4-22.**

Wire Nut Installation

Wire nuts have almost eliminated the need for soldering and taping. Wire nuts are manufactured in a variety of sizes and shapes. For the most effective use of wire nuts, the wires are twisted together clockwise before the wire nut is twisted on and taped. **See Figure 4-23.** When solid and stranded wires are joined together, the solid wire is bent back over the stranded wire. **See Figure 4-24.**

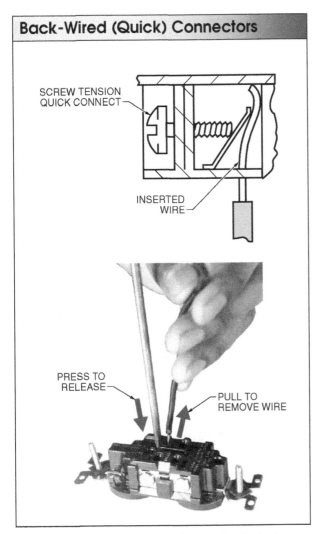

Figure 4-22. *The conductor attached to a back-wired receptacle is held in place by spring tension or screw tension.*

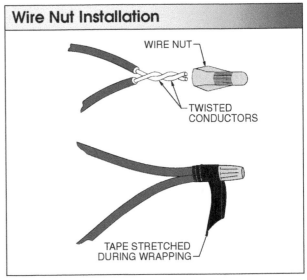

Figure 4-23. *To ensure a proper connection, conductors are twisted together before the wire nut and tape are applied.*

A voltage indicator can be used to check if a receptacle is energized.

To remove a wire from a spring-type quick connector, individuals must insert a screwdriver or a stiff piece of wire into the spring opening next to the connection. Pressing down the spring through the opening releases the wire. Screw-type connectors release a wire as the screw is loosened.

Connecting Stranded and Solid Wires

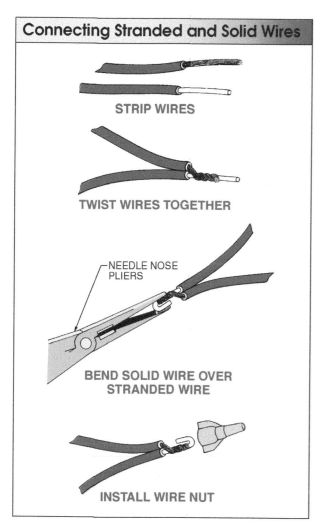

STRIP WIRES

TWIST WIRES TOGETHER

NEEDLE NOSE PLIERS

BEND SOLID WIRE OVER STRANDED WIRE

INSTALL WIRE NUT

Figure 4-24. *The technique used to connect solid and stranded wires together is used to ensure a mechanically and electrically sound connection.*

The size of wire nuts is determined by the size and number of wires to be connected. Shape is an individual preference and is determined by the manufacturer. To ensure safe connections, every wire nut is rated for minimum and maximum wire capacity. Wire nuts are used to connect both conducting wires and grounding wires. Green wire nuts are used only for grounding wires.

Crimp-Type Solderless Connectors

A *crimp-type solderless connector* is an electrical device that is used to join wires together or to serve as terminal ends for screw connections. **See Figure 4-25.** Crimp connectors are manufactured as insulated or noninsulated. Noninsulated crimp connectors are less expensive and are used where there is no danger of shorting the connector to a metal surface. When working with crimp-type solderless connectors, identification can be a problem. To avoid confusion, a wire marker is used to identify each conductor.

Crimp Connector Types

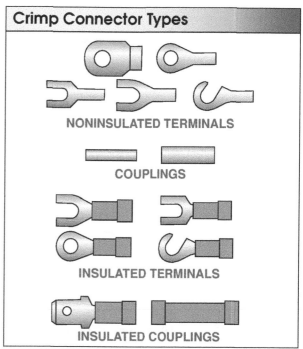

NONINSULATED TERMINALS

COUPLINGS

INSULATED TERMINALS

INSULATED COUPLINGS

Figure 4-25. *Crimp-type solderless connectors are designed for a variety of applications.*

Electrical connector categories are based on shape, function, and smallest size contact in the series. Over 1000 companies worldwide manufacture electrical connectors.

Crimp connectors are secured into place with a multipurpose crimp tool. Crimping tools are also used to cut and strip wire. **See Figure 4-26.**

SOLDERING CONNECTIONS

Soldering is the process of joining metals by using filler metal and heat to make a strong electrical and mechanical connection. The parts to be soldered must be clean, fluxed, and hot enough to melt solder for the solder to properly adhere.

Installing Crimp Connectors

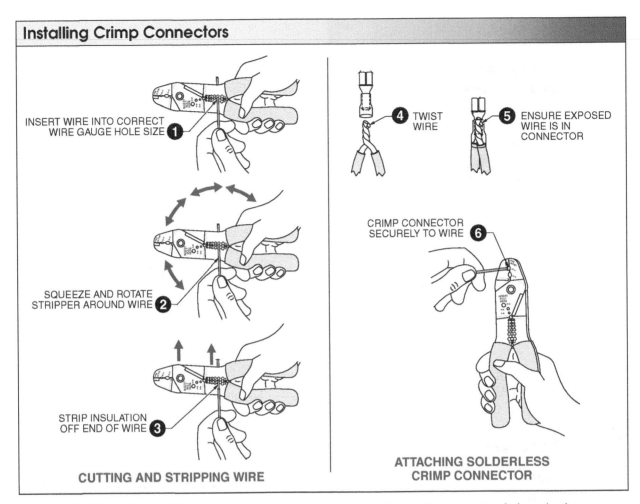

INSERT WIRE INTO CORRECT WIRE GAUGE HOLE SIZE **1**

SQUEEZE AND ROTATE STRIPPER AROUND WIRE **2**

STRIP INSULATION OFF END OF WIRE **3**

CUTTING AND STRIPPING WIRE

4 TWIST WIRE

5 ENSURE EXPOSED WIRE IS IN CONNECTOR

CRIMP CONNECTOR SECURELY TO WIRE **6**

ATTACHING SOLDERLESS CRIMP CONNECTOR

Figure 4-26. *The male stub of the crimping tool is placed on the seam of the crimp connector during crimping.*

In addition to surface dirt, oil, and corrosion, the metal surfaces must be cleaned of all surface oxides. Oxide is formed on metal surfaces by oxygen in the air. For example, when copper is exposed to air long enough, the oxide appears as a green tarnish. Flux is used with solder to remove surface oxides. **See Figure 4-27.** Flux removes the oxide by making the oxide soluble and evaporating the oxide as the flux boils off during heating.

A soldered connection must be mechanically strong before soldering. Heat is applied until the materials are hot, allowing solder to be applied.

Solder must be melted by the heat of the material to be joined, not by the heat of the soldering tool.

Only a small amount of solder should be used. The connection should appear smooth and shiny. **See Figure 4-28.** When the solder appears dull and crackly, the connection is a cold solder joint. A cold solder joint is a soldered joint with poor electrical and mechanical properties. Cold solder joints are typically caused by insufficient heat during soldering or the parts being moved after the solder is applied but before cooling has finished.

Care must be taken not to damage surrounding parts by overheating them. Semiconductor components, such as transistors and ICs, are very sensitive to heat. A heat sink, such as an alligator clip, is used to help prevent heat damage. The heat sink is placed between the soldered connection point and component that requires protection. The heat sink absorbs the heat produced during soldering. **See Figure 4-29.**

Solder

Solder Wire Diameters*		
0.015	0.032	0.093
0.020	0.040	0.125
0.025	0.062	

* in inches

1 LB SPOOL

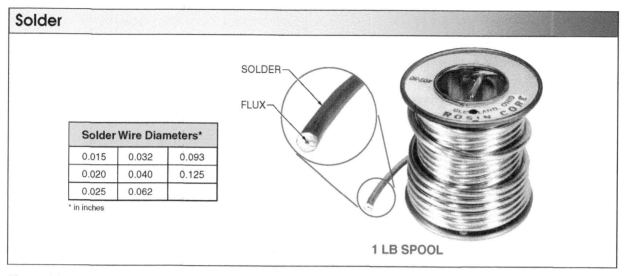

Figure 4-27. *When the solder being used is solid, flux is purchased in a separate container to be brushed on to the connection for soldering.*

Solder Appearance

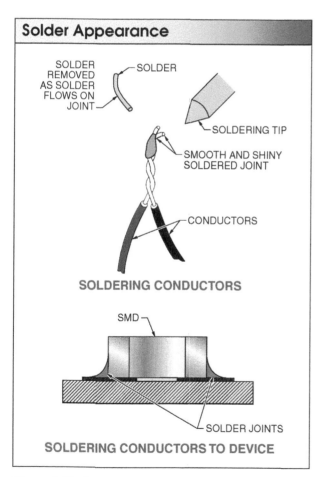

SOLDER REMOVED AS SOLDER FLOWS ON JOINT

SOLDER

SOLDERING TIP

SMOOTH AND SHINY SOLDERED JOINT

CONDUCTORS

SOLDERING CONDUCTORS

SMD

SOLDER JOINTS

SOLDERING CONDUCTORS TO DEVICE

Figure 4-28. *A properly soldered connection looks smooth and shiny while a cold solder connection has a dull grainy texture.*

Heat Sinks

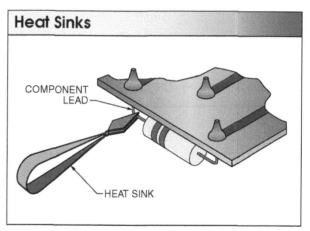

COMPONENT LEAD

HEAT SINK

Figure 4-29. *A heat sink applied to a component lead protects the component from heat damage.*

REPLACING PLUGS

A *plug* is a device at the end of a cord that connects the device to an electrical power supply by means of a receptacle. Plugs sometimes crack, break, or lose their electrical connection when the device is disconnected from the receptacle by pulling on the cord instead of the plug. The plug should be replaced when it goes bad. A broken plug should never be repaired. If the cord is also broken or damaged, both the cord and the plug should be replaced.

Standard Grounded Plugs

Ungrounded equipment can cause shock or fires. Grounded plugs should always be used if included with the original equipment. If a grounded plug is not used, or the ground wire is not connected, the load will work but the circuit will not be safe.

A damaged grounded plug should always be replaced with a new grounded plug. **See Figure 4-30.** To replace a grounded plug, remove the old plug, and apply the following procedure:

1. Insert the cord into the plug cover and pull about 8″ of cord through the plug.
2. Cut the plastic sheathing away from about 6″ of cord. Peel back and cut off any paper wrapping. Be careful not to cut into the insulation around the individual wires.
3. Strip about ¾″ insulation from the wire.
4. Insert wires into the wire pockets. The hot terminal is identified by a bronze screw, the neutral by a silver screw, and the ground terminal by a green screw.
5. Tighten the terminal screws. Care must be taken to ensure that no strands are exposed because most cords use stranded wire. Solder the stranded ends if required.
6. Insert the plug into the cover, and tighten the assembly screw.

Appliance Plugs

An *appliance plug,* also known as a heater plug, is a plug used to power an appliance that produces heat, such as an electric grill, roaster, broiler, waffle iron, or large coffeemaker. Appliance plugs fail when the cord is pulled or overheated. Appliance plugs and their cords should be replaced with new ones if they are bad. **See Figure 4-31.** To replace an appliance (heater) plug, remove the old plug, and apply the following procedure:

1. Strip about ¾″ insulation from the wire and twist the stranded ends into a loop. Do not solder because later, while the appliance is in use, the plug may get hot enough to melt the solder.
2. Wrap the wire around the terminal screws and tighten terminal screws.
3. Screw plug halves together.

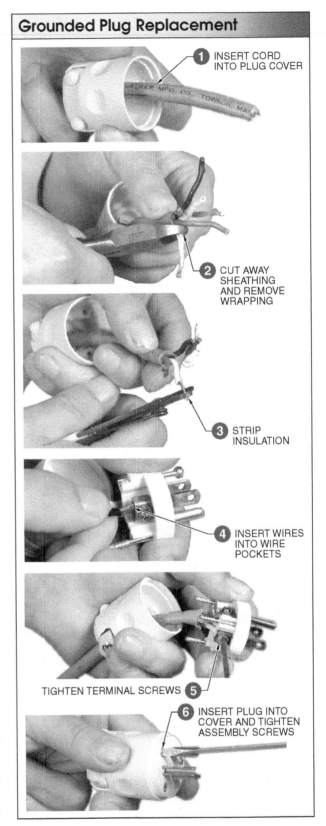

Grounded Plug Replacement

1. INSERT CORD INTO PLUG COVER
2. CUT AWAY SHEATHING AND REMOVE WRAPPING
3. STRIP INSULATION
4. INSERT WIRES INTO WIRE POCKETS
5. TIGHTEN TERMINAL SCREWS
6. INSERT PLUG INTO COVER AND TIGHTEN ASSEMBLY SCREWS

Figure 4-30. *A damaged grounded plug should always be replaced with a new grounded plug.*

Appliance Plug Replacement

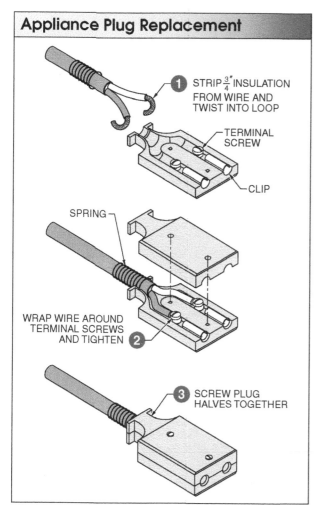

1 STRIP $\frac{3}{4}$" INSULATION FROM WIRE AND TWIST INTO LOOP

TERMINAL SCREW

CLIP

SPRING

WRAP WIRE AROUND TERMINAL SCREWS AND TIGHTEN **2**

3 SCREW PLUG HALVES TOGETHER

Figure 4-31. *Appliance plugs fail and must be replaced when the cord is pulled or overheated.*

LARGE APPLIANCE CONNECTORS

Heavy electrical loads, such as those found on washers, dryers, and ovens require large 120/220 V appliance connectors and receptacles to adequately secure the large conductors used for such equipment. A 120/220 V appliance plug is typically fed by two colored conductors for the 220 V connection and a neutral conductor for the 110 V connection. This is because the heating element of such equipment may require 220 V power, while the lights and controls may require 110 V power.

For 120/220 V loads, an appropriate receptacle with a configuration that matches the plug is also required. Typically, there is no grounding slot. Grounding occurs from a grounding wire in the cable to the outlet box and to the service panel. Receptacles can be box-mounted or surface-mounted. **See Figure 4-32.**

120/220 V Connection Devices

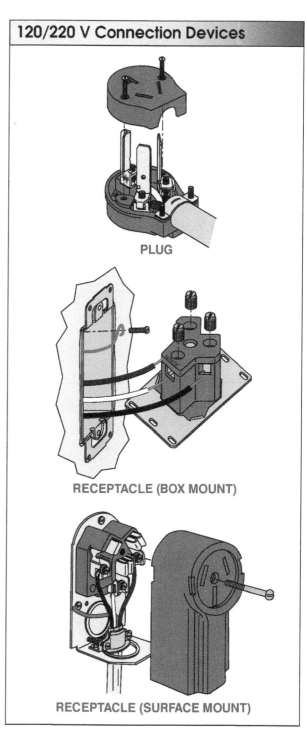

PLUG

RECEPTACLE (BOX MOUNT)

RECEPTACLE (SURFACE MOUNT)

Figure 4-32. *Heavy electrical loads, such as those found on certain appliances, require large 120/220 V appliance connectors and receptacles to adequately secure the large conductors used for such equipment.*

TESTING CONNECTIONS

Resistance measurements using a DMM are useful in isolating problems from loose connections, corrosion, or shorts. With all power OFF, the test leads of a DMM are connected across a connection, splice, load, or circuit. **See Figure 4-33.** The resistance of the connection or circuit being tested is displayed.

When a DMM is set to measure resistance, the measurement should not change with a proper connection. When a varying high-resistance measurement is recorded, there is probably a loose connection. For a connection with a high fixed-resistance value (150.0 Ω or greater) the measurement indicates there is probably a bad connection. A DMM can be left for long-term measurements as long as the batteries are in good condition (typically 1000 hr).

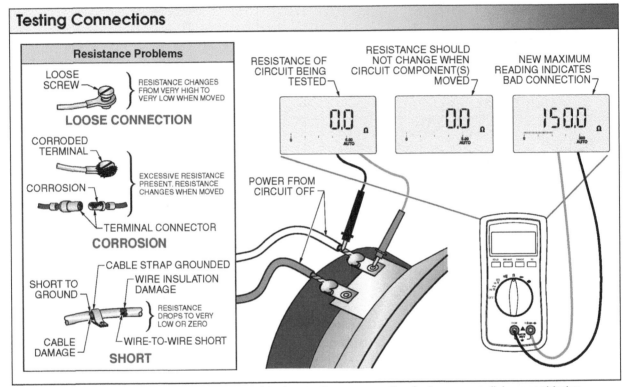

Figure 4-33. *Resistance measurements are useful when testing a connection for improper splicing or soldering.*

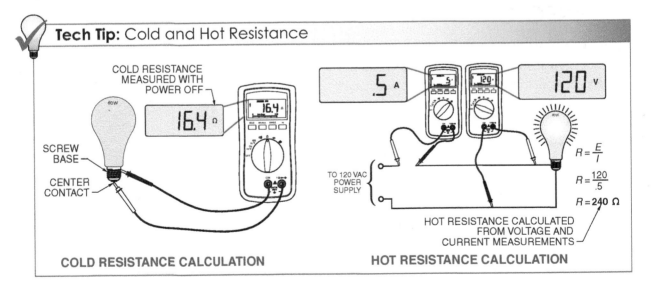

Making Electrical Connections 4

TEST

Name_____ Date _____

Making Electrical Connections

_____ **1.** Solder used in electrical work is usually an alloy of ___.
    A. 60% tin and 40% lead
    B. 60% lead and 40% tin
    C. 50% tin and 50% lead
    D. none of the above

_____ **2.** The amount of ___ in solder determines its melting point.

_____ **3.** The amount of ___ in solder determines its strength.

_____ **4.** Flux removes ___ and other small impurities from metal surfaces.

_____ **5.** ___ solder should never be used on electrical connections.

_____ **6.** The ___ splice is the most commonly used electrical splice.

T   F **7.** Splices are designed to stop the flow of electricity.

T   F **8.** Wire ends are cut off when a solderless connector is used for a splice.

_____ **9.** ___ may be added to solder to ensure proper solder adhesion.

_____ **10.** Plastic tape insulates against voltages up to ___ V.

_____ **11.** Melting solder flows toward the source of ___.

_____ **12.** Taping a splice protects the splice from ___.

_____ **13.** Taping a splice protects personnel against ___.

_____ **14.** A back-wired (quick) connector holds the wire in place by ___ or screw tension.

_____ **15.** Back-wired (quick) connectors are used to secure wires on the backs of switches and ___.

_____ **16.** ___ are used to join large cables together.

_____ 17. Wiring is always connected to electrical equipment with ___ screws.

T   F   18. When solid and stranded wires are spliced together, the solid wire should be bent back over the stranded wire.

T   F   19. Rapid cooling is preferred for developing strong solder joints.

T   F   20. Solder should always be applied on the side opposite from where the heat is applied.

_____ 21. Solder is commercially available in ___ form.
     A. liquid or bar
     B. wire or liquid
     C. bar or wire
     D. none of the above

_____ 22. Cord splices are ___.
     A. weak connections
     B. staggered
     C. soldered for additional strength
     D. all of the above

_____ 23. The ___ splice is strong enough to support long lengths of heavy wire.

_____ 24. ___ core solder contains no flux.

_____ 25. ___ core solder contains flux within the solder.

T   F   26. Solder joints should be dipped in water to stop the cooling process.

T   F   27. Wires on screw terminals should always bend and tighten in a clockwise direction.

T   F   28. Solderless connectors are used extensively to save time when making splices.

_____ 29. A(n) ___ splice may be used to connect into an uncut wire.

_____ 30. A(n) ___ may be used to heat solder joints if electricity is not available.

Taping Splices

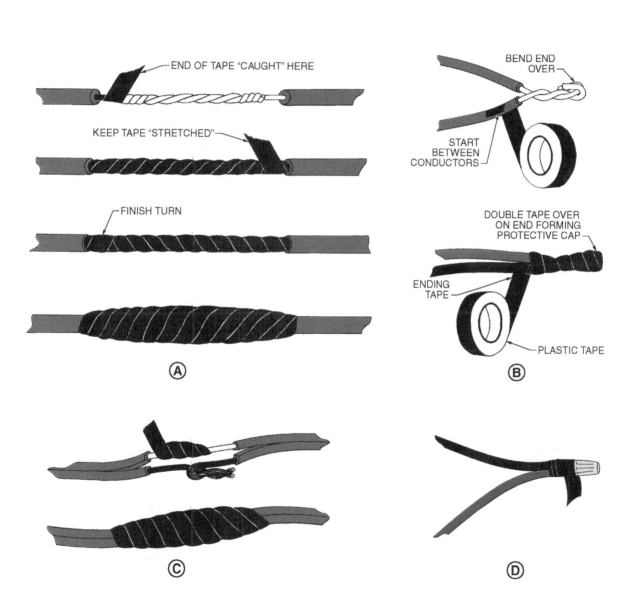

END OF TAPE "CAUGHT" HERE

KEEP TAPE "STRETCHED"

FINISH TURN

Ⓐ

BEND END OVER

START BETWEEN CONDUCTORS

DOUBLE TAPE OVER ON END FORMING PROTECTIVE CAP

ENDING TAPE

PLASTIC TAPE

Ⓑ

Ⓒ

Ⓓ

Pigtail Splice with Wire Nut

_____ **1.** ___ is the first step.

_____ **2.** ___ is the second step.

_____ **3.** ___ is the third step.

_____ **4.** ___ is the fourth step.

_____ **5.** ___ is the last step.

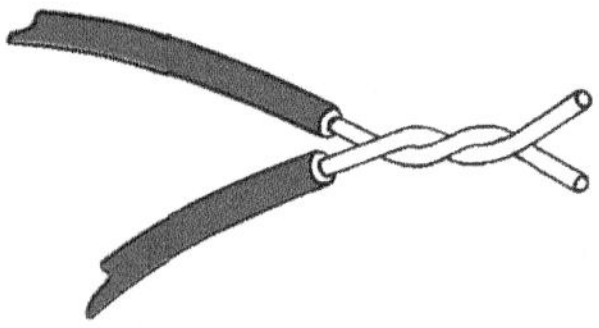

Ⓐ

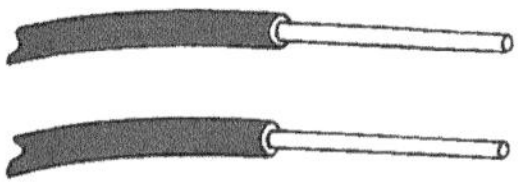

Ⓑ

Ⓒ

Ⓓ

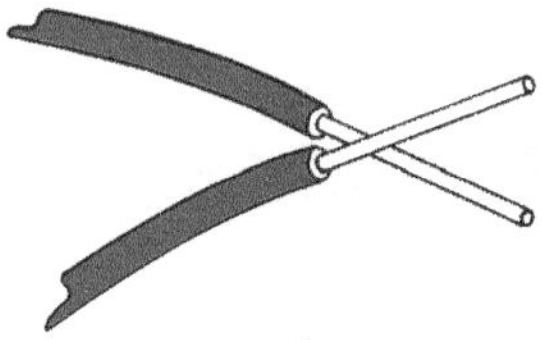

Ⓔ

Pigtail Splice with Tape

_____ **1.** ___ is the first step.

_____ **2.** ___ is the second step.

_____ **3.** ___ is the third step.

_____ **4.** ___ is the fourth step.

_____ **5.** ___ is the fifth step.

_____ **6.** ___ is the last step.

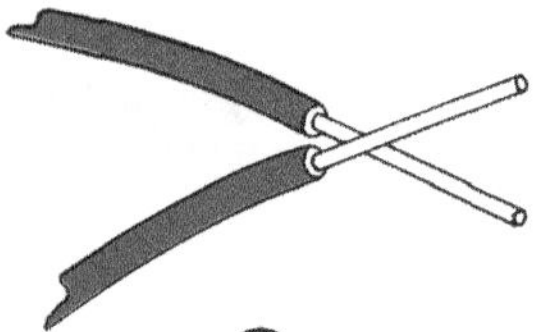

Ⓐ

Ⓑ

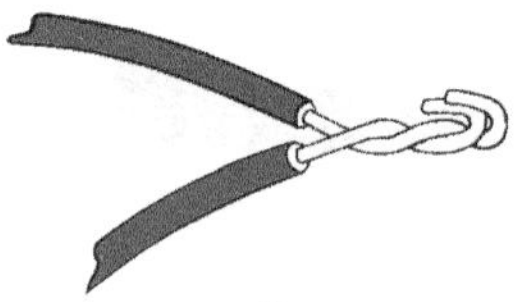

Ⓒ

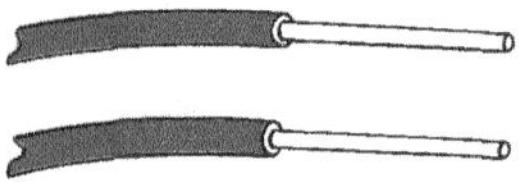

Ⓓ

Ⓔ

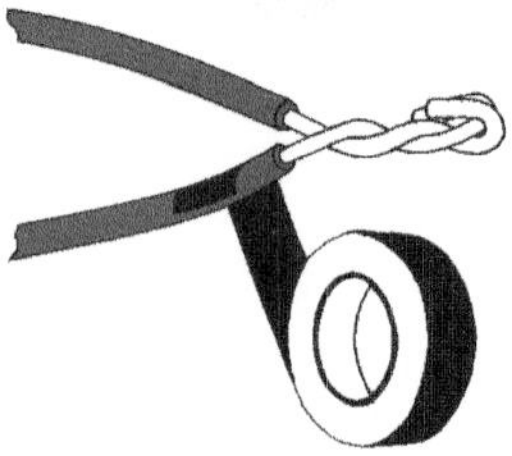

Ⓕ

106

Screw Terminals

_____   **1.** Correct

_____   **2.** Wrong; end not looped

_____   **3.** Wrong; end too long

_____   **4.** Wrong; end too short

_____   **5.** Wrong; wire in counterclockwise direction

_____   **6.** Wrong; too much insulation stripped off wire

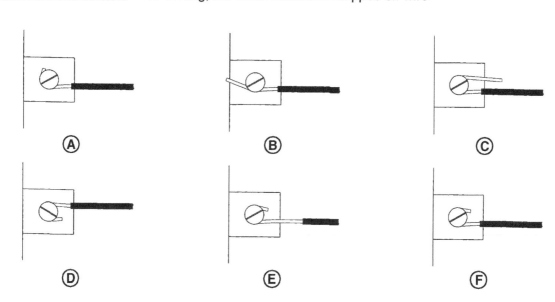

Electrical Tools and Materials

_____ **1.** Tightens clockwise

_____ **2.** Cuts bolts and wire and removes insulation

_____ **3.** Repairs large stranded cables

_____ **4.** Twists wires and insulates

_____ **5.** Common alloy of tin and lead used in solder

_____ **6.** Connection without cutting the main line

_____ **7.** Used to apply heat

_____ **8.** Should be stretched

_____ **9.** Used in emergency only

_____ **10.** Released with a screwdriver or stiff wire

A. Solderless connectors

B. T-tap

C. Soldering iron

D. Tape

E. Portable cord splice

F. Screw terminal

G. Spring-type (quick) connectors

H. Stripper/crimping tool

I. 60/40

J. Cable splice

Switches and Receptacles 5

Important rating information is printed or embossed on switches and receptacles that must be understood prior to use. Chapter 5 provides an overview of the types of switches and receptacles that are available for residential use. Commonly used switches and receptacles are shown, and selection of appropriate switches and receptacles for an application is explained.

SWITCHES

In most residential electrical circuits, individuals should expect to find some type of control device such as a switch. A *switch* is an electrical device used to control loads in a residential electrical circuit. **See Figure 5-1.**

Originally, the switches used to open and close circuits had to be physically changed by hand to either the ON (closed) or OFF (open) condition. Modern technology and the demand for convenience have greatly expanded the role of switches.

Today, switches can be activated by light, heat, water, smoke, and radio waves. Switches can turn lights ON and OFF or vary the intensity (brightness) of lights. Switches are classified as simple or complex depending on the requirements of the user.

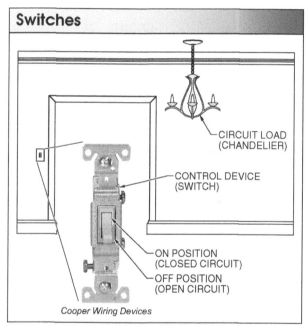

Switches

CIRCUIT LOAD (CHANDELIER)

CONTROL DEVICE (SWITCH)

ON POSITION (CLOSED CIRCUIT)

OFF POSITION (OPEN CIRCUIT)

Cooper Wiring Devices

Figure 5-1. *Switches control loads by opening and closing circuits.*

Switch Markings

When an individual picks up a switch, a great deal of information is found marked on the switch. Information provided can be in the form of color-coding or abbreviations. **See Figure 5-2.** Individuals must pay careful attention to the information because the information determines where and how a switch can be used.

Switch Markings

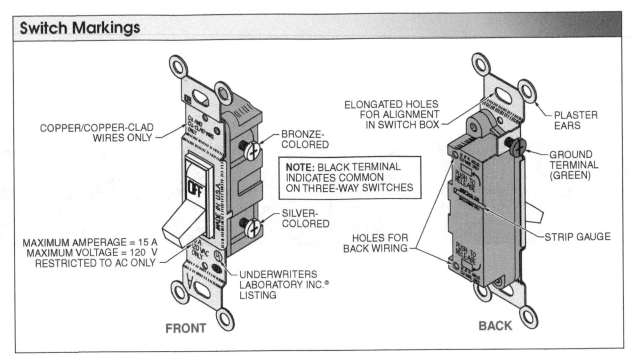

Figure 5-2. *A great deal of information is indicated on the body of a switch.*

Color Coding. Color coding is a technique used to identify switch terminal screws through the use of various colors. Switches use black, steel, and green colors to identify certain screw terminals. Black is used on three-way switches to indicate the screw that is the common terminal for the switch. Power will always come into a switch at the common and leave the switch by the traveler terminal. **See Figure 5-3.** The common screw terminal is typically black or darker than other screws. Green always indicates a ground terminal screw. Switches are constructed with a green ground screw attached to the metallic strap of the switch for grounding.

Printed Information. In addition to color coding, abbreviated words and symbols printed on a switch provide important information. Printed information is typically found on the front of a switch, but some information can be found on the back. **See Figure 5-4.**

UL Label. A *UL label* is a stamped or printed icon that indicates that a device or material has been approved for consumer use by Underwriters Laboratories Inc.® Underwriters Laboratories (UL) was created by the National Board of Fire Underwriters to test electrical materials and devices. Manufacturers, who wish to obtain UL approval, must submit samples to a UL testing laboratory in New York, Chicago, or Santa Clara, California. When a sample passes an extensive testing program, the sample is listed in a UL category as having met the minimum safety requirements.

A UL label indicates the minimum safety requirements and is not meant to serve as a quality comparison between manufacturers. Two switches, both with UL approval, can differ greatly in quality and performance.

CSA Label. A *Canadian Standard Association (CSA) label* is a marking that indicates that extensive tests have been conducted on a device by the Canadian Standards Association. Both CSA and UL labels are found on many devices to indicate that the device is accepted in both Canada and the United States.

Conductor Symbols. A *conductor symbol* is an electrical symbol that represents copper and aluminum respectively. Certain electrical devices are made to work with copper only and some devices with aluminum or copper. When a device is specified CU/AL, the device can be used with copper or aluminum wires.

Switch Color Coding

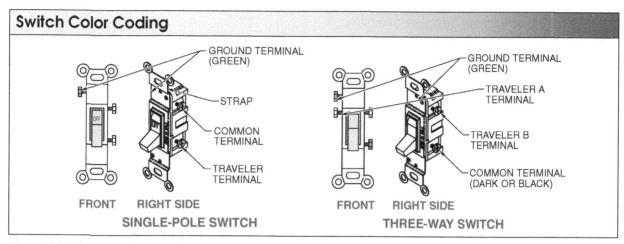

Figure 5-3. *The color of the common and traveler screw terminals are the same on single-pole switches. Three-way switches have a darker or black common terminal.*

Switch Inscribed Information

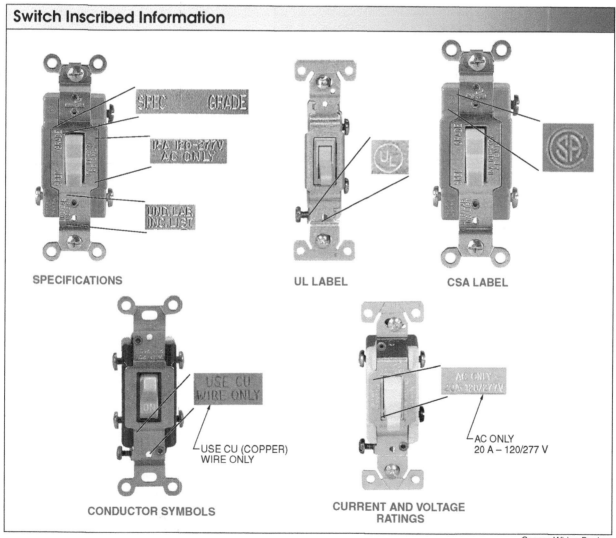

SPECIFICATIONS

UL LABEL

CSA LABEL

CONDUCTOR SYMBOLS

CURRENT AND VOLTAGE RATINGS

Cooper Wiring Devices

Figure 5-4. *Switch specifications are typically found on the front of a switch.*

When aluminum wire is used with copper connectors, the typical result is overheating, but electrical fires due to the chemical reaction between the dissimilar metals are also possible.

Current and Voltage Ratings. Current and voltage ratings are always provided on a switch. The maximum current rating of a switch for residential use is 15 A or 20 A. The maximum voltage rating is typically 120 V. In addition, some switches are specified for alternating current (AC) use only. Failure to follow current and voltage ratings along with other specifications reduces the useful life of a switch.

T Ratings. Switches used to control loads with a tungsten filament (such as standard incandescent lamps) must be marked with the letter T. A *T rating* is special switch information that indicates a switch is capable of handling the severe overloading created by a tungsten load as the switch is closed. **See Figure 5-5.** Tungsten has a very low resistance when cold and increases in resistance as heated. At the moment a switch is closed, the low resistance of a tungsten load causes the current draw to be 8 to 10 times the normal operating current. Once the switch is closed, the tungsten filament (load) heats up and the current flow drops immediately. The process of heating up a tungsten filament takes about 1/240 of a second. Failure to observe the T rating of a switch will reduce the life expectancy of the switch.

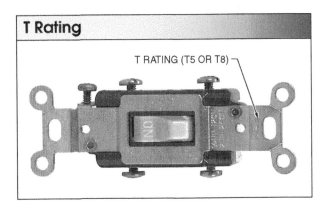

T Rating

T RATING (T5 OR T8)

Figure 5-5. *A T rating is a special rating given to switches that are capable of handling the severe overloading created by a tungsten load as the switch is closed.*

Strip Gauges. Strip gauges are found on the rear portion of a back-wired switch. **See Figure 5-6.** A *strip gauge* is a short groove that indicates how much insulation must be removed from a wire so the wire can be properly inserted into a switch. When a wire has not had enough insulation removed, a proper connection is not made. When too much insulation is removed, bare wire is left exposed allowing a short to possibly occur.

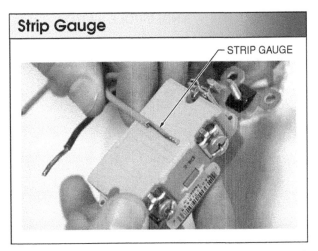

Strip Gauge

STRIP GAUGE

Figure 5-6. *A strip gauge is a marker on the back of a switch that identifies the amount of insulation that must be removed from a wire so the wire can be attached to the switch.*

Types of Switches

Switches are designed for a variety of applications. **See Figure 5-7.** Common switches used in residences are the single-pole, three-way, four-way, and double-pole switches. Some modern residential-use switches are the timer, programmable, and motion sensor switches. A timer switch is set to the times that the switch is turned ON and OFF. A programmable switch allows a sequence of events to be established, such as turning lights ON when no one is home. Motion sensor switches are used as security devices or energy savers. As a security device, motion sensor switches turn ON lights when someone enters an area. As an energy saver, motion sensor switches turn lights OFF when no one is detected in an area.

Switch Types . . .

Cooper Wiring Devices

SINGLE-POLE SWITCH

Standard ON/OFF switch provides control from one location; a three-way switch used with another three-way switch provides control from two locations (has no ON/OFF markings)

Leviton Manufacturing Co., Inc.

THREE-WAY SWITCH

Leviton Manufacturing Co., Inc.

FOUR-WAY SWITCH

Used with two three-way switches to provide control from three or more locations

Cooper Wiring Devices

DOUBLE-POLE SINGLE-THROW SWITCH

Equivalent to two single-pole switches. May open and close two circuits or wires at the same time; Looks identical to a four-way switch except double-pole switch is marked with an ON/OFF

Cooper Wiring Devices

DUPLEX

Useful when space is a problem; duplex switch allows two single pole switches to be placed in a standard switch box

Cooper Wiring Devices

SWITCH/RECEPTACLE COMBINATION

Switch may be wired to control receptacle or receptacle can be hot while the switch controls another circuit

Cooper Wiring Devices

DIMMER

Provides light adjustment from low levels to full brilliancy; Available in single-pole or three-way operation.

Leviton Manufacturing Co., Inc.

KEY

Used where access to the switch is to be restricted

Cooper Wiring Devices

AC PUSHBUTTON

Replaces old and loud snap action switches; Quiet switches are also available with toggle operation

Cooper Wiring Devices

WEATHERPROOF

Suitable for outside use

Leviton Manufacturing Co., Inc.

TIMER

Used to turn ON and OFF lights or appliances at predetermined times

Cooper Wiring Devices

MOTION SENSOR

Allows a circuit to be turned ON and OFF according to any movement within a 400 to 1600 sq ft area

Figure 5-7. *(continued . . .)*

. . . Switch Types

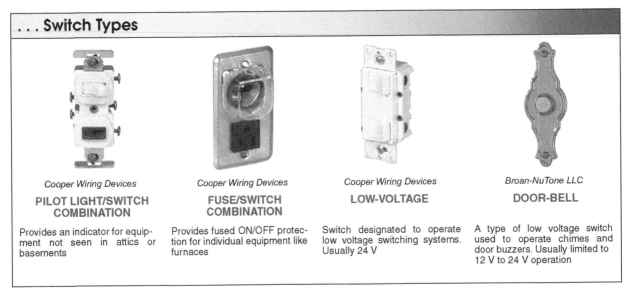

PILOT LIGHT/SWITCH COMBINATION	FUSE/SWITCH COMBINATION	LOW-VOLTAGE	DOOR-BELL
Cooper Wiring Devices	*Cooper Wiring Devices*	*Cooper Wiring Devices*	*Broan-NuTone LLC*
Provides an indicator for equipment not seen in attics or basements	Provides fused ON/OFF protection for individual equipment like furnaces	Switch designated to operate low voltage switching systems. Usually 24 V	A type of low voltage switch used to operate chimes and door buzzers. Usually limited to 12 V to 24 V operation

Figure 5-7. *Common switches used in residences are the single-pole, three-way, four-way, and double-pole switches.*

Single-Pole Switches. A *single-pole switch* is an electrical control device used to turn lights or appliances ON and OFF from a single location. Switches are connected into a circuit in the positive or hot side (leg) of a circuit. Single-pole switches are the easiest type of switches to install. **See Figure 5-8.** Single-pole switches are installed by applying the following procedure:

1. Remove the fuse or turn the circuit breaker to the OFF position.
2. Remove the wall plate and existing switch.
3. Straighten conductors from the box.
4. Trim conductors and strip insulation as required. Bend loops at the end of conductors.
5. Connect looped conductors to the respective screws on the switch.
6. Connect grounding wire from green hex head screw on switch to the metal box.
7. Attach new switch to the box and replace the wall plate.
8. Turn the power ON.

When adding a switch to a light or outlet at the end of a conductor run, a switchloop connection may be made. **See Figure 5-9.** A switchloop connection for adding a light is made by applying the following procedure:

1. Use existing incoming power to connect neutral conductor to the light base.

2. Connect the "hot" incoming conductor to the white conductor. *Note:* The white conductor must be marked with black tape to indicate that it is to be used as a hot feeder conductor.
3. Connect the black conductor and marked white conductor to the switch to complete the circuit.

Replacing Single-Pole Switches

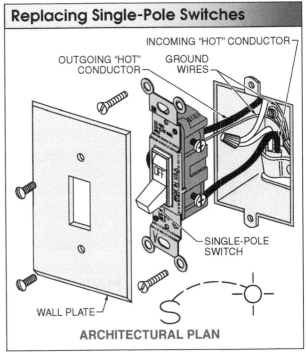

ARCHITECTURAL PLAN

Figure 5-8. *Single-pole switches are the easiest type of switches to install.*

Switchloop Single-Pole Switch Connections

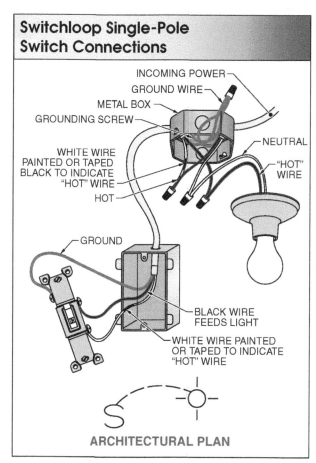

Figure 5-9. *When adding a switch to a light or outlet that is at the end of a run, a "switchloop" connection may be made.*

When armored or nonmetallic cable is used, an identified white conductor can also be used as the feeder conductor to a switch. **See Figure 5-10.** The National Electrical Code (NEC®) allows this exception.

A single-pole or two-way dimmer switch is used if the lamp that is controlled requires only one switch. **See Figure 5-11.** A dimmer switch is a switch that varies lamp brightness by changing the voltage applied to the lamp. Dimmer switches are single-pole switches used to control the light levels in living rooms, dining rooms, and other areas in which variable light level is desired.

A two-way dimmer switch is connected into a circuit in the same manner as a standard two-way switch. The only difference is the dimmer switch includes a knob (or lever) that is used to set the brightness in addition to an ON/OFF position. Most dimmer switches adjust the lamp through the full range of brightness.

Single-Pole Switch Circuits

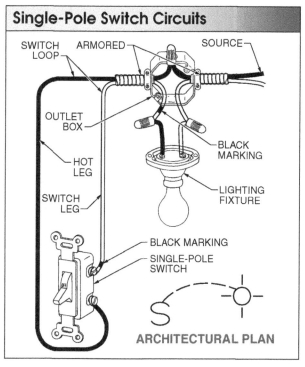

Figure 5-10. *When armored or nonmetallic cable is used, an identified white conductor can also be used as the feeder conductor to a switch.*

Dimmer Switches

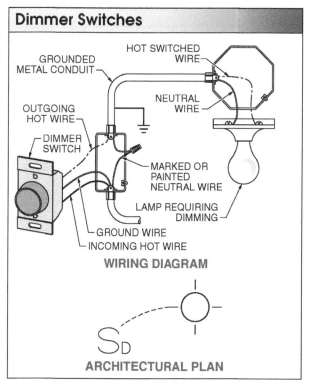

Figure 5-11. *Dimmer switches are used to control light levels in living rooms, dining rooms, and other areas in which variable light level is desired.*

Three-Way Switches. A *three-way switch* is an electrical control device used in pairs to control a light or load from two locations. The term "three-way" is the name given to the switch and in no way describes the operation of the switch. The terminals of a three-way switch are identified as common, traveler A, and traveler B. **See Figure 5-12.** The single terminal at one end of the switch is the common. The common terminal is easily identified because the terminal is darker than the other terminals or is black. The positive or hot wire of a circuit is always connected to the darkened terminal. The remaining two terminals are the connecting points for the two traveler wires that connect paired switches together.

A three-way switch operates by moving a pole between two positions. A three-way switch is really 2 two-position switches. When the handle is down, contact is made with terminal 1; when the handle is up, contact is made with terminal 2. Three-way switches have no ON or OFF markings because the handle indicates the position of the contacts from 1 three-way switch in relation to another three-way switch. The pole position of each switch determines whether or not the circuit is complete (load ON). **See Figure 5-13.**

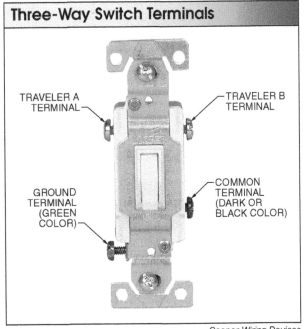

Cooper Wiring Devices

Figure 5-12. *The terminals of a three-way switch are identified as common, traveler A, and traveler B.*

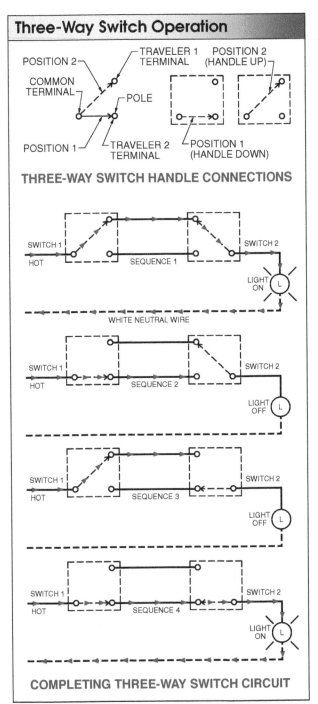

Figure 5-13. *A three-way switch operates by moving a pole between two positions.*

When the same respective traveler of each switch is connected, a conducting path is completed, which allows a load (light) to turn ON. When either switch contact is moved, the light will turn OFF because the conducting path is broken. Because either switch can break the conducting path, the load can be turned OFF from two locations. When the light must be turned back ON, either switch can be moved to allow a flow path to be completed. Two three-way switches are typically used for hallways, stairways, and rooms with two entrances.

A lamp that is used to light a staircase (six steps or more) must have a three-way switch at the top and bottom of the staircase. Rooms that have two entrances (or exits) require 2 three-way switches to control the light(s) from each location. Two three-way switches must be used to control a lamp from two locations. **See Figure 5-14.**

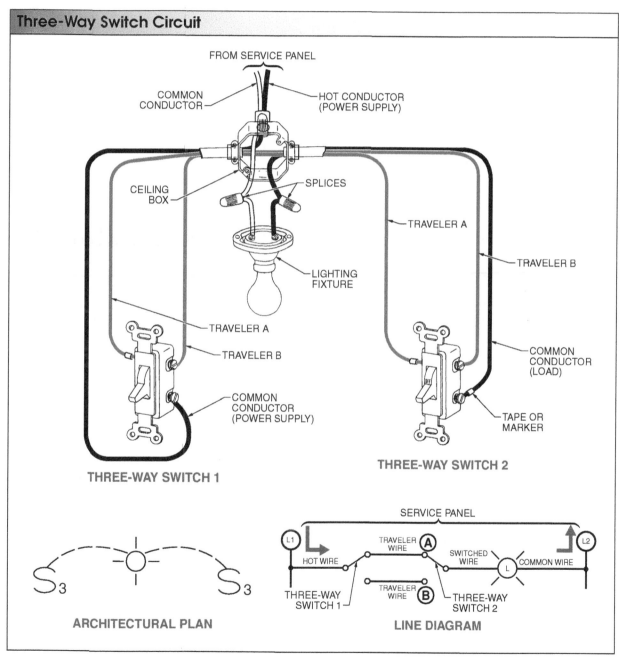

Figure 5-14. *Two three-way switches are used to control a lamp from two locations.*

Three-way switch circuits have the common terminal of 1 three-way switch connected to the hot power supply at all times. The common point of the other three-way switch is connected directly to the load (lamp) at all times. The traveler terminals of both three-way switches are connected together. A black wire is used to connect the hot power line to the first three-way switch. Red wires are used for the travelers. The load wire connecting the second three-way switch to the lamp is typically red or black with colored tape or marker used to distinguish the load wire from others.

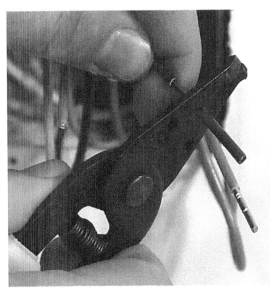

Conductors should be properly stripped of the proper amount of insulation with a wire stripper prior to termination to a switch or receptacle.

Residential circuits typically use a green or bare wire for the ground when nonmetallic cable is used. A ground wire is required to ground all non-current-carrying metal to earth. Metal conduit may be used to maintain a solid ground connection when metal conduit is used. Some applications require a load (lamp) to be controlled from more than two locations.

A three-way dimmer switch is installed by applying the following procedure:

1. Shut OFF power to the circuit.
2. Remove the wall plate and switch.

3. Test the circuit with a DMM or voltage tester to verify that there is no power.
4. Before removing conductors from an existing three-way switch, mark the conductor that leads to the common connection with electrical tape. With dimmer switches, wires are used for connection purposes rather than screws.
5. Loosen terminal screws and separate the wires from existing switch if it is a standard switch.
6. Trim and strip incoming conductors.
7. Use plastic connectors to join the white conductors together.
8. Use plastic connectors to connect each black lead to a black lead on the dimmer.
9. Connect the ground wires.
10. As the dimmer switch is repositioned in the box, fold back the conductors as far as possible to allow room for the dimmer switch, which is larger than a standard switch.

Switches for residential applications can be easily identified by their markings. Two-way (single-pole) single-throw switches and double-pole single-throw switches have marked ON and OFF positions. Three-way (single-pole) double-throw switches and four-way switches do not have ON or OFF markings.

Minimal power is consumed using a solid-state dimmer switch. Solid-state dimmer switches generate less heat than basic switches, are 99% efficient, and rely on electronic switching for control. For example, to control a 100 W lamp, only 1 W of power is required. Dimmer switches must be installed according to the following considerations:

- On a standard three-way switch, a dark-colored screw indicates the common terminal.
- On a three-way dimmer switch, the common lead is marked on the switch body.
- Only use a dimmer switch with a wattage rating high enough to handle the load. When installing a dimmer on a fixture with multiple bulbs, such as a chandelier, multiply the wattage of each bulb by the number of sockets in the fixture to

determine the total wattage. For example, 60 W bulbs × 6 sockets = 360 W total. Most dimmer switches are rated around 600 W and should be good for most small chandeliers. If the wattage is higher than the rating of the dimmer switch, a dimmer switch with a higher rating must be used for the installation.

- Do not use a dimmer switch to control a fan, as it will destroy the fan motor.
- Use only one three-way dimmer switch in a circuit. Other installed switches should be standard three-way switches.

Four-Way Switches. A *four-way switch* is a control device that is used in combination with 2 three-way switches to allow control of a load from three locations. One or more four-way switches are used with 2 three-way switches to provide control of a load from three or more locations. The two types of four-way switches are the through-wired type and crossed-wired type. **See Figure 5-15.** The through-wired type of a four-way switch is the most popular. Any number of four-way switches can be connected into a circuit, but all four-way switches must be connected between 2 three-way switches.

Dimmer switches can be dual functioned to control more than one device, such as a ceiling fan and luminaire, from the same switch.

The positions of the switch contacts in relation to each other determine whether a load (lamp or receptacle) is ON or OFF. To completely understand all combinations of three-location control, electricians should sketch all possible combinations of the contacts. **See Figure 5-16.** With 2 three-way switches and 1 four-way switch, there are six possible combinations.

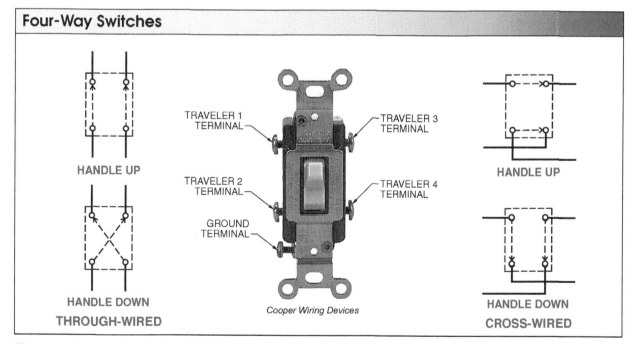

Figure 5-15. *Four-way switches provide multiple switch locations for controlling loads such as lamps.*

Four-Way Switch Locations

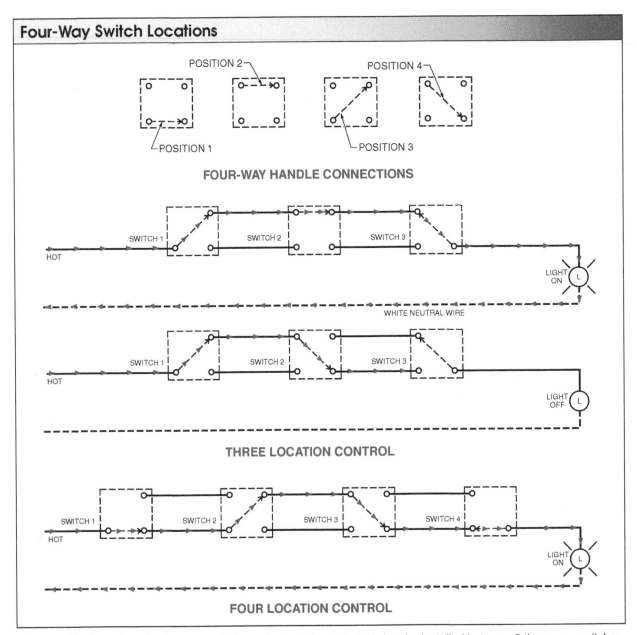

FOUR-WAY HANDLE CONNECTIONS

THREE LOCATION CONTROL

FOUR LOCATION CONTROL

Figure 5-16. *Operation of a four-way switch requires that four-way switches be installed between 2 three-way switches.*

Tech Tip:
Switch Contacts

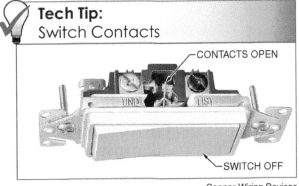

Cooper Wiring Devices

For example, a room that contains three entrances requires a switch to turn a lamp ON from any of the three entrances. A four-way switch is placed between 2 three-way switches to control the lamp from three locations. **See Figure 5-17.**

Two three-way switches and 2 four-way switches can be used to control a light from four locations. The 2 four-way switches must always be wired between 2 three-way switches. Any number of four-way switches can be added to a lamp control circuit to increase the number of control locations.

Four-Way Switch Circuits

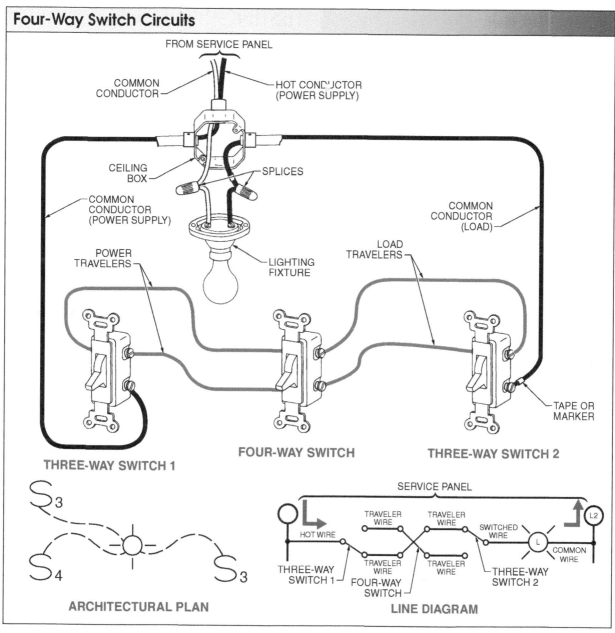

Figure 5-17. *Four-way switches add as many control locations to a circuit as required: 1 four-way equals three locations, 2 four-ways equal four locations, 3 four-ways equal five locations, etc.*

Double-Pole Switches. A *double-pole switch* is a control device that is two switches in one for controlling two separate loads. Double-pole switches are designed to connect or disconnect (open or close) at the same time. **See Figure 5-18.** Double-pole switches are common for 230 V circuits where both conductors are hot conductors. Double-pole switches open or close the dual path to the 230 V load, turning the load ON or OFF. Double-pole switches look similar to four-way switches, however double-pole switches are distinctly marked with an ON and OFF position.

> Use a commercial-grade switch for extra protection or longer service. Commercial-grade switches increase voltage capability from 120 V to 277 V, horsepower capability from ½ HP to 2 HP, and amperage capability from 15 A to 20 A or 30 A.

Double-Pole Switches

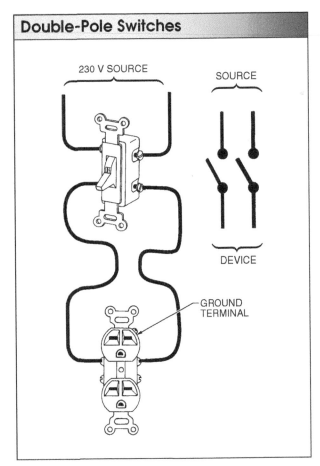

Figure 5-18. *Double-pole switches open two lines or circuits simultaneously.*

TESTING SWITCHES

A continuity tester or test instrument with a continuity test setting is used to test a switch for a complete path for current flow. A closed switch that is operating correctly has continuity, while an open switch does not have continuity. **See Figure 5-19.** A continuity tester or test instrument with a continuity test setting indicates when a switch has a complete path by emitting an audible beeping sound. Some test instruments with a continuity test setting display the resistance of a device while beeping. Indication of a complete path is used to determine the condition of a switch as open.

A quad box, or four-gang, is the largest switch box that is typically used with residential construction. Larger boxes are available but are rarely used for residential applications for aesthetic reasons. A four-gang connection can be used at the main entrance of a residence to control entry lights, outdoor lights, ceiling fans, room lights, spotlights, hallway lights, and closet lights from one location. A four-gang arrangement can be separated into two double-gang arrangements either for ease in wiring or for memorization of which circuit is controlled by the switch. All switches should be clearly marked when installed in a quad box.

Testing Switches

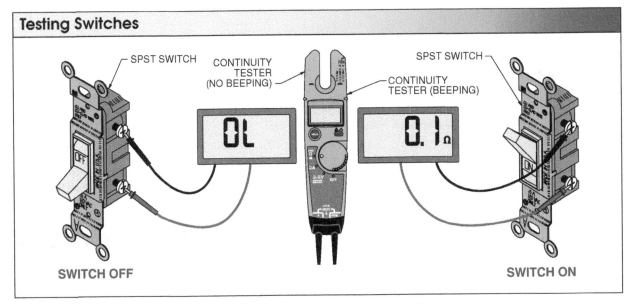

Figure 5-19. *Switches should be tested prior to installation using continuity testing.*

RECEPTACLES

Receptacles are often called convenience outlets. A *receptacle* is a contact device installed for the connection of plugs and flexible cords to supply current to portable electrical equipment.

Receptacle Markings

Receptacles, like switches, convey important information by colors and symbols. In addition, the shapes and positions of the openings (slots) in a receptacle also determine proper use. **See Figure 5-20.**

Shape and Position. The shape and position of the connection slots are used to differentiate between a 15 A, 125 V receptacle and a 15 A, 250 V receptacle. A 125 V receptacle has connection slots that are vertical and typically different lengths. A 250 V receptacle has connection slots that are horizontal and the same size. **See Figure 5-21.**

A 125 V receptacle with different size connection slots is a polarized receptacle. A *polarized receptacle* is a receptacle where the size of the connection slots determines the plug connection. The short connection slot is the hot connection and the long connection slot is the common (neutral) connection. Because the slots are designated and a plug can only be connected one way, the receptacle is a polarized receptacle. **See Figure 5-22.**

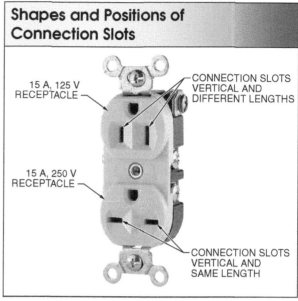

Shapes and Positions of Connection Slots

15 A, 125 V RECEPTACLE

CONNECTION SLOTS VERTICAL AND DIFFERENT LENGTHS

15 A, 250 V RECEPTACLE

CONNECTION SLOTS VERTICAL AND SAME LENGTH

Cooper Wiring Devices

Figure 5-21. *The shape and position of the connection slots are used to differentiate between various amperage and voltage rated receptacles.*

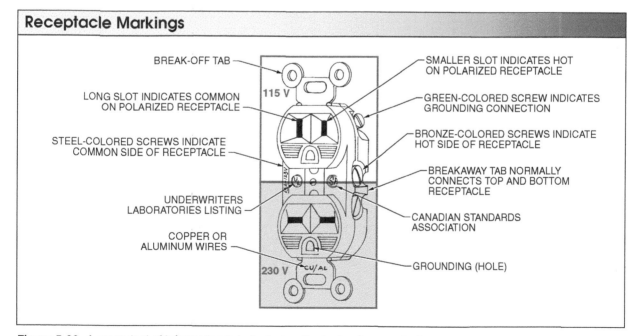

Receptacle Markings

BREAK-OFF TAB

LONG SLOT INDICATES COMMON ON POLARIZED RECEPTACLE

STEEL-COLORED SCREWS INDICATE COMMON SIDE OF RECEPTACLE

UNDERWRITERS LABORATORIES LISTING

COPPER OR ALUMINUM WIRES

115 V

230 V

SMALLER SLOT INDICATES HOT ON POLARIZED RECEPTACLE

GREEN-COLORED SCREW INDICATES GROUNDING CONNECTION

BRONZE-COLORED SCREWS INDICATE HOT SIDE OF RECEPTACLE

BREAKAWAY TAB NORMALLY CONNECTS TOP AND BOTTOM RECEPTACLE

CANADIAN STANDARDS ASSOCIATION

GROUNDING (HOLE)

Figure 5-20. *A great deal of information is indicated on the body of a receptacle.*

Polarized Receptacles

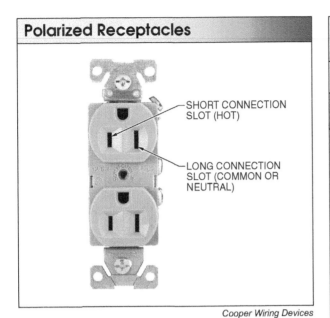

Cooper Wiring Devices

Figure 5-22. *Because the slots of a polarized receptacle are designated, a plug can only be connected one way.*

Ratings determine the number of contacts and the configuration in which the slots are positioned in the receptacle. **See Figure 5-23.** There are a large number of configurations and diagrams for receptacles encountered in residential construction work. **See Appendix.**

Lew Electric Fittings Co.

Floor-mount boxes can be used when routing telecommunications or electrical conductors to specially designed locations such as a home office.

Receptacle Connection Slot Configurations

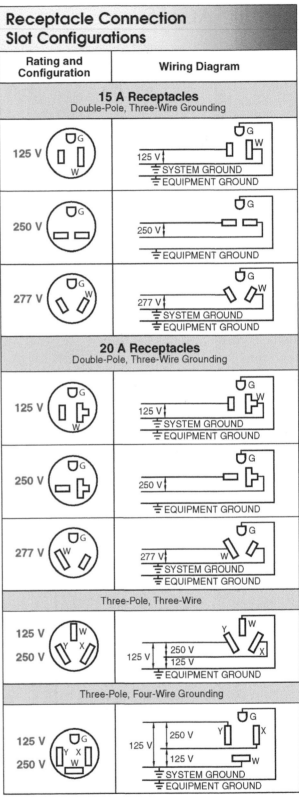

Figure 5-23. *Receptacle amperage and voltage ratings determine the number of connection slots and the configuration in which the slots are positioned.*

Receptacles are designed for straight blade or locking type plugs. Locking type receptacles are more common in commercial and industrial applications than in residential work. **See Figure 5-24.**

Straight and Locking Blade Receptacles

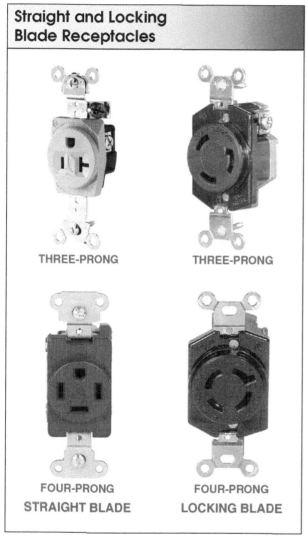

THREE-PRONG

THREE-PRONG

FOUR-PRONG
STRAIGHT BLADE

FOUR-PRONG
LOCKING BLADE

Cooper Wiring Devices

Figure 5-24. *Straight blade receptacles are typically used for residential applications while locking blade receptacles are typically used for commercial and industrial applications.*

Color Coding. Steel, bronze, and green colors are typically used on receptacle terminal screws. A silver-colored screw indicates the neutral terminal, a bronze-colored screw indicates the hot terminal, and a green colored screw indicates the ground terminal.

Printed Information. The markers "UL", "CSA", "CU/AL", "15 A", and "125 V" have the same meaning for receptacles as they do for switches: Underwriter Laboratories, Canadian Standards Association, copper/aluminum wire connections, maximum current, and maximum voltage rating.

> *Each piece of rating information is important when purchasing or installing receptacles.*

Types of Receptacles

Receptacles are found in a variety of shapes and sizes depending on the application. Typical receptacles found in residences are standard, isolated ground, split-wired, and GFCI. **See Figure 5-25.**

Standard Receptacles. Standard receptacles include a long neutral slot, a short hot slot, and a U-shaped ground hole. **See Figure 5-26.** Wires are attached to the receptacle at screw terminals or push-in fittings. A connecting tab between the two hot and two neutral screw terminals provides an electrical connection between the terminals. The electrical connection allows for both terminals to be powered when one wire is connected to either terminal screw.

Receptacles are marked with ratings for maximum voltage and amperes. Standard receptacles are marked 15 A, 125 or 20 A, 125 V (maximum values). Receptacles marked 20 A, 125 V are typically required for laundry and kitchen circuits. Other 125 V receptacles found in a residence are 15 A.

Receptacles marked CU are used with solid copper wire only. Receptacles marked CU-CLAD are used with copper-coated aluminum wire. Receptacles marked CO/ALR are used with solid aluminum wire.

⚠️ **DANGER**

Receptacles require a grounding jumper if not automatically grounded.

Receptacle Types . . .

STANDARD DUPLEX

A grounded receptacle commonly used for 15 A, 115 V appliances in the home

STANDARD DUPLEX

A grounded receptacle commonly used for 20 A, 115 V appliances in the home

COMBINATION HIGH/LOW VOLTAGE

Used to provide a 230 V circuit and a 115 V circuit in one receptacle

SINGLE GROUNDED 115 V

Maximum load is 15 A

30 A

Used on clothes dryers for 120/240 V operation; maximum load is 30 A

TWIST-LOCK

Used with twist-lock plugs in areas where positive connection is to be maintained

SINGLE GROUNDED 115 V

Maximum load is 20 A

50 A

Used on electric ranges to supply 120/240 V operation; maximum load is 50 A

NOTE: Four-wire can be required by local code

50 A SURFACE MOUNT

Used on electric devices to supply 120/240 V operation; maximum load is 50 A

Figure 5-25. *(continued . . .)*

... Receptacle Types

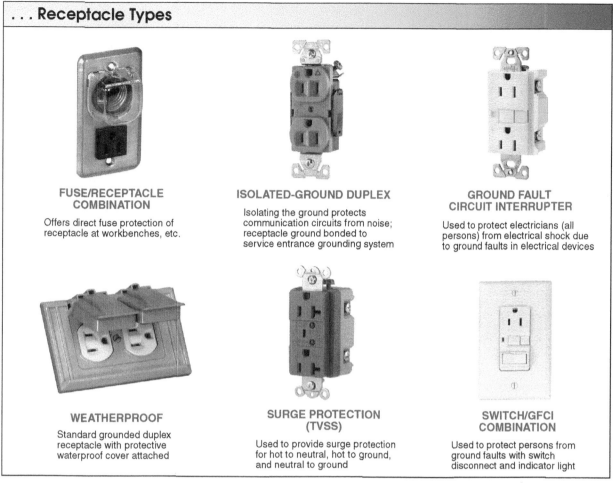

FUSE/RECEPTACLE COMBINATION

Offers direct fuse protection of receptacle at workbenches, etc.

ISOLATED-GROUND DUPLEX

Isolating the ground protects communication circuits from noise; receptacle ground bonded to service entrance grounding system

GROUND FAULT CIRCUIT INTERRUPTER

Used to protect electricians (all persons) from electrical shock due to ground faults in electrical devices

WEATHERPROOF

Standard grounded duplex receptacle with protective waterproof cover attached

SURGE PROTECTION (TVSS)

Used to provide surge protection for hot to neutral, hot to ground, and neutral to ground

SWITCH/GFCI COMBINATION

Used to protect persons from ground faults with switch disconnect and indicator light

Cooper Wiring Devices

Figure 5-25. *Receptacles are found in a variety of shapes and sizes depending on the application.*

Standard Receptacles

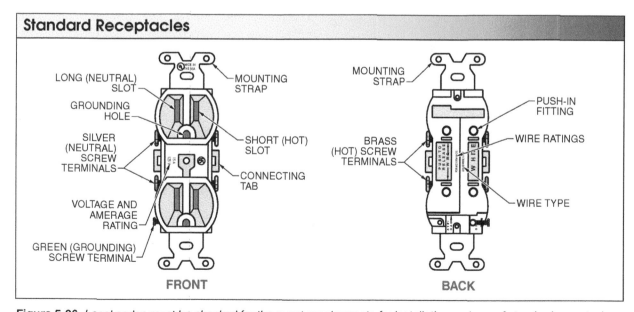

LONG (NEUTRAL) SLOT

GROUNDING HOLE

SILVER (NEUTRAL) SCREW TERMINALS

VOLTAGE AND AMERAGE RATING

GREEN (GROUNDING) SCREW TERMINAL

MOUNTING STRAP

SHORT (HOT) SLOT

CONNECTING TAB

FRONT

MOUNTING STRAP

BRASS (HOT) SCREW TERMINALS

PUSH-IN FITTING

WIRE RATINGS

WIRE TYPE

BACK

Figure 5-26. *Local codes must be checked for the exact requirements for installation and use of standard receptacles.*

When wiring a 120 V duplex receptacle, the black or red (hot) wire is connected to the brass-colored screw, the white wire is connected to the steel-colored (common) screw, and the green or bare wire is connected to the green screw. **See Figure 5-27.**

Standard Receptacle Wiring

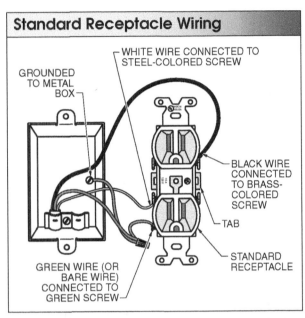

Figure 5-27. *Most local codes require that a standard receptacle be installed as a polarized receptacle.*

Tamper-Resistant Receptacles. Electrical safety has been one of the primary issues addressed in the NEC®. This is particularly true when it comes to safety for children. In 2008, the NEC® increased rules on child safety by requiring tamper-resistant receptacles in all new construction of dwelling units. This includes single dwelling units providing complete and independent living facilities for one or more persons.

These living units provide for living, sleeping, cooking, and sanitation. All 125 V, 15 A and 20 A receptacles shall be tamper resistant in these dwellings. UL-listed devices must have a "TR" (tamper resistant) marking clearly visible when installed.

Tamper-resistant receptacles have an internal mechanism that limits access to electrically live components within the receptacles. Tamper-resistant receptacles resemble standard receptacles but include automatic shutters, which admit plugs but block other objects such as hairpins, keys, and paper clips.

See Figure 5-28. The objective of all technologies used in tamper-resistant receptacles is to prevent the insertion of foreign objects into the receptacles.

Tamper-Resistant Receptacles

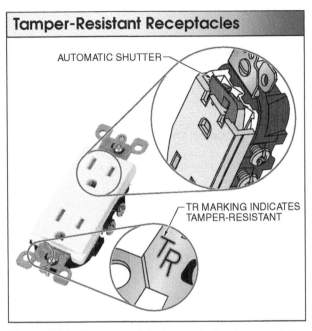

Figure 5-28. *Tamper-resistant receptacles have an internal mechanism that limits access to electrically live components within the receptacles.*

Split-Wired Receptacles. A *split-wired receptacle* is a standard receptacle that has had the tab between the two brass-colored (hot) terminal screws removed. The tab between the two steel-colored (neutral) terminals has not been removed. Split-wired receptacles are used to provide a standard and switched circuit or two separate circuits at the same duplex outlet. **See Figure 5-29.**

Split-Wired Receptacles

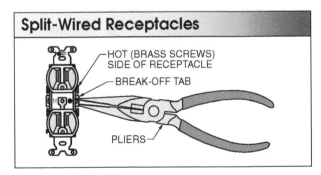

Figure 5-29. *Split-wired receptacles provide a standard-wired receptacle and a switched-circuit receptacle or two separate circuits at the same duplex receptacle.*

Isolated-Ground Receptacles. In a standard receptacle, the receptacle ground is connected to the common grounding system when the receptacle is installed in a metal outlet box. A common grounding system normally includes all metal wiring, boxes, conduit, water pipes, and the non-current-carrying metal parts of most electrical equipment. The receptacle ground becomes part of the larger grounding system when a piece of electrical equipment is plugged into the receptacle.

The common grounding system acts as a large antenna and conducts electrical noise. This electrical noise causes interference in computers, medical, security, military, and communication equipment.

An isolated-ground receptacle is used to minimize problems in sensitive applications or areas of high electrical noise. An *isolated-ground receptacle* is a special receptacle that minimizes electrical noise by providing a separate grounding path for each connected device. Isolated-ground receptacles are identified by an orange color. **See Figure 5-30.** A separate ground conductor is run with the circuit conductors in an isolated grounding system.

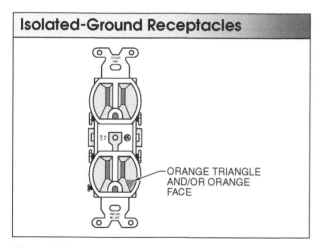

Isolated-Ground Receptacles

Figure 5-30. *Isolated-ground receptacles minimize noise by providing a separate grounding path for each receptacle.*

GFCI Receptacles. A ground fault circuit interrupter (GFCI) is a fast-acting receptacle that detects low levels of leakage current to ground and opens the circuit in response to the leakage (ground fault). A *ground fault* is any current above the level that is required for a dangerous shock. GFCIs provide greater protection than standard or isolated-ground receptacles. **See Figure 5-31.**

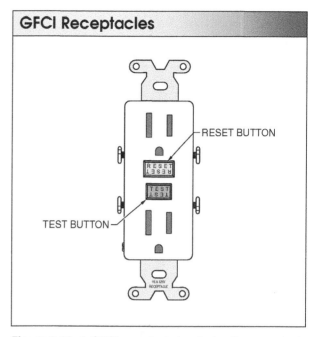

GFCI Receptacles

RESET BUTTON

TEST BUTTON

Figure 5-31. *A GFCI receptacle is a fast-acting receptacle that detects low levels of leakage current to ground and opens the circuit in response to the leakage.*

TESTING RECEPTACLES

A test light or test instrument with a voltage setting is used to test a receptacle for voltage. A receptacle that is operating correctly has 120 V between the hot slot and the common slot or ground, while an open switch does not have continuity. **See Figure 5-32.** Indication of correct voltage is used to determine the condition of a receptacle.

Testing receptacles is also possible with a receptacle tester. A *receptacle tester* is a test instrument that is plugged into a standard receptacle to determine if the receptacle is properly wired and energized. **See Figure 5-33.** Some receptacle tester models include a ground fault circuit interrupter or ground fault interrupter test button that allows the receptacle tester to be used on GFCI or GFI receptacles. **See Figure 5-34.**

> *Cover receptacle terminal screws with several layers of electrical tape prior to installation in a metal box to prevent electrical shock or a short circuit to the system.*

Testing Standard Receptacles— Test Lights

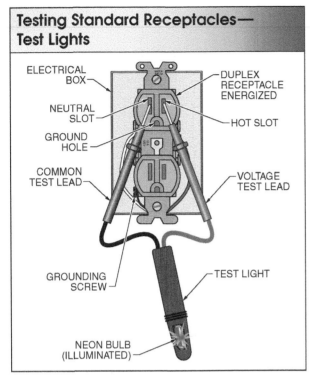

Figure 5-32. *A test light illuminates when a standard receptacle is energized.*

Testing Standard Receptacles— Receptacle Testers

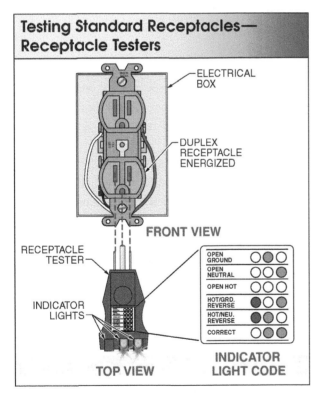

Figure 5-33. *Receptacle testers are plugged into a standard receptacle to determine if the receptacle is properly wired and energized.*

Testing GFCI Receptacles— GFCI Testers

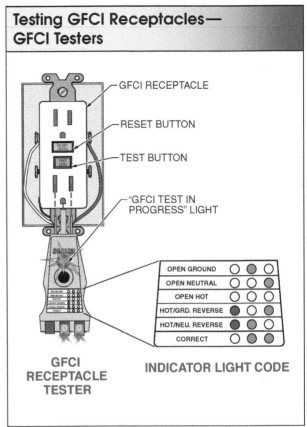

Figure 5-34. *GFCI receptacle testers are plugged into a GFCI receptacle to determine if the receptacle is properly grounded, wired, and energized.*

SWITCH AND RECEPTACLE COVERS

To properly trim out a switch or receptacle, the device must be properly covered. **See Figure 5-35.** Switches can be combined with receptacles and trimmed out by using the correct cover plate configuration. Switches and receptacles can also be located outside. For outdoor applications, weatherproof covers are required.

To avoid having excess conductors in the back of a receptacle box, a box that is larger than required should be used. The largest size single-gang, nonmetallic receptacle box has a volume of 22.5 cu in.

Switch and Receptacle Covers

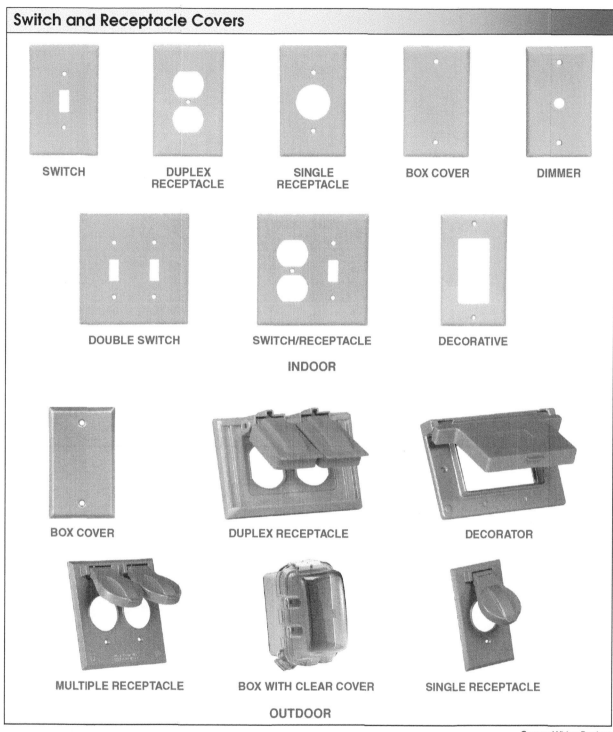

SWITCH **DUPLEX RECEPTACLE** **SINGLE RECEPTACLE** **BOX COVER** **DIMMER**

DOUBLE SWITCH **SWITCH/RECEPTACLE** **DECORATIVE**

INDOOR

BOX COVER **DUPLEX RECEPTACLE** **DECORATOR**

MULTIPLE RECEPTACLE **BOX WITH CLEAR COVER** **SINGLE RECEPTACLE**

OUTDOOR

Cooper Wiring Devices

Figure 5-35. *Switches and receptacles must be properly covered according to the NEC®.*

Switches and Receptacles 5

TEST

Name_____ **Date** _____

Switches and Receptacles

_____ **1.** The ___ screw on a three-way switch indicates the common pivot point for the switch.

_____ **2.** Printed information is generally found on the ___ side of a switch.

_____ **3.** The UL label indicates that an item meets ___ and NEC® safety requirements.

_____ **4.** The letters ___ on a device indicate copper or aluminum wires can be used.

_____ **5.** If a device is not specified as copper or aluminum, ___ should be used.

T   F **6.** Current and voltage ratings are always provided on a switch.

_____ **7.** Switches that are used to operate devices with a tungsten filament must be marked with the letter ___.

_____ **8.** ___ switches open two circuit conductors at the same time.

_____ **9.** A single-pole switch is connected in the ___ leg of a circuit.

_____ **10.** A(n) ___ on the back of a switch or receptacle indicates what length of insulation must be removed from a wire.

T   F **11.** As a general rule, the neutral conductor should not be switched or used as one leg in a switch loop.

T   F **12.** Three-way switches do not have ON and OFF markings.

_____ **13.** Receptacles are also known as ___ outlets.

_____ **14.** Receptacles are rated by their ___.

        A. physical size   **C.** type of material
        B. color   **D.** amperage and voltage

_____ **15.** The ___ tests electrical devices in Canada.

_____ **16.** ___ switches are used to control a light or device from two locations.

T   F   **17.** One or more four-way switches may be used with 2 three-way switches to provide control of a device from three or more locations.

T   F   **18.** Receptacles are designed as straight blade or locking blade.

_____ **19.** A(n) ___ plug is identified by the size of the openings, which determines which blade opening is the hot and which opening is the neutral.

_____ **20.** Silver-colored screws on a receptacle indicate the ___ terminal.

_____ **21.** Bronze-colored screws on a receptacle indicate the ___ terminal.

_____ **22.** Green-colored screws on a receptacle indicate the ___ terminal.

_____ **23.** A(n) ___ plug and receptacle may be used where a positive connection is to be maintained.

_____ **24.** A(n) ___ switch may be used where access to the switch is to be restricted.

_____ **25.** A(n) ___ switch provides light adjustment from low levels to full brilliancy.

_____ **26.** A(n) ___ switch is used to turn ON and OFF devices at predetermined times.

_____ **27.** A(n) ___ switch may be used where space is a problem.

_____ **28.** A(n) ___ switch is used to turn circuits ON and OFF according to movement within a specified distance.

_____ **29.** Receptacles marked CO/ALR are used with ___ wire only.

_____ **30.** Color coding, abbreviations, and ___ are used to provide information about switches.

Receptacles

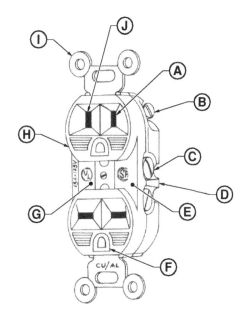

Light Controlled by Single-Pole Switch-Conduit

Switches

_____	1.	Canadian Standards Association	_____	7.	Strip gauge
_____	2.	Copper/copper-clad wires only	_____	8.	Maximum amperage/voltage
_____	3.	Colored screw	_____	9.	Elongated holes
_____	4.	Plain screw	_____	10.	Back wiring holes
_____	5.	Plaster ears	_____	11.	Underwriters Laboratories
_____	6.	Captive mounting screw			

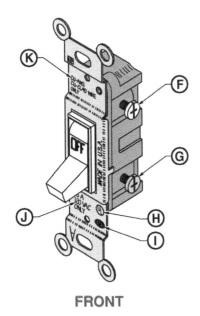

FRONT

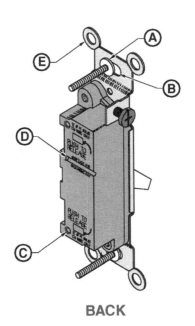

BACK

15 A Receptacles, 2-Pole, 3-Wire

_____	1.	125 V
_____	2.	250 V
_____	3.	277 V

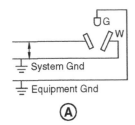

Ⓐ

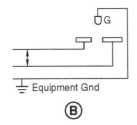

Ⓑ

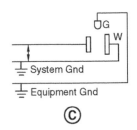

Ⓒ

20 A Receptacles, 2-Pole, 3-Wire

_____ **1.** 125 V

_____ **2.** 250 V

_____ **3.** 277 V

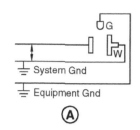
(A)

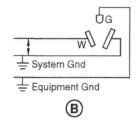
(B)

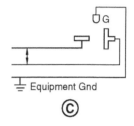
(C)

30 A Receptacles, 2-Pole, 3-Wire

_____ **1.** 125 V

_____ **2.** 250 V

_____ **3.** 277 V

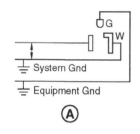
(A)

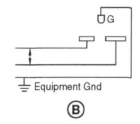
(B)

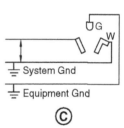
(C)

50 A Receptacles, 2-Pole, 3-Wire

_____ **1.** 125 V

_____ **2.** 250 V

_____ **3.** 277 V

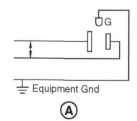
(A)

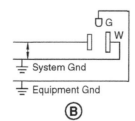
(B)

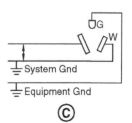
(C)

Working with Nonmetallic-Sheathed Cable 6

Chapter 6 provides an overview of the characteristics of nonmetallic-sheathed cable and the procedures for preparing and routing the cable. The safe, economical, and effective installation of nonmetallic-sheathed cable is shown. The mounting of common types of boxes is covered in detail.

NONMETALLIC-SHEATHED CABLE

A *nonmetallic-sheathed cable* is an electrical conductor (cable) that has a set of insulated electrical conductors held together and protected by a strong plastic jacket. **See Figure 6-1.** The plastic jacket can contain any number of conductors; however, the typical cable contains two or three insulated conductors and a separate bare ground wire. Nonmetallic-sheathed cable is very popular in residential wiring because it is inexpensive and easy to install.

Nonmetallic-Sheathed Cable

BLACK WIRE

14/2 W/GR TYPE NM

GROUND WIRE (CAN BE BARE)

PLASTIC JACKET   WHITE WIRE

Figure 6-1. *The electrical conductors in nonmetallic-sheathed cable are protected by a rugged plastic jacket.*

INFORMATION PRINTED ON NONMETALLIC-SHEATHED CABLE

Important information is printed on the outside jacket of nonmetallic-sheathed cable. The printed information typically includes size of conductors (wire), number of conductors (with or without ground), and type of jacket.

Numbers

A considerable amount of information is printed on the outside jacket of nonmetallic-sheathed cable. The numbers, which are printed first, provide information about the size of the wire and the number of conductors in the cable, such as cable that is marked 14/2 or 12/3. The first number (in this case 14 or 12) indicates the size of the wire. The smaller the number, the larger the wire size. The second number (in this case 2 or 3) refers to the number of current-carrying conductors in the cable, not counting the ground wire. **See Figure 6-2.** To determine the load carrying capacity of any wire, the National Electrical Code® (NEC®) should be referred to for wire size and ampacity.

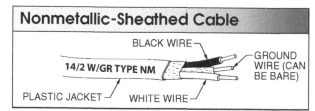

Over 185 companies produce nonmetallic-sheathed cable.

Nonmetallic-Sheathed Cable Numbers

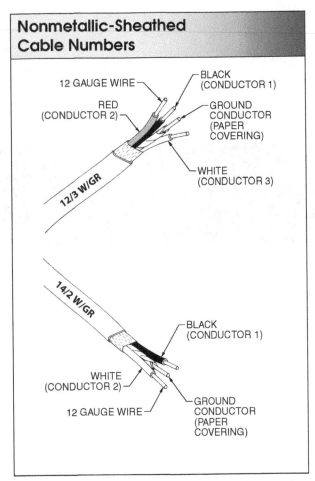

Figure 6-2. *Nonmetallic-sheathed cable is manufactured in various wire sizes and with a specified number of conductors.*

Letters

The letters NM and NMC indicate specific types of nonmetallic cable and what characteristics the cable has. *Type NM cable* is a nonmetallic-sheathed cable that has the conductors enclosed within a nonmetallic jacket and is the type typically used for dry interior wiring. *Type NMC cable* is a nonmetallic-sheathed cable that has the conductors enclosed within a corrosion-resistant, nonmetallic jacket. Article 334 of the NEC® contains detailed information on NM and NMC cable.

The letters WITH GR (with ground) indicate that a separate grounding conductor is present. **See Figure 6-3.** When a nonmetallic-sheathed cable does not clearly specify WITH GR, the cable will not contain a grounding wire.

Nonmetallic-Sheathed Cable Letters

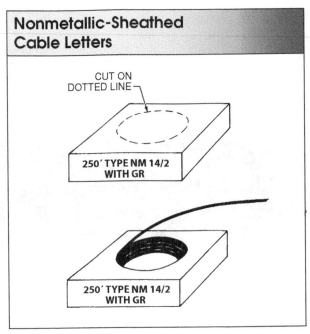

Figure 6-3. *The identification letters and numbers of a nonmetallic-sheathed cable are found on the jacket of the cable and on the packaging that the cable is sold in.*

BUYING NONMETALLIC-SHEATHED CABLE

Nonmetallic-sheathed cable is found at electrical supply houses, hardware stores, and in some large retail stores with home repair centers. Nonmetallic-sheathed cable is typically sold in boxed rolls of 250 ft. **See Figure 6-4.** For small jobs, smaller quantities are available at hardware stores and at some electrical supply stores.

Nonmetallic-Sheathed Cable Packaging

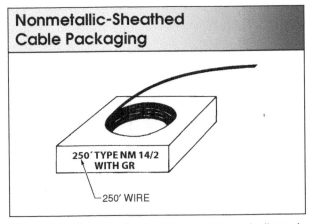

Figure 6-4. *Nonmetallic-sheathed cable is typically packaged in rolls of 250'.*

When purchasing nonmetallic-sheathed cable, electricians must specify the length and type of cable required. For example, when an application requires 100′ of cable that has two conductors of 14 gauge wire with a separate ground wire, the cable is specified as 100′ of 14/2 WITH GR Type NM or NMC.

PREPARING NONMETALLIC-SHEATHED CABLE

Nonmetallic-sheathed cable must be prepared. Typically the cable must be cut to length, have the outer jacket removed to a distance from the end, and have a specified amount of insulation removed from the individual conductors.

Plastic Jacket Removal

Nonmetallic-sheathed cable is easy to install and requires few tools to perform a quality job. Lineman side cutting pliers are capable of cutting NM cable quickly to length. **See Figure 6-5.**

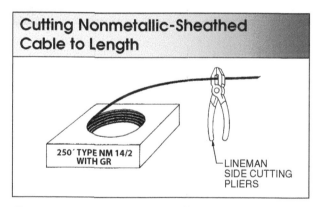

Cutting Nonmetallic-Sheathed Cable to Length

250′ TYPE NM 14/2 WITH GR

LINEMAN SIDE CUTTING PLIERS

Figure 6-5. *Lineman side cutting pliers easily cut through nonmetallic-sheathed cable.*

When NM cable has been cut to length, the ends of the plastic jacket must be removed before the cable is inserted into outlet or junction boxes. To open the plastic jacket of an NM cable, a knife is used to cut through the center of the jacket where the ground wire is located. **See Figure 6-6.** The bare ground wire will guide the path of the knife blade. At least 8″ of the jacket should be removed from each end of the cable.

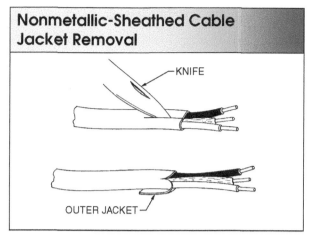

Nonmetallic-Sheathed Cable Jacket Removal

KNIFE

OUTER JACKET

Figure 6-6. *The plastic jacket of nonmetallic-sheathed cable can easily be removed with a pocketknife.*

When great lengths of cable must be prepared, cable rippers are used to save time. **See Figure 6-7.** Cable rippers have a small, razor-sharp blade designed to penetrate a short distance into the cable to remove only the outer plastic cover.

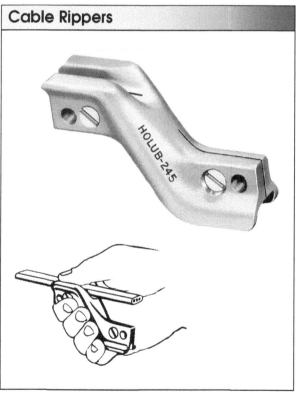

Cable Rippers

HOLUB-245

ITT Holub Industries

Figure 6-7. *Cable rippers are useful where large quantities of cable are being used.* Note: *There is a small blade (not shown) that cuts the jacket.*

Cable Conductor Insulation Removal

When the outer jacket is cut away, insulation from the individual conductors (wires) can be removed. Various multipurpose wire-stripping tools are used to remove insulation from individual conductors. **See Figure 6-8.**

CAUTION

Never use a pocketknife to remove insulation from individual wires.

ROUGHING-IN NONMETALLIC CABLE

Roughing-in is a phrase that refers to the placement of electrical boxes and wires before wall coverings and ceilings are installed. Roughing-in must be performed in such a way that the entire electrical system

(wires and boxes) can be easily traced after walls and ceilings are in place. All rough-in work must be carefully checked by the Authority having Jurisdiction (AHJ). Errors in the roughing-in process are compounded once the wall coverings are in place.

When roughing-in the electrical system of a residence, the integrity of the wood studs and joists must be taken into consideration. Care should be taken to not drill too many large-diameter holes. Consult local codes prior to drilling and cutting studs and joists.

Receptacle Boxes

A *receptacle box* is an electrical device designed to house electrical components and protect wiring connections. Depending on the specific application, number of wires, and size of wires, receptacle boxes can have various shapes, sizes, and standard accessories.

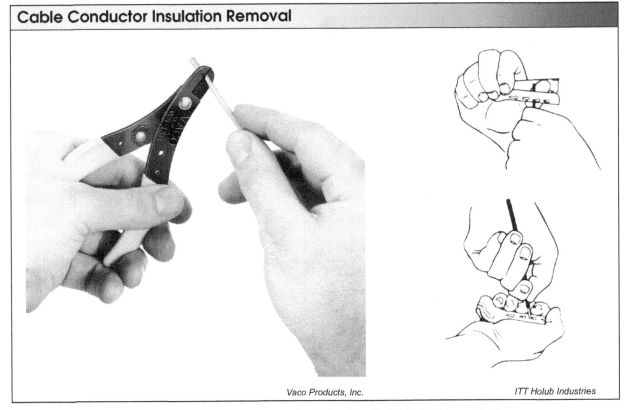

Cable Conductor Insulation Removal

Vaco Products, Inc.   ITT Holub Industries

Figure 6-8. *Wire-stripping tools can be used to remove insulation from individual cable conductors.*

Uses for Receptacle Boxes. Receptacle boxes and junction boxes must be installed at every point in the electrical system where NM cable is spliced or terminated. receptacle boxes are installed so that every point in the system is accessible for future repairs or additions. Boxes must never be completely covered to the point that a box becomes inaccessible.

Shapes and Sizes of Receptacle Boxes. Typically three shapes of receptacle boxes are used in residential wiring. The three boxes are square, octagonal, and rectangular. **See Figure 6-9.** Square, octagonal, and rectangular boxes come in various widths and depths, and with various knockout arrangements.

Never use a nonmetallic box for hanging lights or ceiling fans. A metal ceiling-mount box has large attachment screws and will be more secure than a nonmetallic box.

Boxes come in various sizes as well as various shapes. Rectangular boxes for switches are available in one-, two-, three-, or four-gang box sizes.

Box Shapes

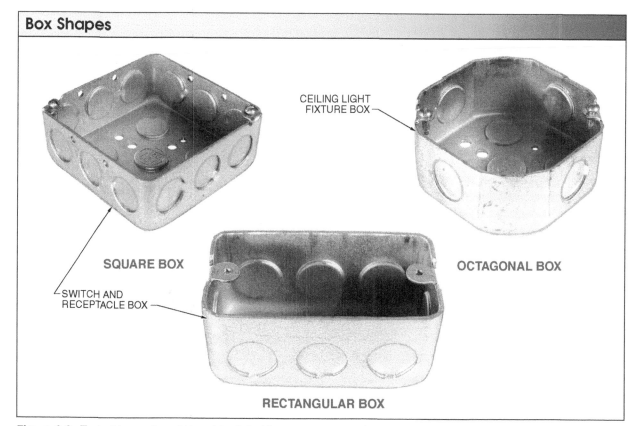

SQUARE BOX

SWITCH AND RECEPTACLE BOX

CEILING LIGHT FIXTURE BOX

OCTAGONAL BOX

RECTANGULAR BOX

Figure 6-9. *Typical boxes found in residential wiring are square and rectangular (for switches and receptacles) and octagonal (for ceiling fixtures).*

Per the NEC®, luminaire outlet boxes must provide adequate space for conductors.

Although there is no hard-and-fast rule regarding the shape of a box, octagonal boxes are typically used in ceilings for lighting fixtures. Square and rectangular boxes are typically used for switches and receptacles. Any of the three are used for junction boxes.

The size of an outlet box is based on the number and the size of the wires entering the box. More wires and larger size wires require a larger or deeper box. Manufacturer specification sheets must be consulted when ordering boxes to find the appropriate volume (cubic inch capacity) box. Additional information can be obtained from the NEC®.

A *knockout* is a round indentation punched into the metal of a box and held in place by unpunched narrow strips of metal. Knockouts (abbreviated KO) are located in the sides and bottoms of outlet boxes and provide ready access from any direction. **See Figure 6-10.** Knockouts must be removed to attach fittings for wiring.

Mounting Boxes. Boxes are typically secured by nails. The nails may be part of the box or driven through straps in the box. **See Figure 6-11.** When

a box must be mounted between studs and joists, hangers are used. Almost all ceiling outlet boxes are mounted with bar hangers.

Removing Box Knockouts

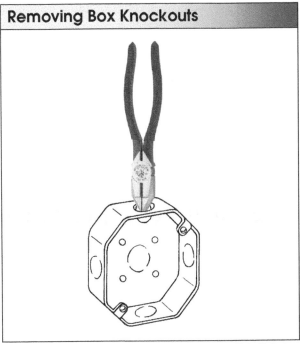

Klein Tools, Inc.

Figure 6-10. *Box knockouts must be removed to attach fittings for wiring.*

Mounting Plastic Boxes

When mounting plastic boxes, the box should be positioned against the stud so that the front of the box will be flush with the finished surface of the wall. **See Figure 6-12.** For example, if the finished wallboard is ½″ drywall, the box should be positioned ½″ past the face of the stud. When the wallboard will be covered by tile, the thickness of the tile must be taken into account.

Too many conductors in a box are known as a "cable fill violation." Refer to NEC® Article 314.16, Number of Conductors in Outlet, Device, and Junction Boxes and Conduit Bodies, for proper box size. Boxes must have enough volume to account for the receptacle and incoming and outgoing conductors. Nonmetallic boxes are available in sizes of 16, 18, 20.3, and 22.5 cu in. and have the volume stamped on the inside of the box. Check the manufacturer's catalog for the volume of metal boxes.

Mounting Boxes

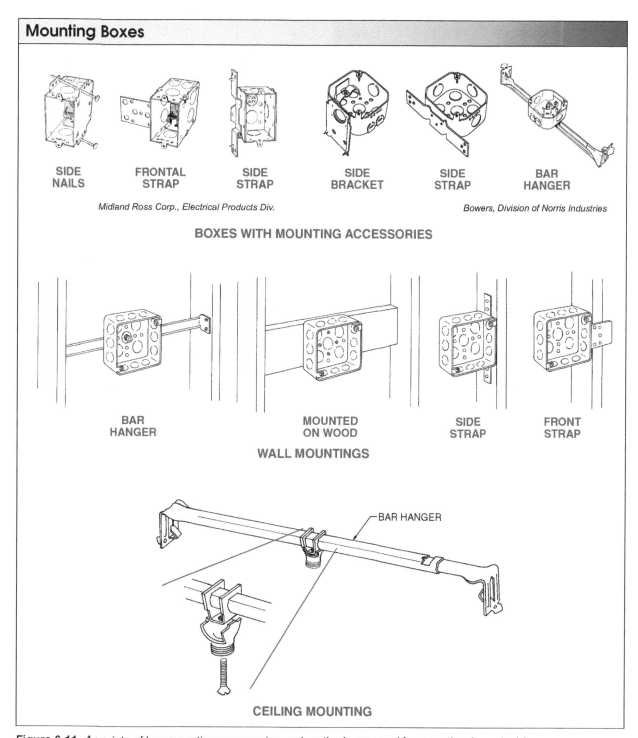

BOXES WITH MOUNTING ACCESSORIES

SIDE NAILS — FRONTAL STRAP — SIDE STRAP — SIDE BRACKET — SIDE STRAP — BAR HANGER

Midland Ross Corp., Electrical Products Div. *Bowers, Division of Norris Industries*

WALL MOUNTINGS

BAR HANGER — MOUNTED ON WOOD — SIDE STRAP — FRONT STRAP

BAR HANGER

CEILING MOUNTING

Figure 6-11. *A variety of box mounting accessories and methods are used for mounting (securing) boxes.*

Knockout Holes. In each plastic box there is an area where knockout holes can be created. The knockouts can be removed before or after a box is mounted. **See Figure 6-13.** Any sharp edges that might penetrate the plastic sheathing that surrounds the NM cable should be removed. Burrs or sharp edges can typically be removed by rotating a screwdriver around the knockout hole. **See Figure 6-14.**

Plastic Box Depth Positioning

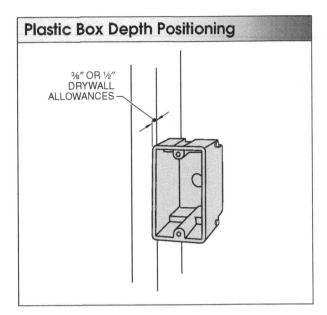

³⁄₈″ OR ½″ DRYWALL ALLOWANCES

Figure 6-12. *When mounting plastic boxes, the box should be positioned against the stud so that the front of the box is flush with the finished surface of the wall.*

Plastic Box Knockouts

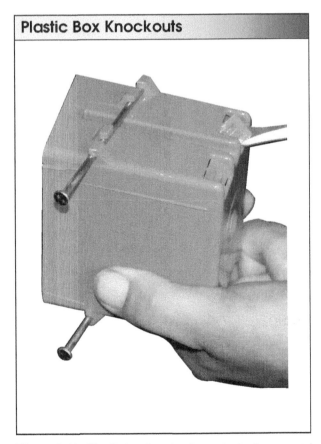

Figure 6-13. *Plastic box knockouts are typically removed with a screwdriver.*

Deburring Plastic Box Knockout Holes

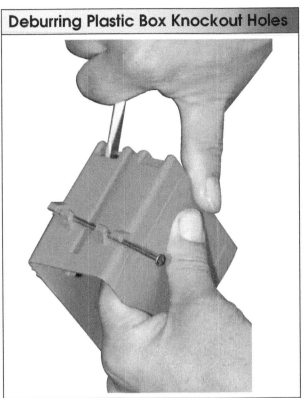

Figure 6-14. *Burrs and sharp edges can be removed from knockout holes by rotating a screwdriver around the hole.*

Receptacle and switch boxes must be installed to allow enough room for the drywall to be flush with the front section of the box.

When securing cables to a box, at least 8″ of wire must extend beyond the bottom of the box to allow enough wire for making connections to receptacles and switches. **See Figure 6-15.** When securing nonmetallic cables to studs and joists, a wide range of staples are available. **See Figure 6-16.**

Determining Length of Conductors to Plastic Box

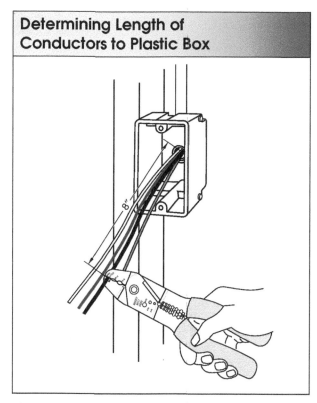

Figure 6-15. *At least 8″ of wire must extend beyond the bottom of a box to allow for connections to receptacles and switches.*

Mounting Plastic Boxes between Studs. To mount a plastic box between studs a cross block is used. Cross blocks are installed with the top edge 46″ above the floor so that when the box is attached, the face of the box is flush with the finished wall. The box is attached to the cross block with nails or screws. **See Figure 6-17.**

Drilling Holes

There is some difference of opinion on whether holes should be drilled before or after boxes are mounted. For the beginner, it is best to mount the boxes first according to the plan and then drill the holes. Mounting the box first reduces the number of unnecessary holes and saves time.

Large-diameter holes (½″ and larger) can be cut into wooden joists and studs by using a power drill and either a spade bit, auger bit, self-feed wood borer, or hole saw attachment.

NM Cable Staples

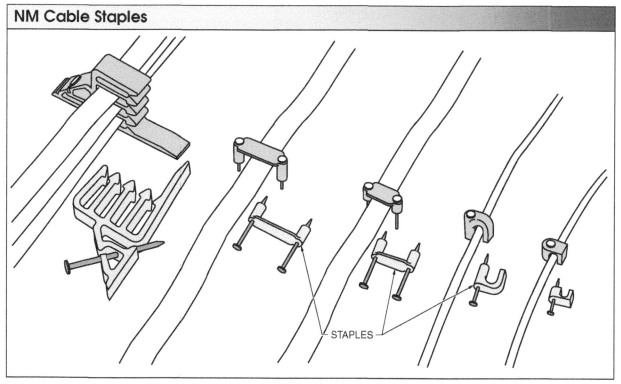

STAPLES

Figure 6-16. *NM cables must be secured with staples at various positions of the cable run.*

Mounting Plastic Boxes between Studs

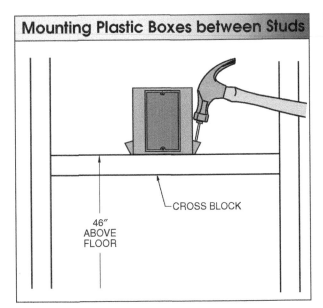

Figure 6-17. *Plastic boxes are mounted to cross blocks when located between studs.*

Holes in studs or floor joists can be made most effectively with an offset drill. **See Figure 6-18.** Electricians use offset drills almost exclusively because offset drills easily fit between studs and joists. When an offset drill is not available, a heavy-duty electric drill with a spade power bit is substituted. **See Figure 6-19.** All holes should be drilled as straight as possible to aid in pulling wires through multiple studs. **See Figure 6-20.**

Offset Drills

Milwaukee Electric Tool Corp.

Figure 6-18. *Offset or right angle drills are ideal for drilling in tight spots between studs.*

Heavy-Duty Electric Drills

Milwaukee Electric Tool Corp.

Figure 6-19. *A heavy-duty electric drill with a hole saw is used for drilling holes when a right angle drill is not available.*

Drilling Holes in Studs

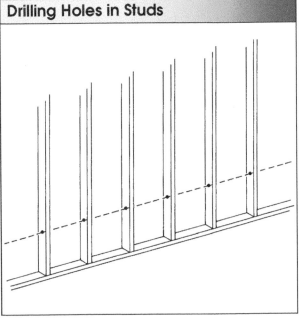

Figure 6-20. *Holes in studs should be drilled at the same height to aid in running cables.*

Installing Cable Runs

When all holes have been properly drilled, NM cable can be installed quickly and easily. To obtain the proper length of cable required, the cable should be pulled into position before cutting. Positioning the cable ensures that enough, but not an excessive amount of, cable is pulled. **See Figure 6-21.**

Pulling NM Cable to Determine Length

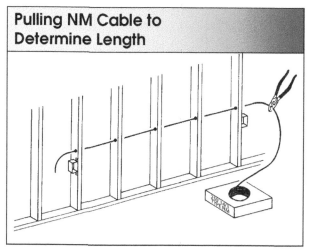

Figure 6-21. *Cable should be pulled to the proper positions and cut to length to avoid waste.*

Cable and conductors should be stripped before being installed in a box. **See Figure 6-22.** Stripping the wire before installing the wire into a box saves time later when outlets and switches are being installed.

Stripping Cable and Conductors

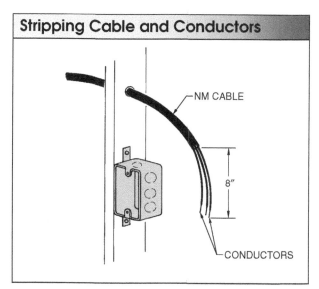

NM CABLE

8″

CONDUCTORS

Figure 6-22. *Cable jacket and conductor insulation should be removed from the wire before it is installed in the box.*

Stripping cable is much easier to perform from outside a box than from inside a box. When stripping the cable, leave at least 8″ of wire for rough-in purposes. It is easier to cut off excess wire later than to try to add wire.

Wiring through Metal Studs. Metal studs typically have holes cut in them for the purpose of routing conduit and cable through them. Prior to routing any cable or conduit through the holes, plastic bushings must be inserted in each hole to prevent damage to cable and nonmetallic conduit. **See Figure 6-23.**

Working with Metal Studs

METAL STUDS — CABLE

PLASTIC BUSHINGS

Figure 6-23. *Plastic bushings must be inserted in each hole in metal studs to prevent damage to cable and nonmetallic conduit.*

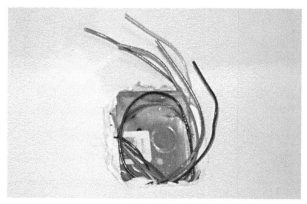

Conductors must be routed through a box with enough length out of the box to provide for adequate stripping of insulation and termination of devices.

Routing through Studs and Joists. Running nonmetallic cable through stud walls and through floor joists is the most popular way of routing. **See Figure 6-24.** Typically, holes are drilled through the center of the studs and joists. This practice reduces the possibility of a nail penetrating deep enough to puncture the cable. When cable is run through drilled holes, no additional support is required.

In some instances, studs may have to be notched instead of drilled. **See Figure 6-25.** When notches are substituted for holes, the cable run must be protected from nails by a steel plate at least 1/16″ thick.

Routing Cable through Notches in Studs

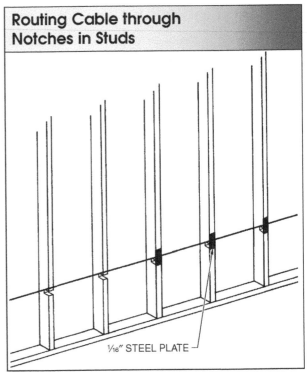

1/16″ STEEL PLATE

Figure 6-25. *When notches are substituted for holes, a cable must be protected by a minimum of a 1/16″ steel plate.*

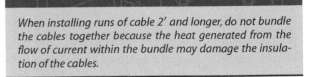

When installing runs of cable 2′ and longer, do not bundle the cables together because the heat generated from the flow of current within the bundle may damage the insulation of the cables.

Routing Cables through Studs and Joists

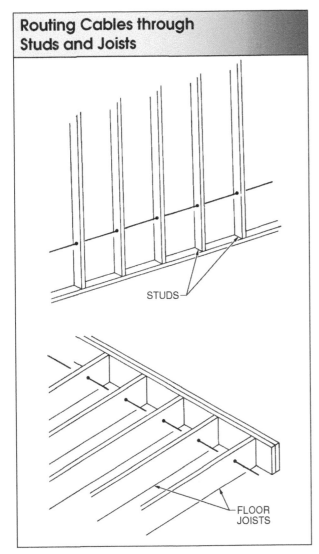

STUDS

FLOOR JOISTS

Figure 6-24. *Generally cable is run through holes in studs or floor joists.*

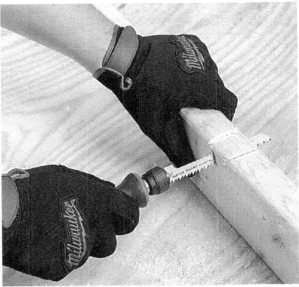

Milwaukee Electric Tool Corp.

A jab saw can be used to notch studs when routing cables.

Routing around Corners. Routing cables through solid corners can sometimes be complicated. Various methods are used to overcome the problem of solid corners.

One method is to drill holes from each side at an angle to accommodate a cable run. **See Figure 6-26.**

Drilling holes from each side will save wire but may be time-consuming if the holes do not line up properly. Another method is to notch the corner studs, using steel plates for protection. A steel plate is a common, inexpensive item available at most electrical supply distributors and is easily installed with a hammer.

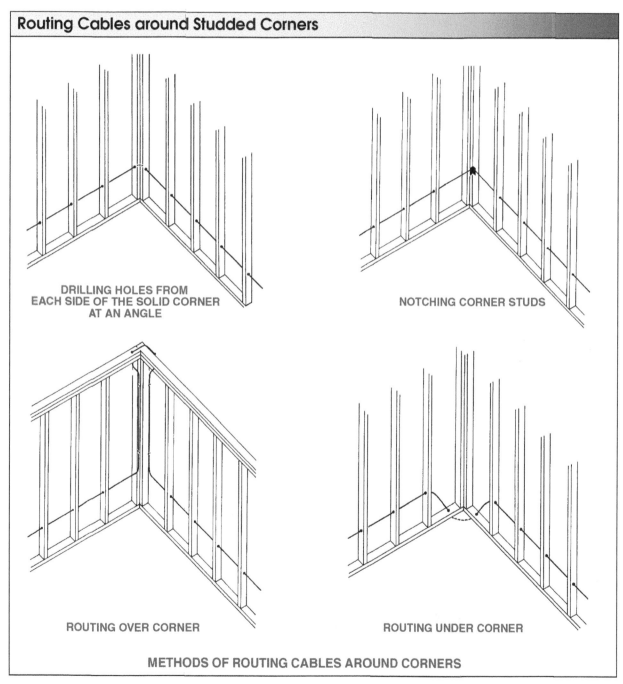

Routing Cables around Studded Corners

DRILLING HOLES FROM EACH SIDE OF THE SOLID CORNER AT AN ANGLE

NOTCHING CORNER STUDS

ROUTING OVER CORNER

ROUTING UNDER CORNER

METHODS OF ROUTING CABLES AROUND CORNERS

Figure 6-26. *Cables can be run through holes or notches in corner studs, or run over or under the corner.*

Routing over or under a corner is another method of routing cable. At first glance, the looping under method appears to be the most economical method, but looping under requires extra time. The extra time is due to someone being required to go into the basement or crawl space to send the wire back up through the floor. The over method is accomplished rapidly by one electrician without leaving the room.

Routing through Masonry Walls. Type NM cable may be fished through masonry walls if no moisture is present and the wall is above grade. **See Figure 6-27.** When moisture is present, type NMC cable must be used. Neither NM nor NMC cable can be embedded in poured cement, concrete, or aggregate.

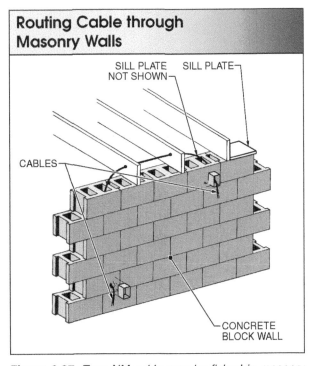

Figure 6-27. *Type NM cable may be fished in masonry walls when no moisture is present and the cable run is above grade; when moisture is present, type NMC must be used.*

Using a hammer or piece of wood cut to the proper length as a height gauge ensures that all holes drilled through studs for cable runs are at the same height.

Securing Cable

NM and NMC cable must be supported or secured (stapled) every 4½′ of cable run and within 12″ of a box. NM and NMC cables can also be secured using guard strips.

Cable Run Stapling. When cable has been pulled into position, the cable must be securely fastened with special staples. Nonmetallic cable staples typically have a plastic strap that reduces the possibility of damage to the cable. **See Figure 6-28.** The NEC® requires that all cables be secured near an outlet box.

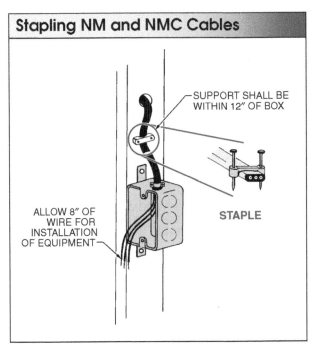

Figure 6-28. *NM and NMC cable is firmly secured using staples designed for nonmetallic cable.*

When cable is not supported by holes or notches, but rests along the sides of joists, studs, or rafters, the cable must be stapled every 4½′. **See Figure 6-29.** Cables must also be stapled under floor joists in such a way that the cables do not present problems later when the basement ceiling is being installed.

Placing Guard Strips. Nonmetallic cables must be protected when routed across the face of roof rafters, studs within 7′ of the floor, or attics accessible by stairs. A quick method of protection is the use of wooden guard strips. **See Figure 6-30.** Wooden guard strips must be at least as thick as the cable being protected.

Cable Stapling Distance

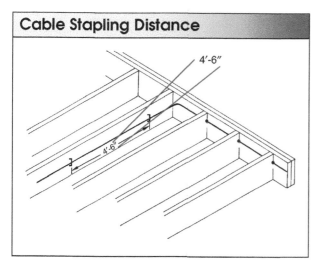

Figure 6-29. *Horizontal cable runs not supported by holes must be stapled every 4½'. Cable runs on the sides of joists, studs, or rafters require no additional protection.*

Placing Guard Strips

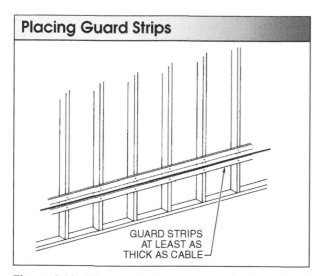

GUARD STRIPS AT LEAST AS THICK AS CABLE

Figure 6-30. *When cable is run across the face of roof rafters, across studding within 7' of the floor, or in attics accessible by stairs, it must be protected by guard strips at least as high as the cable.*

Securing Ground Wires. When a cable is secured, a good mechanical and electrical ground must be established to provide electrical protection. **See Figure 6-31.** The three widely accepted methods of proper residential grounding are component, pigtail, and clip.

Securing Ground Wires

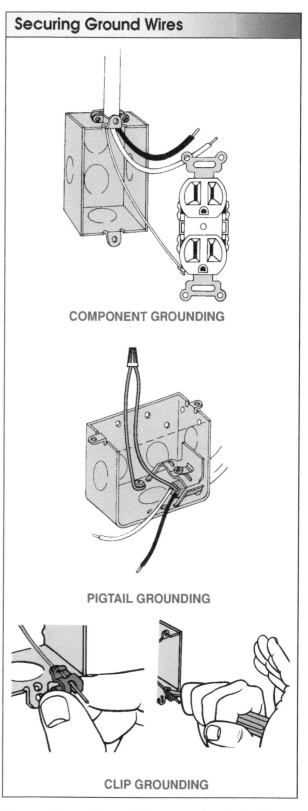

COMPONENT GROUNDING

PIGTAIL GROUNDING

CLIP GROUNDING

Figure 6-31. *The three widely accepted methods of proper residential grounding are component grounding, pigtail grounding, and clip grounding.*

Component grounding is a grounding method where the ground wire is attached directly to an electrical component such as a receptacle. Component grounding requires that the grounding wire be attached before the electrical component is permanently mounted.

Pigtail grounding is a grounding method where two grounding wires are used to connect an electrical device to a grounding screw in the box and then to system ground. The box ground wire is secured to a threaded hole in the bottom of the box. Once secured, the cable ground wire is pigtailed to the box ground wire.

Clip grounding, like pigtail grounding, can be secured as the cable is put in place. *Clip grounding* is a grounding method where a grounding clip is merely slipped over the grounding wire from the electrical device. The grounding wire and grounding clip are secured with pressure using a screwdriver. Clip grounding is the method most often used when there are many ground connections to be made to reduce installer fatigue.

Ground wires are installed to help protect equipment and wiring from electrical power surges. The main cause of these surges is lightning strikes. The magnetic field created by a lightning strike can induce voltage into both current-carrying conductors, such as those found in appliances and electronics, and noncurrent-carrying conductors, such as water pipes. If the current from the surge cannot find a path from the appliance to ground, it will take a path through electronic equipment or a phone line, or arc through the air.

Securing Cables in Outlet Boxes

Cables are secured to outlet boxes by clamps. The three types of clamps typically available for nonmetallic cables are saddle, straight, and cable connectors. **See Figure 6-32.** Saddle clamps and straight clamps are part of an outlet box when manufactured. Cable connectors are installed by electricians. **See Figure 6-33.** Locknuts are installed so that the points of the nut point inward to dig firmly into the metal box.

Tech Tip: Cord-and-Plug-Connected Equipment

CORD-AND-PLUG-CONNECTED EQUIPMENT

RECEPTACLE OUTLET INSTALLED WHEREVER CORD-AND-PLUG-CONNECTED EQUIPMENT IS TO BE USED
• 210.50(B)

6' MAXIMUM
•210.50(C)

6' MAXIMUM
•210.50(C)

Box Cable Clamps

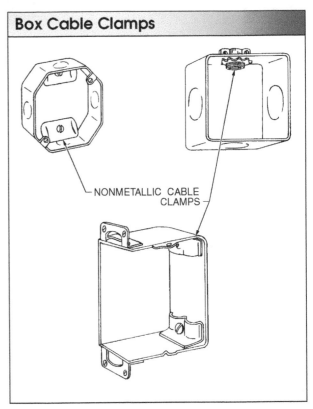

NONMETALLIC CABLE CLAMPS

Figure 6-32. *Various types of cable clamps are used to secure nonmetallic cable to outlet and junction boxes.*

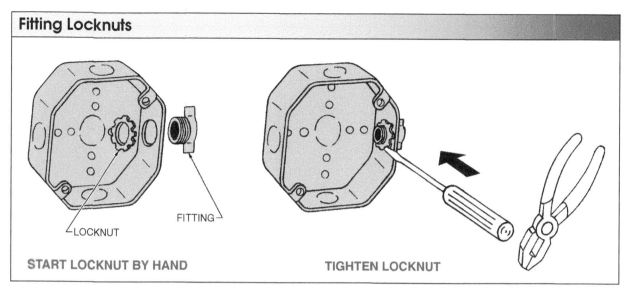

Fitting Locknuts

LOCKNUT

FITTING

START LOCKNUT BY HAND

TIGHTEN LOCKNUT

Figure 6-33. *The points of the locknut dig firmly into the metal box when tightened.*

When inserting cable through clamps, care must be taken not to damage any insulation by scraping the cable on the rough surfaces and edges. Care must also be taken when tightening cable clamps. Too much pressure can cause the clamp to penetrate the insulation and cause a short circuit.

Never use a screwdriver when removing staples from a cable. Use electrician's pliers or a nail puller.

Folding Back Wires

Many wires have been needlessly damaged because they were left exposed. The very last roughing-in task is to tuck the wires neatly into the outlet boxes. **See Figure 6-34.** The time taken to fold back wires helps to avoid damage to the wires as wall coverings are being placed and other construction is being performed. Prior to being folded back into a box, wires must be properly connected and spliced together with connection devices such as wire nuts or splice caps. Wire nuts and splice caps must be the correct size to provide a sound connection. Boxes must be sized to accommodate not only wires, but also wire nuts and splice caps. Incorrectly sized connection devices can cause connections to fail or loosen.

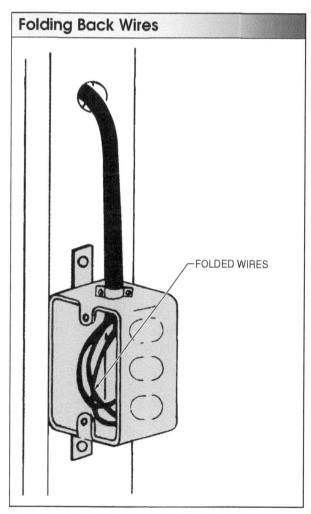

Folding Back Wires

FOLDED WIRES

Figure 6-34. *Wires that are neatly folded into a box avoid being damaged as other construction is performed.*

Name_____   Date _____

Working with Nonmetallic-Sheathed Cable

_____ **1.** NMC is protected by a strong ___ jacket.

_____ **2.** NMC usually contains two or three insulated conductors and a bare ___ wire.

_____ **3.** The first number identifying nonmetallic-sheathed cable provides the ___ size.

_____ **4.** The second number identifying nonmetallic-sheathed cable provides the number of ___ conductors.

_____ **5.** The letters NM and NMC indicate ___ of nonmetallic-sheathed cable.

_____ **6.** Type NMC cable is ___ -resistant and fungus-resistant.

_____ **7.** Type ___ cable is typically used for interior (dry) wiring.

_____ **8.** The letters ___ on nonmetallic-sheathed cable stand for "with ground."

_____ **9.** Nonmetallic cable is typically sold boxed in ___′ rolls.

_____ **10.** An outlet box or junction box is installed when NM cable is ___ or terminated.

T   F   **11.** Boxes in walls can be completely covered for appearance.

T   F   **12.** The size of an outlet box is generally based on the number and size of wires entering the box.

_____ **13.** The three shapes of outlet boxes are ___, ___, and ___.
   A. square; round; rectangular
   B. square; hexagonal; octagonal
   C. square; rectangular; octagonal
   D. none of the above

T   F   **14.** Only square boxes are used for junction boxes.

_____ 15. ___ are round indentations punched into a metal box and held in place by narrow strips of metal.

_____ 16. Almost all ceiling outlet boxes are mounted with ___.

T   F   17. NM cable should be cut before being pulled through a wall.

T   F   18. NM cable should be stripped before being installed in a box.

_____ 19. When stripping NM and NMC cable, at least ___" should be left outside the box.

_____ 20. Cable routed through notches in stud walls is protected by a(n) ___" steel plate.

T   F   21. Types NM and NMC cable can be embedded in concrete.

T   F   22. Component grounding is the direct attachment of the component grounding wire to a box.

T   F   23. Cables are secured in outlet boxes by clamps.

_____ 24. Cable must be supported within ___" of a box.

T   F   25. Type NM cable may be fished in masonry walls when no moisture is present.

T   F   26. Type NM cable can be routed through the corners of stud walls by notching the corner studs.

_____ 27. ___ refers to the placement of electrical boxes and wires before wall coverings and ceilings are installed.

T   F   28. Pocketknives should not be used to remove insulation from wires.

T   F   29. Cable rippers are used to remove the outer cover of cables efficiently.

T   F   30. Outlet boxes are generally secured by nails.

Mounting Square Boxes

_____ **1.** Brace   _____ **3.** Bar hanger

_____ **2.** Side strap   _____ **4.** Front strap

  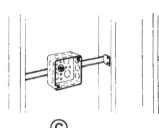 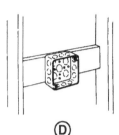

Ⓐ   Ⓑ   Ⓒ   Ⓓ

Cables

_____ **1.** Clamp

_____ **2.** Stripping

_____ **3.** NMC

_____ **4.** Octagonal box

_____ **5.** Notching

_____ **6.** NM

_____ **7.** Pigtail ground

_____ **8.** Knockout

_____ **9.** Ground

_____ **10.** Guard strip

A. Corrosion-resistant cable

B. Provides electrical protection

C. Secures cable in outlet box

D. Used for lighting outlets

E. Round indentation

F. Cable for dry interior wiring

G. Must be at least as thick as cable

H. Removing insulation

I. Ground wire secured to box

J. Requires $\frac{1}{16}''$ steel plate for cable protection

Cables in Walls and Floors

_____ **1.** Cable through drilled studs

_____ **2.** Cable through drilled floor joists

_____ **3.** Cable protected by guard strips within 7′ of floor, in attics, or across roof rafters

_____ **4.** Cable through notched studs protected by $\frac{1}{16}$″ steel plate

_____ **5.** Cable supported every 4½′ on sides of joists, studs, or rafters

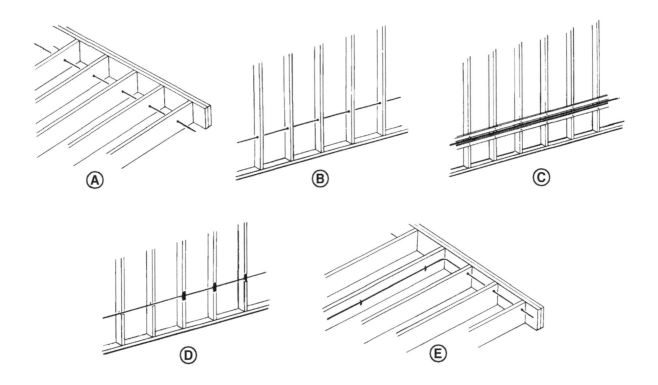

Electrical Boxes

_____   **1.** Octagonal box, side strap

_____   **2.** Rectangular box, side nails

_____   **3.** Rectangular box, frontal strap

_____   **4.** Octagonal box, side bracket

_____   **5.** Octagonal box, bar hanger

_____   **6.** Rectangular box, side strap

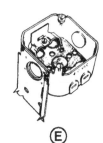

Ⓐ

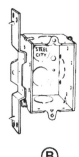

Ⓑ

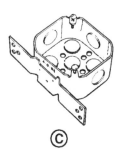

Ⓒ

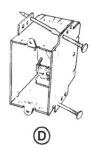

Ⓓ

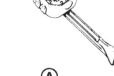

Ⓔ

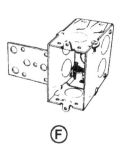

Ⓕ

161

Grounding

_____ **1.** Component grounding

_____ **2.** Clip grounding

_____ **3.** Pigtail grounding

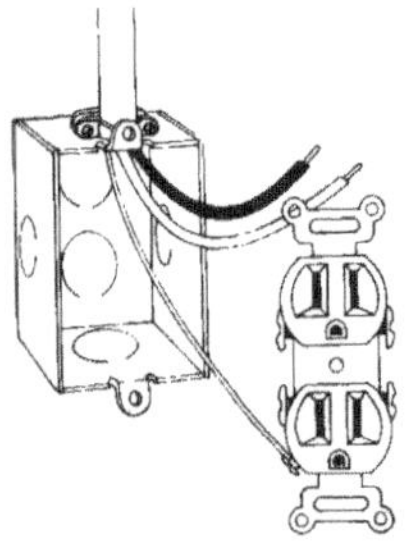

(A)

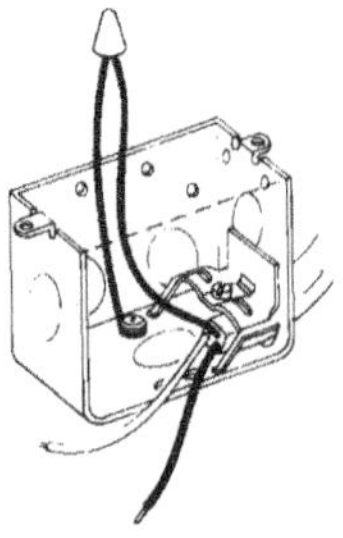

(B)

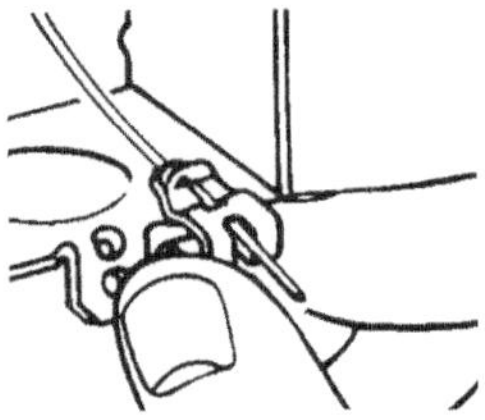

(C)

NM and NMC

_____ **1.** #14 wire, two conductor, with ground, corrosion-resistant

_____ **2.** #14 wire, three conductor, with ground

_____ **3.** #12 wire, two conductor, with ground

_____ **4.** #12 wire, three conductor, with ground

_____ **5.** #14 wire, two conductor

_____ **6.** #12 wire, two conductor, with ground, corrosion-resistant

14/3 W/GR TYPE NM   14/2   14/2 W/GR TYPE NMC
(A)   (B)   (C)

12/3 W/GR TYPE NM   12/2 W/GR TYPE NM   12/2 W/GR TYPE NMC
(D)   (E)   (F)

Creating Electrical Systems from Electrical Plans

Chapter 7 provides an overview of how complete electrical circuits are installed. Combining electrical components and wires into circuits is the most rewarding part of electrical construction. Various symbols and schematics are used to design systems to fit specific needs. Chapter 7 provides information on how electrical plans are developed into actual system wiring, how various switching circuits are wired, and how switches, receptacles, and lights are installed.

ROOM LAYOUTS

The ultimate objective of learning to read and follow electrical plans is the finished product, an electrical system that is ready to use. All the knowledge of electrical components and installation is expressed in the finished electrical system. The use of standard plans and private plans result in a full-scale house plan and electrical system that meets the needs of the occupants.

Component Layout

Electrical components are installed in locations typically shown on the floor plans. An isometric drawing (construction layout) can help to clarify the approximate location of all switches, receptacles, and light

fixtures. **See Figure 7-1 through Figure 7-15.** Component layouts are directly coordinated to the actual construction layout which includes the stud walls and electrical boxes.

Electrical Layouts

An electrical layout is a drawing that indicates how the component parts of a circuit will be connected to one another and where the wires will be run. Careful planning of the electrical layout results in substantial savings by eliminating long runs of wire. Electrical layouts are used by individuals to determine where to place electrical boxes and how many boxes are required. Electrical layouts are helpful in determining a bill of material and approximate costs of an electrical project.

Wire runs shown on actual construction drawings should indicate the most economical use of wire.

Wire runs are laid out on a wiring layout in a very smooth and definite pattern to make the drawing easier to follow. In many cases, wire runs shown at right angles on the wiring layout could have been run diagonally in the room to conserve wire. When any cable runs are routed on a job site, shortening the wire runs always results in lower installation costs.

Bedroom Wiring

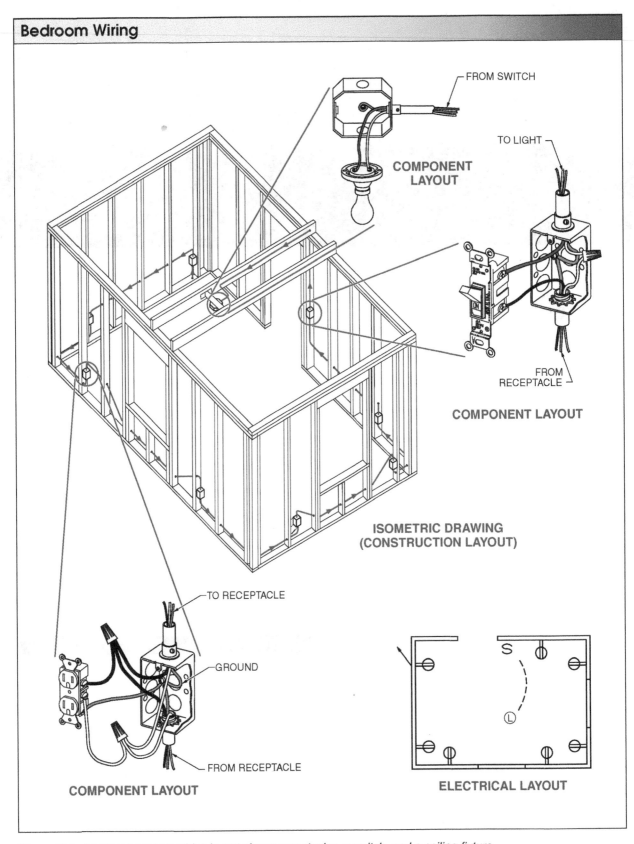

FROM SWITCH

COMPONENT LAYOUT

TO LIGHT

COMPONENT LAYOUT

FROM RECEPTACLE

ISOMETRIC DRAWING (CONSTRUCTION LAYOUT)

TO RECEPTACLE

GROUND

FROM RECEPTACLE

COMPONENT LAYOUT

ELECTRICAL LAYOUT

Figure 7-1. *A typical bedroom wiring layout shows receptacles, a switch, and a ceiling fixture.*

Exploded Views

Exploded views or component layouts provide key points for the construction of an electrical circuit by providing detailed representations of how each component is actually wired into the system. When component wiring is duplicated, the component can be given a letter designation to indicate the device is wired exactly the same way.

Wiring Methods

Although it is assumed that conduit is being used in most construction layouts, all circuits can be wired with metallic sheathed cable or nonmetallic cable.

When performing remodeling work, it is important to locate the studs behind a finished wall without damaging the wall. This can be done by either using a battery-operated stud finder or by drilling a ¼" hole and inserting a wire bent to a 90° angle and then rotating the wire until you feel the stud.

Use of Room Designs

Wiring layouts provide enough information to successfully wire each room. Wiring layouts must be studied carefully to determine how each component will be connected. All residential electrical systems vary, but all systems have items in common. The design of an electrical circuit for one room might require only minor changes to be used for another room.

The minimum capacity branch circuit for a dwelling unit is 15 A. A branch circuit can be any size required to supply one load or multiple loads. Typical branch circuits in dwellings supply combination or single loads. For example, lighting and multiple items connected to receptacles are combination loads, while HVAC units, food waste disposals, trash compactors, and other similar appliances are single loads.

A 15 A branch circuit commonly has 10 outlets to supply power to lighting fixtures and general purpose receptacles. A 20 A branch circuit has 13 outlets to supply power to lighting fixtures and general purpose receptacles. The number of receptacles permitted on a branch circuit is determined by dividing the rating of the overcurrent protection device (OCPD) by 1.5 A. The NEC® does not limit the number of receptacles in a branch circuit. **See Figure 7-16.**

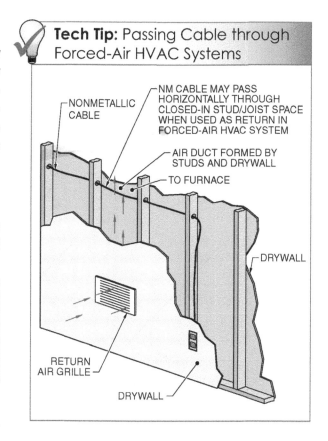

Tech Tip: Passing Cable through Forced-Air HVAC Systems

NONMETALLIC CABLE

NM CABLE MAY PASS HORIZONTALLY THROUGH CLOSED-IN STUD/JOIST SPACE WHEN USED AS RETURN IN FORCED-AIR HVAC SYSTEM

AIR DUCT FORMED BY STUDS AND DRYWALL

TO FURNACE

DRYWALL

RETURN AIR GRILLE

DRYWALL

Per the NEC®, metal and nonmetallic raceways must be continuous between all cabinets, boxes, fittings, or other enclosures and outlets.

Bedroom and Closet Wiring

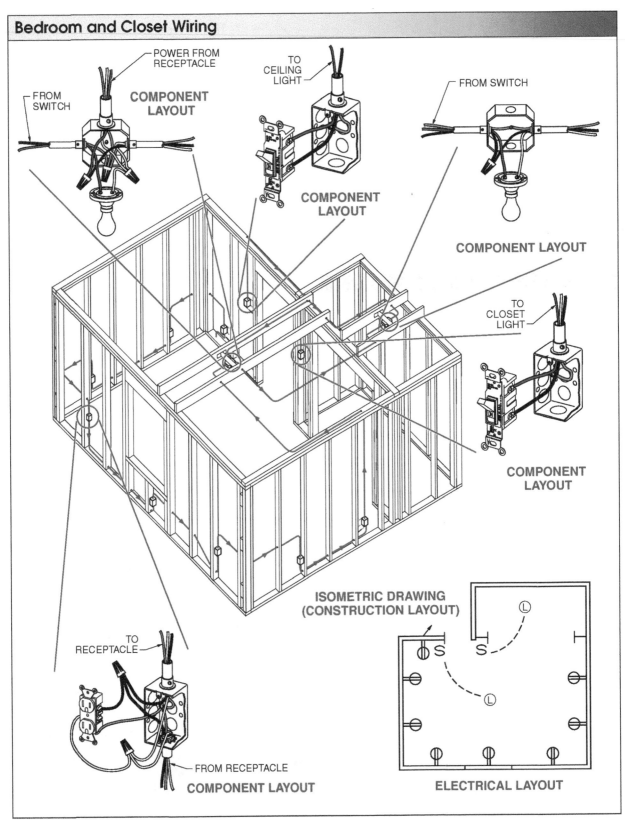

POWER FROM RECEPTACLE

FROM SWITCH

COMPONENT LAYOUT

TO CEILING LIGHT

COMPONENT LAYOUT

FROM SWITCH

COMPONENT LAYOUT

TO CLOSET LIGHT

COMPONENT LAYOUT

TO RECEPTACLE

FROM RECEPTACLE

COMPONENT LAYOUT

ISOMETRIC DRAWING (CONSTRUCTION LAYOUT)

ELECTRICAL LAYOUT

Figure 7-2. *A typical bedroom with closet wiring layout shows receptacles, two ceiling lights, and switches for the two ceiling lights.*

Bathroom Wiring

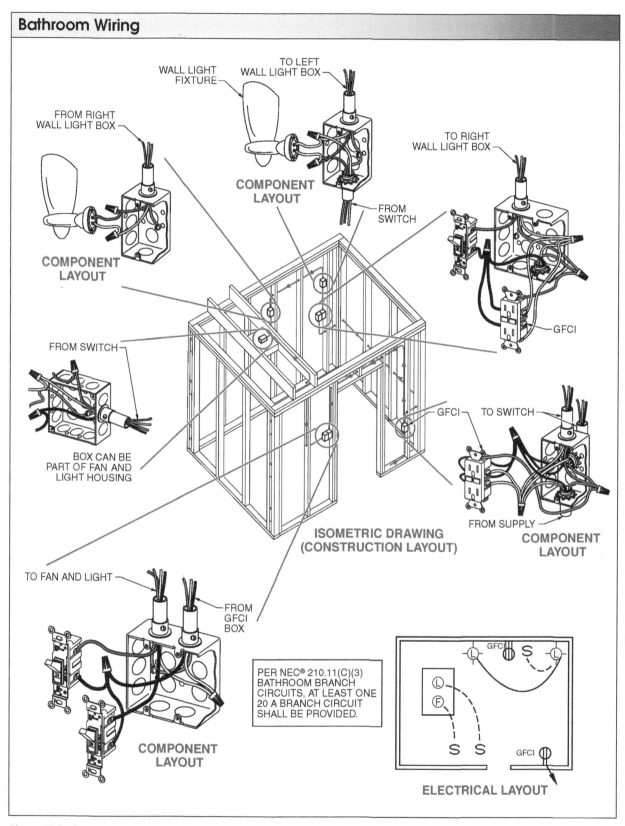

WALL LIGHT FIXTURE

TO LEFT WALL LIGHT BOX

FROM RIGHT WALL LIGHT BOX

COMPONENT LAYOUT

COMPONENT LAYOUT

TO RIGHT WALL LIGHT BOX

FROM SWITCH

GFCI

FROM SWITCH

BOX CAN BE PART OF FAN AND LIGHT HOUSING

GFCI

TO SWITCH

FROM SUPPLY

COMPONENT LAYOUT

ISOMETRIC DRAWING (CONSTRUCTION LAYOUT)

TO FAN AND LIGHT

FROM GFCI BOX

PER NEC® 210.11(C)(3) BATHROOM BRANCH CIRCUITS, AT LEAST ONE 20 A BRANCH CIRCUIT SHALL BE PROVIDED.

COMPONENT LAYOUT

GFCI

S

L

L

F

S   S

GFCI

ELECTRICAL LAYOUT

Figure 7-3. *A typical bathroom wiring layout shows two lighting fixtures, a combined fan and ceiling light, and two GFCI receptacles.*

Hallway and Light Wiring

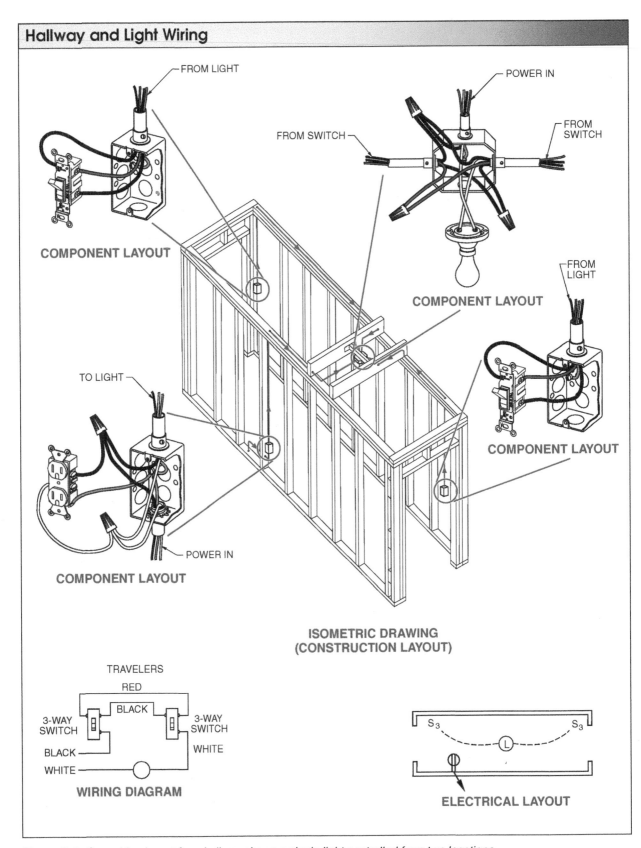

FROM LIGHT

COMPONENT LAYOUT

POWER IN

FROM SWITCH

FROM SWITCH

COMPONENT LAYOUT

FROM LIGHT

COMPONENT LAYOUT

TO LIGHT

POWER IN

COMPONENT LAYOUT

ISOMETRIC DRAWING
(CONSTRUCTION LAYOUT)

TRAVELERS

RED

BLACK

3-WAY
SWITCH

3-WAY
SWITCH

BLACK

WHITE

WHITE

WIRING DIAGRAM

S_3

S_3

L

ELECTRICAL LAYOUT

Figure 7-4. *One wiring layout for a hallway shows a single light controlled from two locations.*

Hallway and Two-Light Wiring—Two Switches

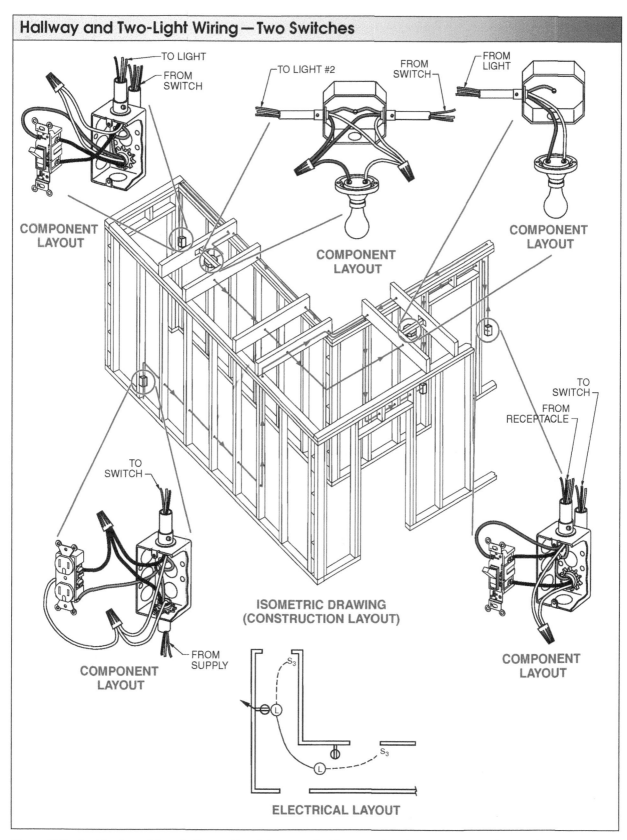

TO LIGHT

FROM SWITCH

COMPONENT LAYOUT

TO LIGHT #2

FROM SWITCH

FROM LIGHT

COMPONENT LAYOUT

COMPONENT LAYOUT

TO SWITCH

FROM RECEPTACLE

TO SWITCH

COMPONENT LAYOUT

FROM SUPPLY

ISOMETRIC DRAWING (CONSTRUCTION LAYOUT)

COMPONENT LAYOUT

S_3

S_3

ELECTRICAL LAYOUT

Figure 7-5. *A wiring layout for a hallway with a light controlled from two locations shows use of two switches.*

Hallway and Two-Light Wiring—Three Switches

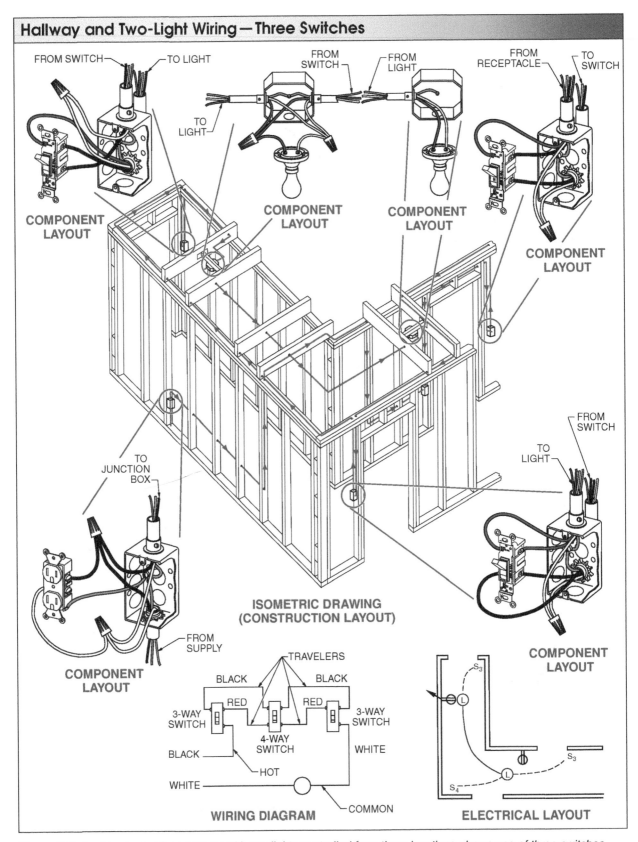

FROM SWITCH — TO LIGHT

COMPONENT LAYOUT

FROM SWITCH — FROM LIGHT

TO LIGHT

COMPONENT LAYOUT

COMPONENT LAYOUT

FROM RECEPTACLE — TO SWITCH

COMPONENT LAYOUT

TO JUNCTION BOX

COMPONENT LAYOUT

FROM SUPPLY

ISOMETRIC DRAWING (CONSTRUCTION LAYOUT)

FROM SWITCH

TO LIGHT

COMPONENT LAYOUT

TRAVELERS

BLACK   BLACK

RED   RED

3-WAY SWITCH   3-WAY SWITCH

4-WAY SWITCH

BLACK   WHITE

HOT

WHITE

COMMON

WIRING DIAGRAM

S_3

L

L

S_3

S_4

ELECTRICAL LAYOUT

Figure 7-6. *A wiring layout for a hallway with two lights controlled from three locations shows use of three switches.*

Living Room Wiring

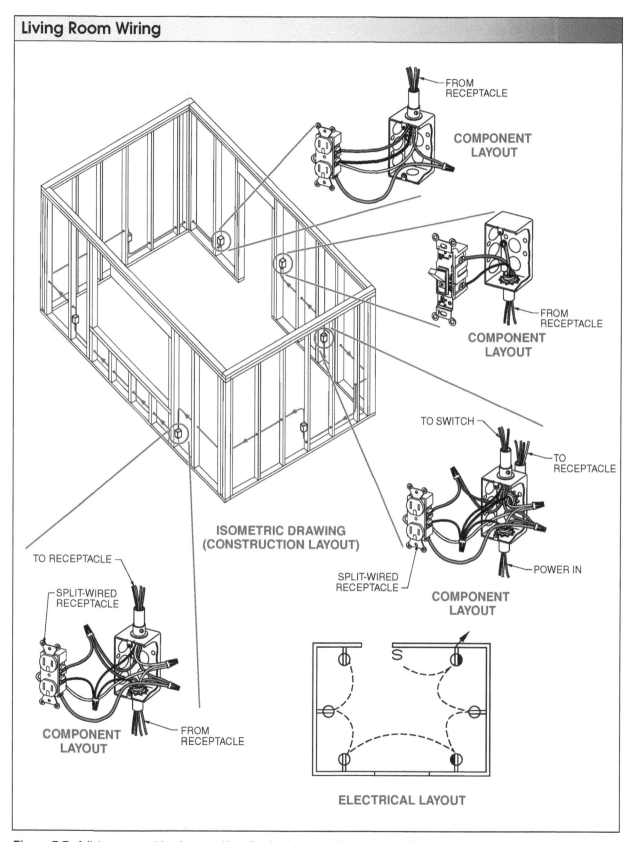

Figure 7-7. *A living room wiring layout with split-wired receptacles controlled from one location uses one switch.*

Living Room Wiring—Two Switches

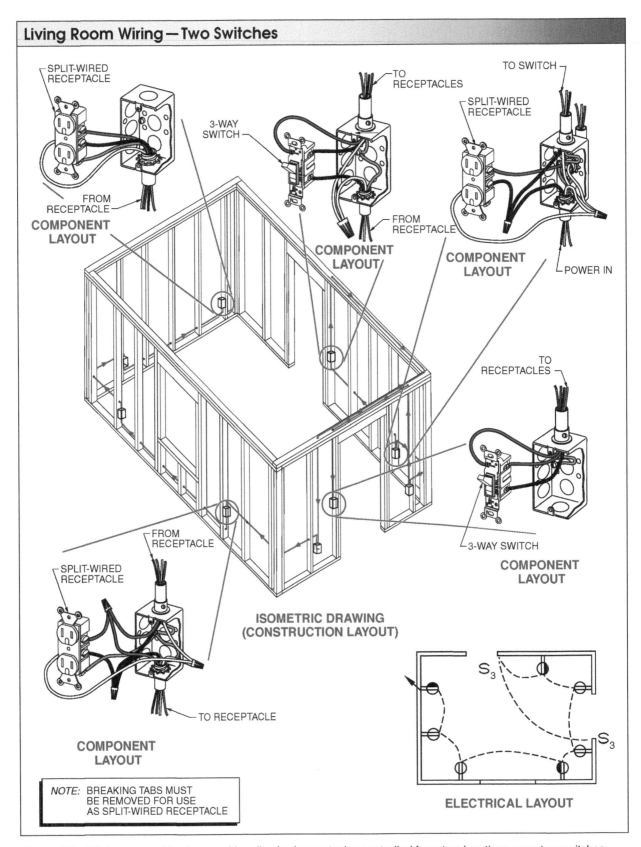

SPLIT-WIRED RECEPTACLE

FROM RECEPTACLE

COMPONENT LAYOUT

TO RECEPTACLES

3-WAY SWITCH

FROM RECEPTACLE

COMPONENT LAYOUT

TO SWITCH

SPLIT-WIRED RECEPTACLE

COMPONENT LAYOUT

POWER IN

ISOMETRIC DRAWING (CONSTRUCTION LAYOUT)

TO RECEPTACLES

3-WAY SWITCH

COMPONENT LAYOUT

FROM RECEPTACLE

SPLIT-WIRED RECEPTACLE

TO RECEPTACLE

COMPONENT LAYOUT

S_3

S_3

ELECTRICAL LAYOUT

NOTE: BREAKING TABS MUST BE REMOVED FOR USE AS SPLIT-WIRED RECEPTACLE

Figure 7-8. *A living room wiring layout with split-wired receptacles controlled from two locations uses two switches.*

Garage Wiring

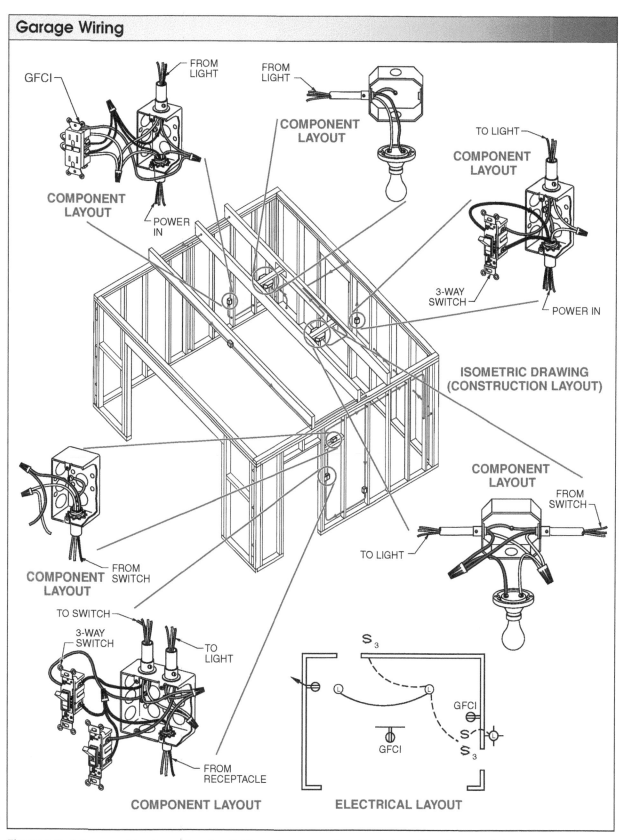

Figure 7-9. *A garage wiring layout has two lights and uses one 1-way switch and two 3-way switches.*

Outdoor Garage Lighting Wiring

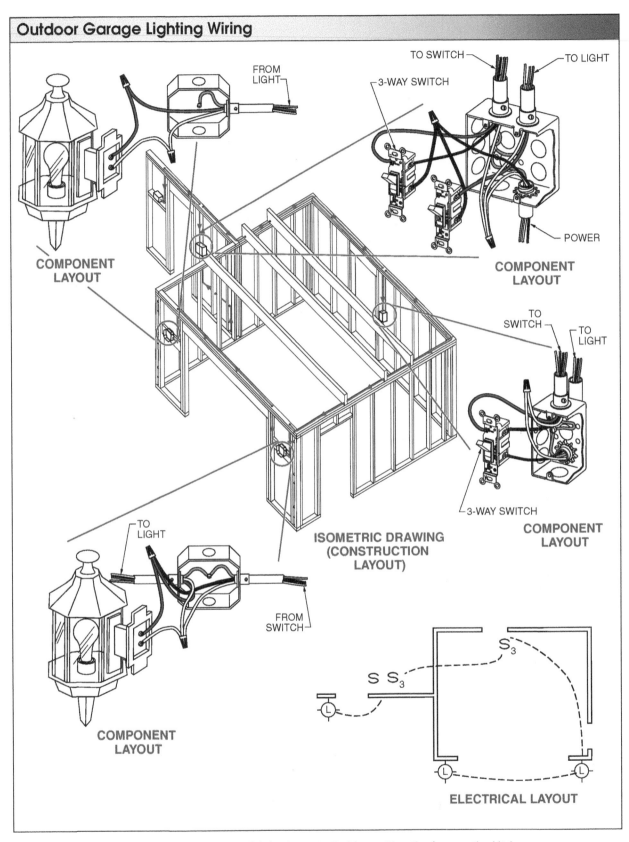

Figure 7-10. *In this wiring layout, the outdoor lighting is controlled from either the foyer or the kitchen area.*

Kitchen Wiring

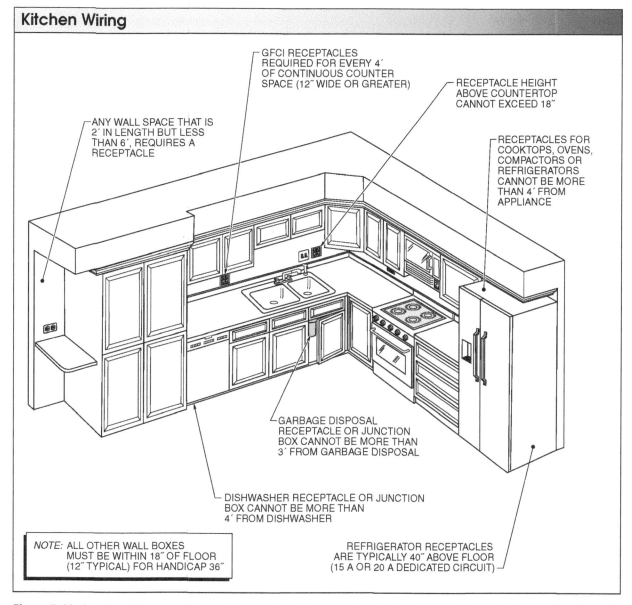

GFCI RECEPTACLES REQUIRED FOR EVERY 4′ OF CONTINUOUS COUNTER SPACE (12″ WIDE OR GREATER)

RECEPTACLE HEIGHT ABOVE COUNTERTOP CANNOT EXCEED 18″

ANY WALL SPACE THAT IS 2′ IN LENGTH BUT LESS THAN 6′, REQUIRES A RECEPTACLE

RECEPTACLES FOR COOKTOPS, OVENS, COMPACTORS OR REFRIGERATORS CANNOT BE MORE THAN 4′ FROM APPLIANCE

GARBAGE DISPOSAL RECEPTACLE OR JUNCTION BOX CANNOT BE MORE THAN 3′ FROM GARBAGE DISPOSAL

DISHWASHER RECEPTACLE OR JUNCTION BOX CANNOT BE MORE THAN 4′ FROM DISHWASHER

NOTE: ALL OTHER WALL BOXES MUST BE WITHIN 18″ OF FLOOR (12″ TYPICAL) FOR HANDICAP 36″

REFRIGERATOR RECEPTACLES ARE TYPICALLY 40″ ABOVE FLOOR (15 A OR 20 A DEDICATED CIRCUIT)

Figure 7-11. *A pictorial kitchen layout shows the locations of all electrical appliances and receptacles specified by building and electrical codes.*

When installing GFCI receptacles near a kitchen countertop, consideration must be given to the height of the backsplash, which is installed to prevent water from getting between the countertop and the wall, at the back of the wall countertop space. Typical backsplash height is 4″ but can vary from one house to another. Check with the builder or check the blueprints before installing receptacles near a kitchen counter.

Leviton Manufacturing Co., Inc.
GFCI receptacles are available in a variety of colors to help coordinate with kitchen and bathroom decor.

Kitchen Wiring — 230 V Receptacles

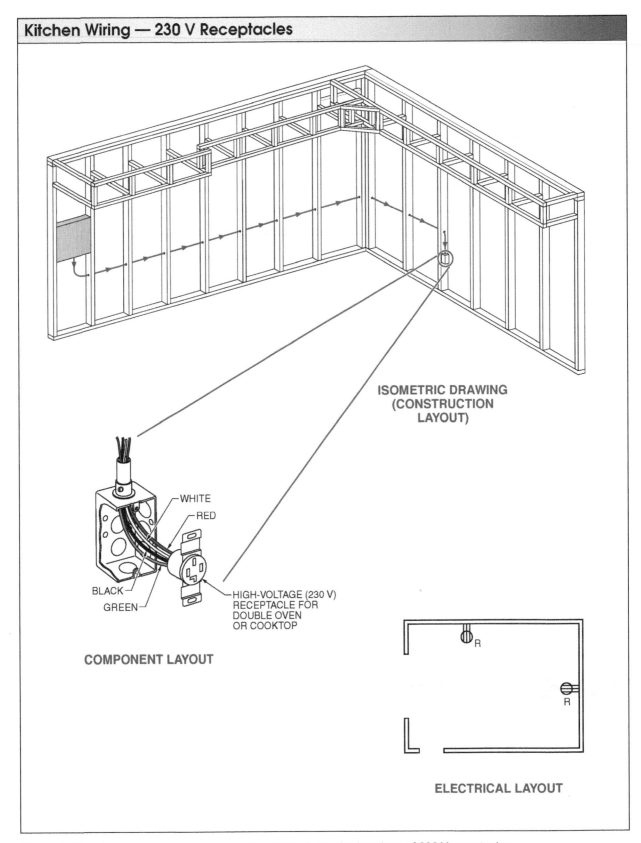

Figure 7-12. *A kitchen oven and cooktop wiring layout shows the locations of 230 V receptacles.*

Kitchen Wiring — GFCIs

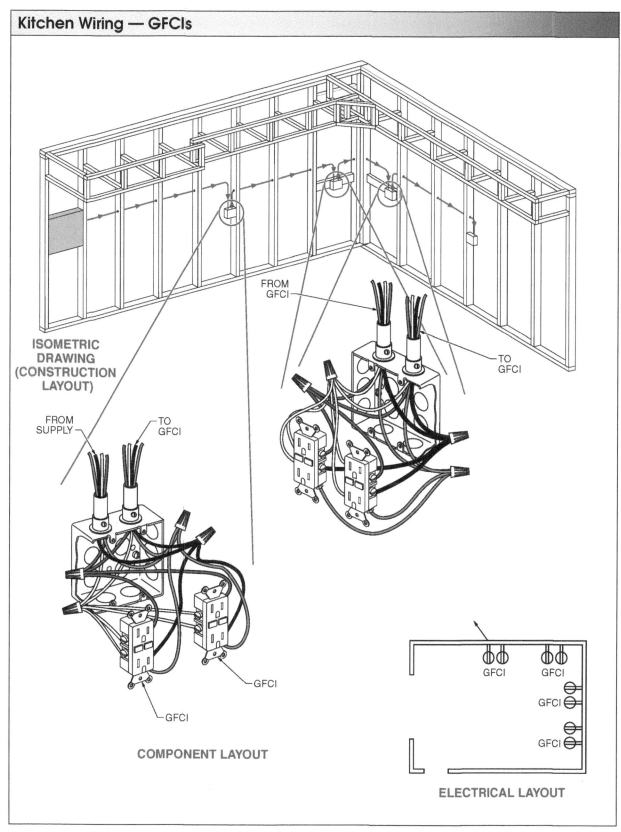

ISOMETRIC
DRAWING
(CONSTRUCTION
LAYOUT)

FROM
GFCI

TO
GFCI

FROM
SUPPLY

TO
GFCI

GFCI

GFCI

COMPONENT LAYOUT

GFCI   GFCI

GFCI

GFCI

ELECTRICAL LAYOUT

Figure 7-13. *A kitchen wiring layout shows the ground fault circuit interrupters (GFCIs).*

Kitchen Wiring — Lighting and Exhaust

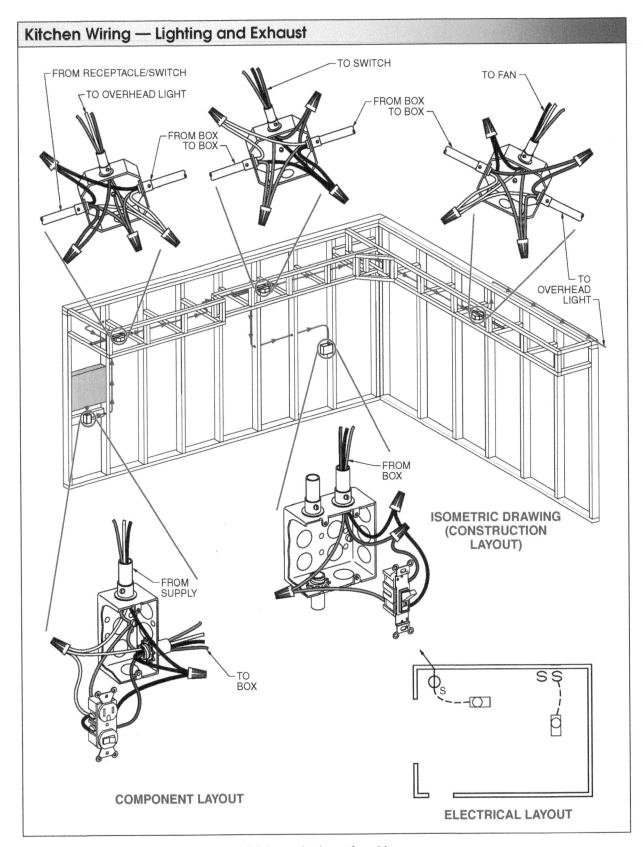

FROM RECEPTACLE/SWITCH

TO OVERHEAD LIGHT

FROM BOX TO BOX

TO SWITCH

FROM BOX TO BOX

TO FAN

TO OVERHEAD LIGHT

FROM BOX

ISOMETRIC DRAWING (CONSTRUCTION LAYOUT)

FROM SUPPLY

TO BOX

COMPONENT LAYOUT

ELECTRICAL LAYOUT

Figure 7-14. *A kitchen wiring layout shows lighting and exhaust fan wiring.*

Kitchen Wiring — Dishwasher, Garbage Disposal, and Refrigerator

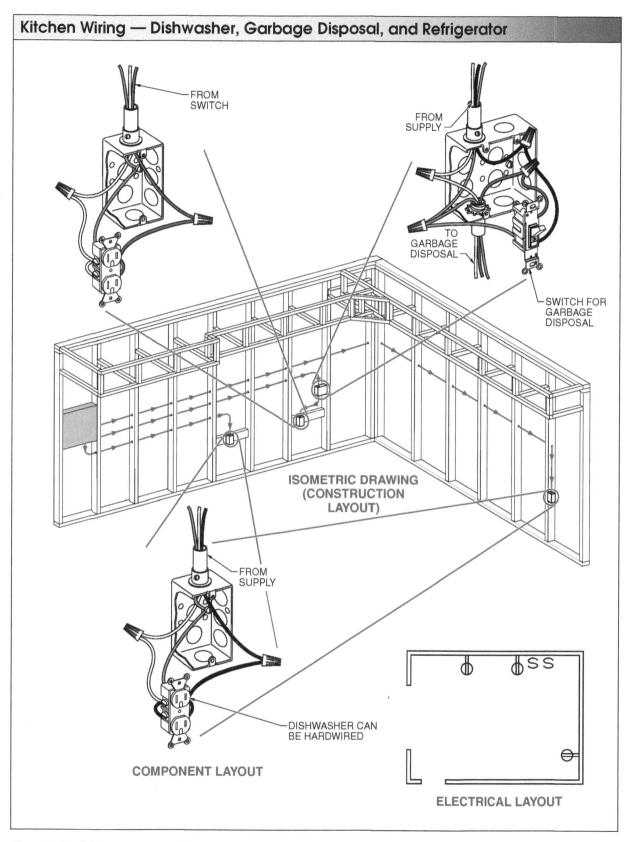

Figure 7-15. *A kitchen wiring layout shows dishwasher and garbage disposal wiring.*

Branch Circuits

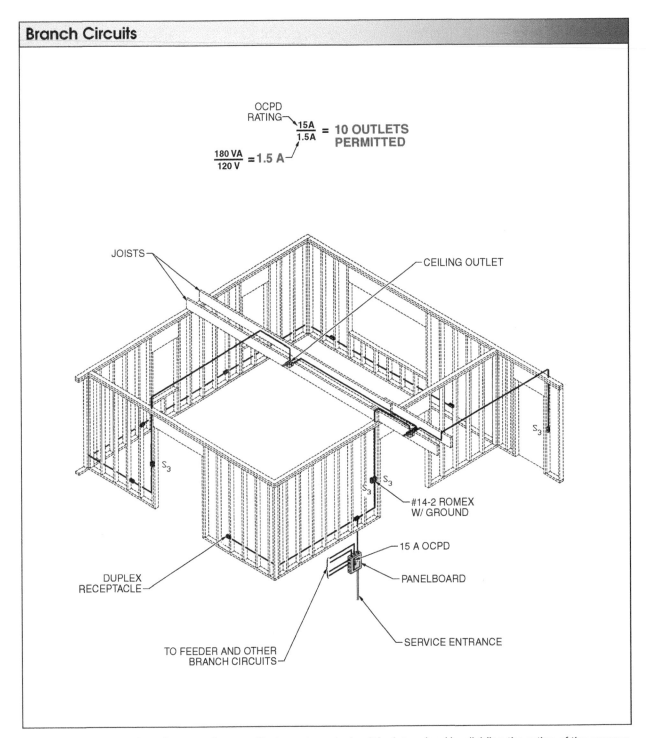

Figure 7-16. *The number of receptacles permitted on a branch circuit is determined by dividing the rating of the overcurrent protection device (OCPD) by 1.5 A.*

Lighting Fixtures

Although lighting fixtures differ greatly in outward appearance, most fixtures are wired to the electrical system of a residence in the same way. **See Figure 7-17.**

Most lighting fixtures will have black and white wires from the manufacturer. The black and white wires are typically connected to the corresponding black and white wires found in the cable or conduit run.

Lighting Fixtures

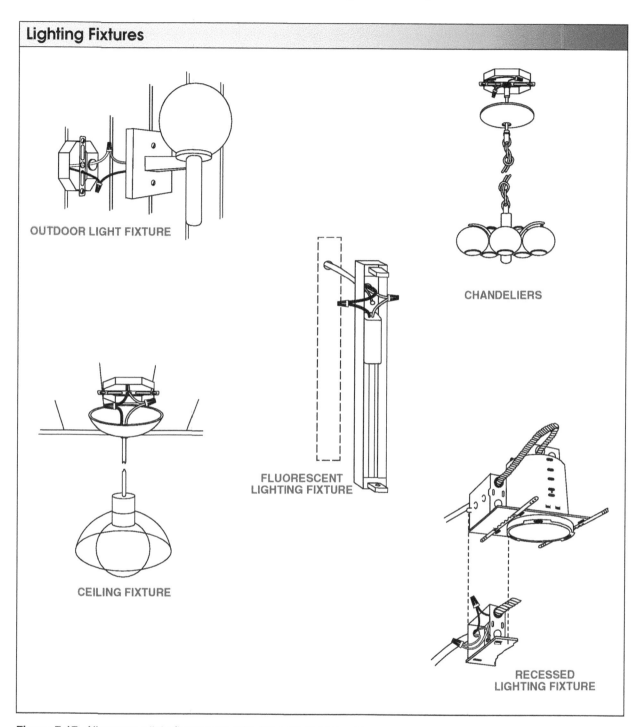

OUTDOOR LIGHT FIXTURE

CHANDELIERS

FLUORESCENT
LIGHTING FIXTURE

CEILING FIXTURE

RECESSED
LIGHTING FIXTURE

Figure 7-17. *All common light fixtures are wired in a similar manner.*

Creating Electrical Systems from Electrical Plans

7

TEST

Name_____ **Date** _____

Wiring Electrical Systems

Connect the components as shown on the wiring diagrams and schematics. Use the following to indicate wires:

BLACK	WHITE	RED
————————————	- - - - - - - - - - - - -	+++++++++++++++++

1.

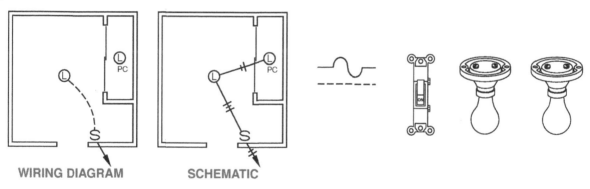

WIRING DIAGRAM   SCHEMATIC

2.

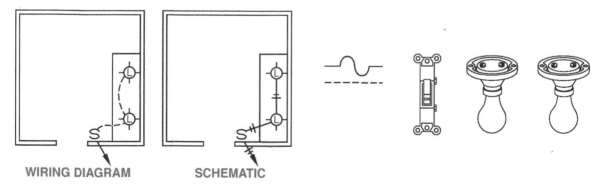

WIRING DIAGRAM   SCHEMATIC

3.

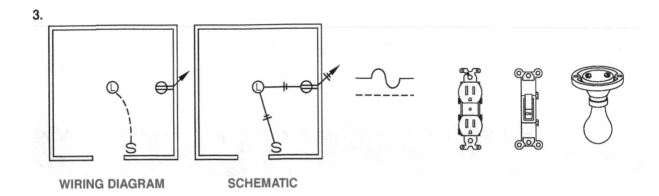

WIRING DIAGRAM SCHEMATIC

4.

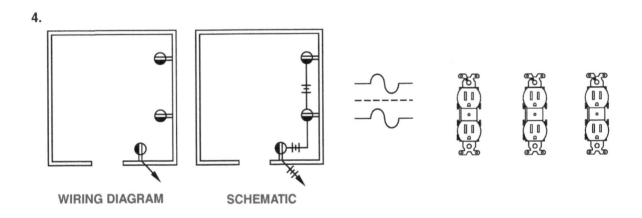

WIRING DIAGRAM SCHEMATIC

5.

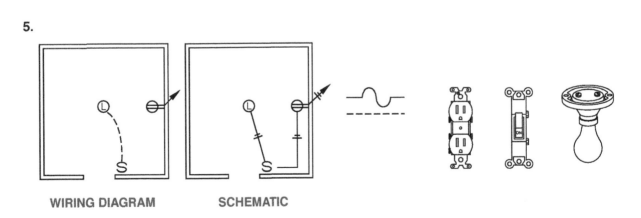

WIRING DIAGRAM SCHEMATIC

184

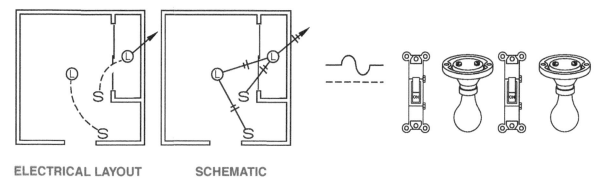

ELECTRICAL LAYOUT   SCHEMATIC

7.

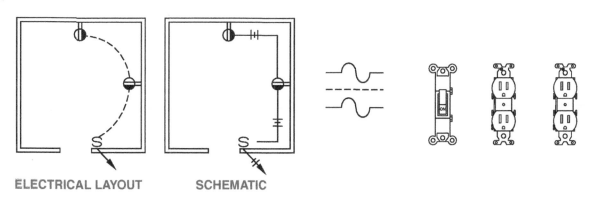

ELECTRICAL LAYOUT   SCHEMATIC

8.

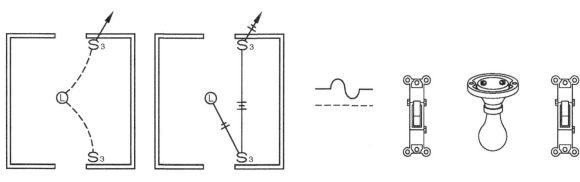

ELECTRICAL LAYOUT   SCHEMATIC

9.

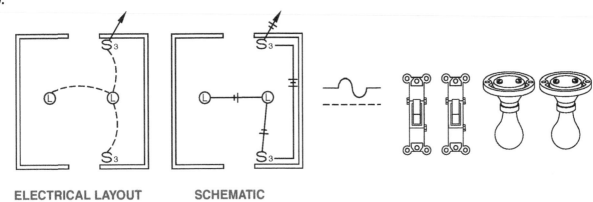

ELECTRICAL LAYOUT   SCHEMATIC

10.

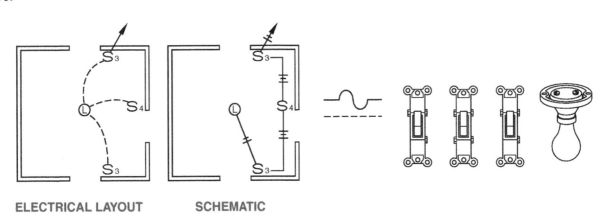

ELECTRICAL LAYOUT   SCHEMATIC

Working with Metallic-Sheathed Cable

Chapter 8 provides an overview of how to install metallic-sheathed cable. Metallic-sheathed cable is similar to nonmetallic-sheathed cable but provides more protection and requires more attention when cutting. The proper installation of metallic-sheathed cable results in an electrical system that meets most local residential electrical codes.

METALLIC-SHEATHED CABLE

Metallic-sheathed cable is a type of cable that consists of two or more individually insulated wires protected by a flexible metal outer jacket. **See Figure 8-1.**

Metallic-sheathed cable is often referred to as armored cable or BX. Armored cable provides additional protection to wires by using a metal outer jacket. The trade name BX was used to denote the armored cable produced by one specific manufacturer. The BX trade name became so popular that even armored cable produced by other manufacturers was called BX.

Armored cable has a typical construction. **See Figure 8-2.** In addition to insulation, each current-carrying conductor has a paper wrapping. The paper provides additional protection within the cable but is removed from the conductors when the cable is being installed into boxes.

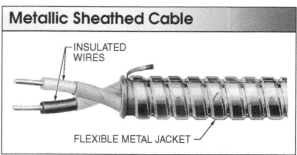

Metallic Sheathed Cable

TRW Crescent Wire and Cable

Figure 8-1. *Metallic-sheathed cable consists of two or more insulated wires protected by a flexible metal jacket.*

A *bonding wire* is an uninsulated conductor in armored cable that is used for grounding. The bonding wire is in contact with the flexible metal outer jacket to assure a proper conducting (ground) path along the entire length of an armored cable. Bonding wires are typically made of aluminum or copper.

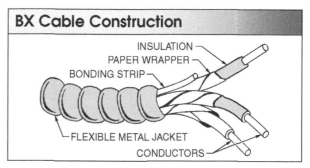

BX Cable Construction

Figure 8-2. *BX cable has a bonding strip, and each insulated conductor of the cable is wrapped with paper for extra protection.*

PURCHASING ARMORED CABLE

As with nonmetallic cable, armored cable is purchased with a specific number of current-carrying conductors. Armored cable is typically sold in coils of 250′. **See Figure 8-3.** When armored cable is ordered, the cable is typically specified as 12/3 or 14/2 to indicate the wire size (8, 10, 12, or 14) and the number of current carrying conductors (2, 3, 4, or 5) required. When designating armored cable, the phrase "with ground" is not required because the outer metal jacket and bonding wire automatically establish a conductive grounding path.

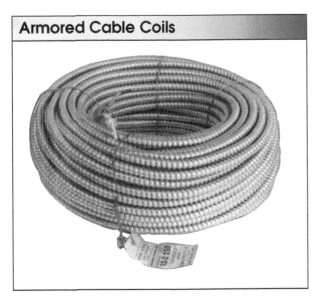

Figure 8-3. *Armored cable (BX) is typically sold in coils of 250′.*

Types of Armored Cable

The NEC® recognizes three types of armored cable for residential work: AC, ACT, and ACL. AC and ACT armored cables are used in dry locations. Both AC and ACT may be fished through the air voids of masonry walls when the walls are not exposed to excessive moisture. Both AC and ACT may be used for under-plaster electrical extensions. ACT armored cable is used for either exposed work or concealed work.

ACL armored cable is embedded in concrete or masonry, run underground, or used where gasoline or oil is present. **See Figure 8-4.** ACL armored cable is lead-covered to provide additional protection for masonry and underground applications. Article 332 of the NEC® contains detailed information on ACL armored cabling.

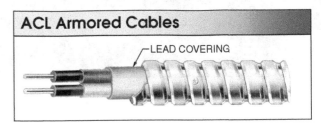

Figure 8-4. *ACL armored cable is covered with a layer of lead.*

Article 320.10 and 320.12 of the NEC® covers uses permitted and uses not permitted for armored cable, Type AC, respectively.

PREPARING ARMORED CABLE

Installing armored cable requires many of the same techniques and equipment used when working with nonmetallic cable. However, armored cable requires more attention when cutting and splicing because of the sharp metal edges.

Cutting Armored Cable

Most electricians use a hacksaw to cut through the metal jacket of armored cable. **See Figure 8-5.** The first step is to cut through the outer armor (one of the convolutions) at a 45° angle about 6″ to 8″ from an end. To avoid damage to the insulation of conductors, care must be taken not to cut too deeply into the jacket of the cable. When the armored cable is cut, the cable is separated by twisting the two sections apart. The armored cable must be carefully flexed until the cable breaks.

Cutting Armored Cable — Hacksaw

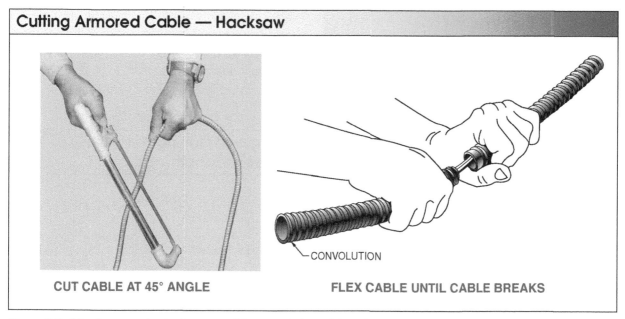

CUT CABLE AT 45° ANGLE

CONVOLUTION

FLEX CABLE UNTIL CABLE BREAKS

Figure 8-5. *A hacksaw must cut across one of the convolutions of the cable at a 45° angle for the cable to separate properly.*

Twisting the cable can open the convolutions enough so that tin snips or cable cutters can be inserted. **See Figure 8-6.** Armored cable cutters are used to avoid damaging the insulation on the conductors. The use of protective gloves helps protect hands from cuts and scratches. When the outer armor and conductors are cut, a second trimming cut may be required to remove any of the metal outer jacket that is bending outward.

A center cut is used to separate the armored cable while side cuts prepare each end for mounting the cable to electrical boxes.

Installing Anti-Short Bushings

Whenever armored cable is cut, an unavoidable sharp edge remains on the ends of the metal armor. To avoid any damage to the insulation of the conductors, anti-short bushings must be installed. **See Figure 8-7.**

Type AC cable is also available with aluminum covering and is half the weight of Type AC steel cable. Standard sizes are 10 AWG, 12 AWG, and 14 AWG.

Cutting Armored Cable — Cable Cutters

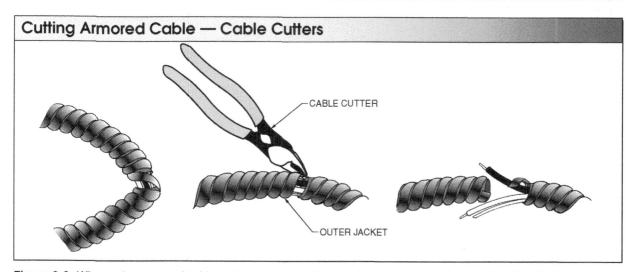

CABLE CUTTER

OUTER JACKET

Figure 8-6. *When using armored cable cutters, care must be exercised not to pinch the conductor insulation.*

Anti-Short Bushings

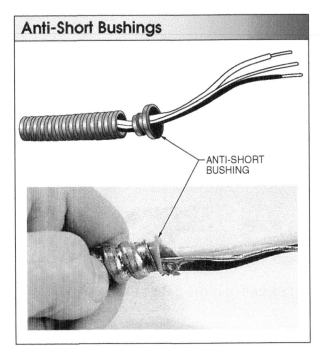

Figure 8-7. *To protect conductors from the sharp edges of cut armored cable, plastic anti-short bushings must be used.*

An *anti-short bushing* is a plastic or heavy fiber paper device used to protect the conductors of armored cable. The bushing covers the sharp edges at the ends of the armor to reduce the possibility of damage to conductor insulation.

> ⚠️ **CAUTION**
>
> *Always verify that an anti-short bushing has been installed on the end of armored cable before installing the cable.*

ROUGHING-IN ARMORED CABLE

To rough-in armored cable, the cable must be pulled, cut, secured, and grounded properly.

Pulling Armored Cables

When roughing-in armored cable into studded walls, the cable must be pulled into position first and then cut to length. **See Figure 8-8.** Pulling the armored cable into position, then cutting to length, reduces

waste. To avoid twisting and kinking the cable, it should be unwound from the center of the coil. After the cable has been pulled into position, three cuts are made in the armor about 6″ to 8″ apart. The entire cable is cut through with a center cut.

Pulling and Cutting Armored Cables to Length

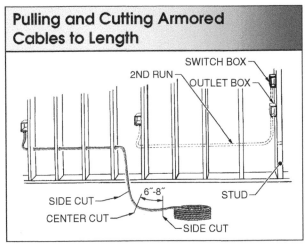

Figure 8-8. *To install armored cable, three cuts must be made to the armored jacket. A center cut is used to separate the cable, and two side cuts prepare each end for mounting to electrical boxes.*

The two side cuts are cut through the metal armor (sheath) only; the conductors and insulation are left intact. One end of the cable is used to terminate the run previously started, and the other end is used to start the next run. When wires are concealed in walls, cable runs can be shortened by allowing the cable to run wild, thus saving an amount of cable. **See Figure 8-9.**

Tech Tip: AC Cable Installation Requirements

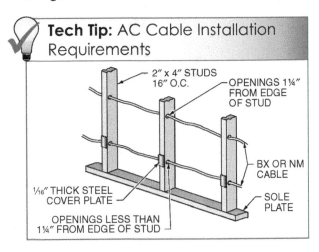

Armored Cable Running Wild

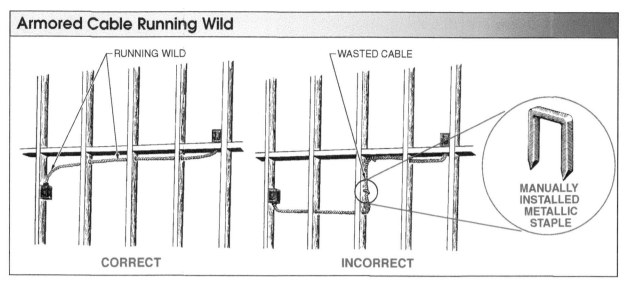

Figure 8-9. *The length of an armored cable that is concealed in a wall can be shortened by allowing the cable to run wild.*

To avoid breaking armored cable, sharp bends should not be made. **See Figure 8-10.** Typically a bend in armored cable must have a minimum radius at least 5 times the diameter of the cable. Armored cable installations have a typical appearance when work is completed. **See Figure 8-11.**

Bending Armored Cable

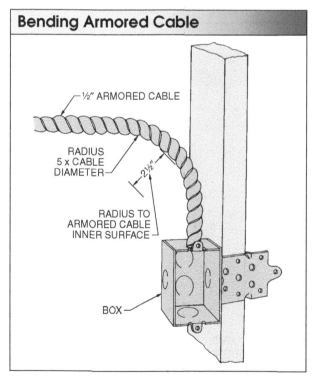

Figure 8-10. *Armored cable cannot be bent tighter than the recommended bending radius without causing damage to conductors.*

Tech Tip: AC Cable Support Requirements

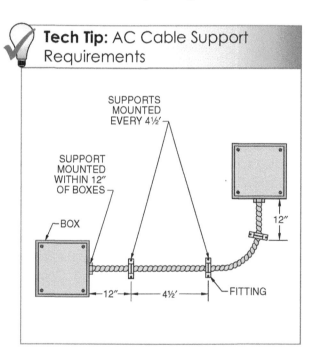

The outer sheathing and terminal connectors of Type AC cable can be used as a grounding path for the equipment it is connected to. Type AC cable is also available with a separate equipment grounding conductor (EGC). The bonding strip is not considered an EGC and can be removed with a cable cutter at the ends of the cable sheathing. Type AC cable is available with three current-carrying conductors and either with or without an EGC.

Armored Cable Installation

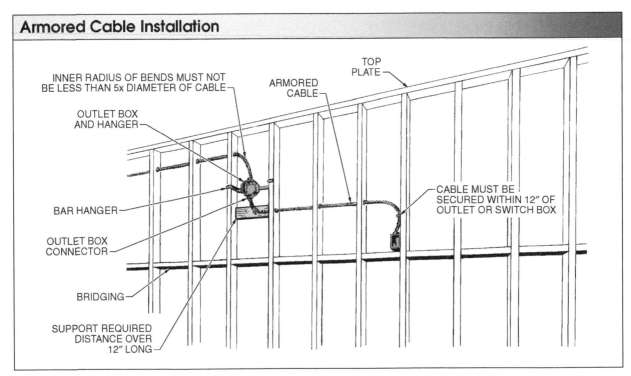

INNER RADIUS OF BENDS MUST NOT BE LESS THAN 5x DIAMETER OF CABLE

OUTLET BOX AND HANGER

BAR HANGER

OUTLET BOX CONNECTOR

BRIDGING

SUPPORT REQUIRED DISTANCE OVER 12" LONG

TOP PLATE

ARMORED CABLE

CABLE MUST BE SECURED WITHIN 12" OF OUTLET OR SWITCH BOX

Figure 8-11. *Armored cable installations have a typical appearance when work is completed.*

Securing Armored Cables

Armored cable is secured in place with cable clamps or cable connectors. **See Figure 8-12.** Cable clamps are typically part of a box, while cable connectors are installed before a box is wired.

A right angle connector (90° connector) is used where the radius bend of the armored cable would be so tight that the cable would snap if forced into the bend. **See Figure 8-13.** Once it is cut, armored cable must be properly grounded.

To make a proper ground, a set screw connector is secured over the end of the cable. The cable is then secured to the metal box with a locknut connector. **See Figure 8-14.**

Grounding Armored Cables

Grounding armored cable requires that the metal jacket and bonding wire be firmly secured. **See Figure 8-15.** A typical method used to ground armored cable is to secure the bonding wire of the cable between the connector and box. Other methods of grounding armored cable are the same as with grounding nonmetallic cable.

Armored Cable Box Clamps and Connectors

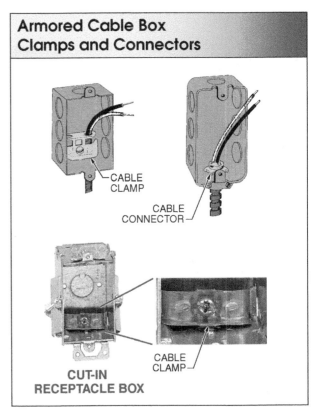

CABLE CLAMP

CABLE CONNECTOR

CUT-IN RECEPTACLE BOX

CABLE CLAMP

Figure 8-12. *Armored cables can be secured to a box with cable clamps, which are part of the box, or cable connectors, which are installed at wiring.*

Armored Cable Box Connectors

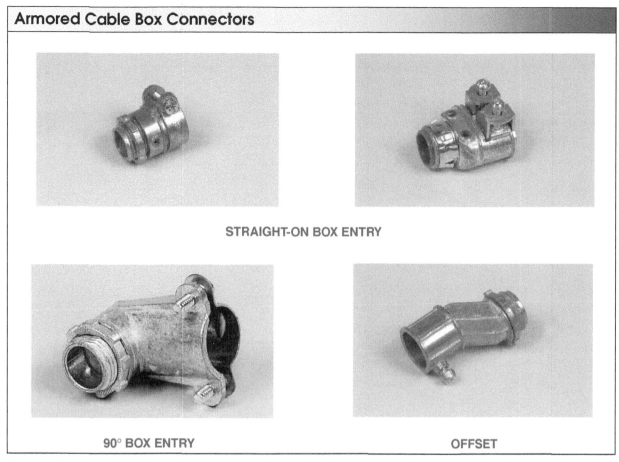

Figure 8-13. *Armored cable connectors are used for straight-on box entry and for 90° box entry.*

Armored Cable Grounding

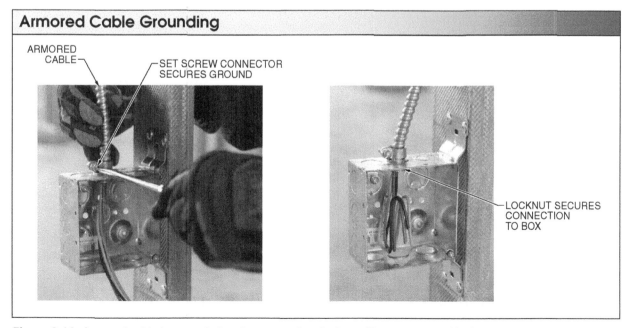

Figure 8-14. *Armored cable is grounded and connected to the box with set screw and locknut connectors.*

Because armored cable is flexible and provides protection to conductors, it is typically installed in limited-space applications where the conductors would normally be subject to vibration or inadvertent contact.

See Figure 8-16. For example, armored cable can be installed under kitchen sinks and in laundry areas, garages, workshops, attics, and any other location that could have exposed conductors.

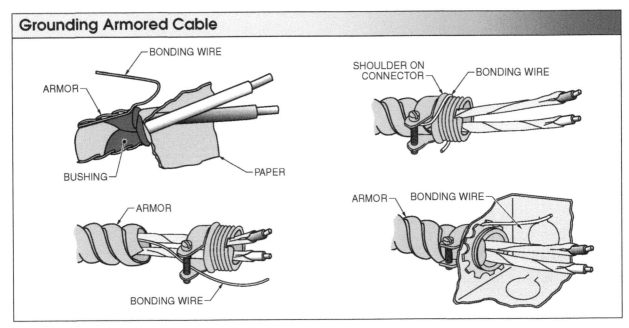

Figure 8-15. *The bonding wire of armored cable is secured between the box connector and the box to create a proper grounding path.*

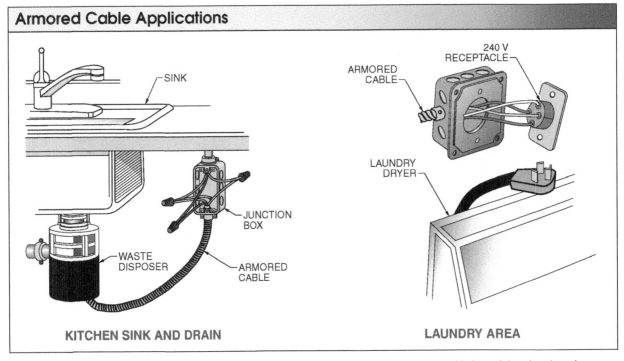

Figure 8-16. *Armored cable is typically installed in residential applications such as under a kitchen sink or in a laundry area.*

Working with Metallic-Sheathed Cable

Name_____ Date _____

Working with Metallic-Sheathed Cable

T	F	**1.**	Armored cable is also known as BX.
_____		**2.**	Each current-carrying conductor of armored cable has a(n) ___ wrapping over the insulation.
_____		**3.**	The ___ wire used for grounding in armored cable can be Cu or Al.
T	F	**4.**	The 3 in the armored cable designation 12/3 refers to the gauge of the conductors.
_____		**5.**	The 14 in the armored cable designation 14/2 refers to wire ___.
T	F	**6.**	Type ACL cable may be embedded in concrete.
T	F	**7.**	Type ACT cable may be used for exposed work.
T	F	**8.**	Type ACL cable may be embedded in masonry.
_____		**9.**	___ cable is lead-covered.
_____		**10.**	___ cable is only for use in dry, concealed locations.
_____		**11.**	Types AC and ACT cable may be fished through air voids of masonry walls when the walls are not exposed to excessive ___.
T	F	**12.**	Type ACL cable may be run underground.
_____		**13.**	When gasoline or oil is present, ___ type of cable must be used.
_____		**14.**	When cutting the jacket of armored cable with a hacksaw, the cut should be made at a(n) ___° angle.
_____		**15.**	An anti-short bushing for armored cable is made of ___ or heavy fiber paper.
T	F	**16.**	Armored cable should be cut to length before being pulled into position.
_____		**17.**	___ are placed on the cut ends of armored cable to avoid any damage to conductor insulation.

_____ **18.** A bend (radius) in armored cable should be at least ___ times the diameter of the cable.

_____ **19.** Armored cable must be secured within ___″ of receptacles or switch boxes.

T   F   **20.** Cable clamps are typically part of a box.

T   F   **21.** Cable connectors are typically part of the box.

_____ **22.** Grounding armored cable requires that the metal jacket and the ___ be firmly secured.

T   F   **23.** The paper wrapper of armored cable should be removed from the individual conductors contained within boxes.

T   F   **24.** Armored cable is typically sold in 100′ coils.

_____ **25.** The ___ recognizes three types of armored cable.

T   F   **26.** AC and ACT cables are used for underplaster extensions.

_____ **27.** To avoid twisting and kinking, armored cable should be removed from the ___ of the coil.

_____ **28.** A ___ cut is used to separate armored cable.

_____ **29.** Metallic ___ are used to fasten armored cable to studs and joists.

T   F   **30.** Armored cable connectors may be used for straight-on or 90° entry into a box.

Armored Cable

_____ 1. Fished in masonry walls   **A.** Metallic cable

_____ 2. Secures metallic cable   **B.** Bonding wire

_____ 3. Exposed or concealed work   **C.** 14/2 or 12/3

_____ 4. BX   **D.** AC

_____ 5. Wire size and number of conductors   **E.** ACL

_____ 6. Protects conductor insulation   **F.** ACT

_____ 7. Continuous electrical path   **G.** Anti-short bushing

_____ 8. Copper or aluminum   **H.** Metallic cable bend

_____ 9. Embedded in concrete   **I.** Cable connector

_____ 10. At least 5 times the cable diameter

Metallic-Sheathed Cable Components

_____ **1.** Conductors

_____ **2.** Insulation

_____ **3.** Flexible metallic jacket

_____ **4.** Paper wrapper

_____ **5.** Bonding wire

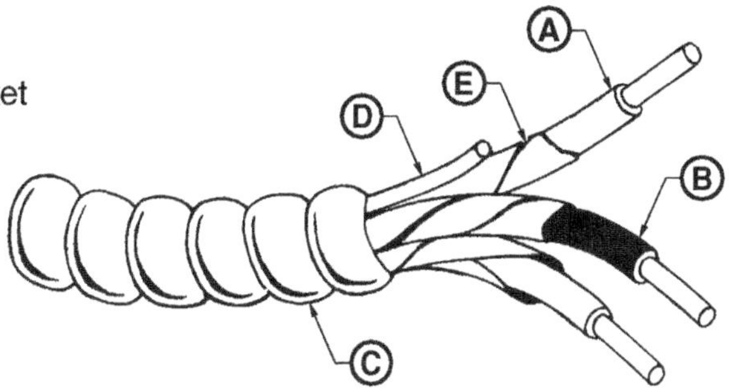

Armored Cable Connectors and Clamps

_____ **1.** Straight armored cable connector

_____ **2.** Armored cable clamp in place

_____ **3.** Armored cable connector in place

_____ **4.** 90° armored cable connector

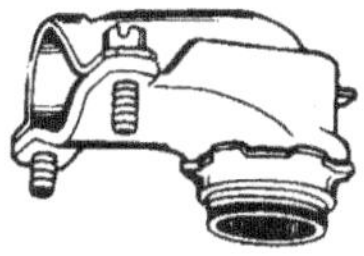

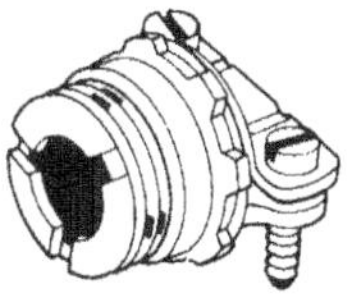

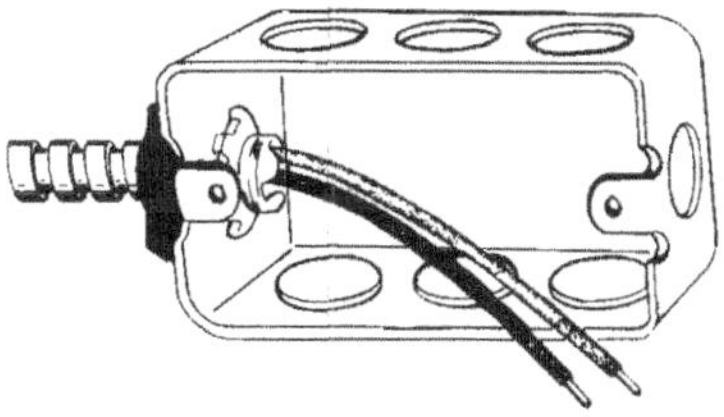

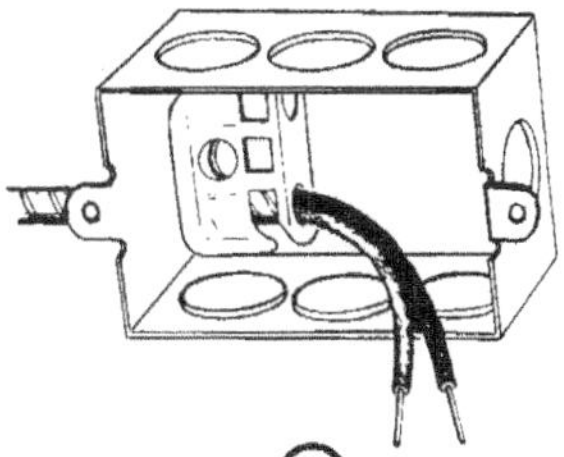

Metallic-Sheathed Cable Parts

_____ **1.** Armored cable

_____ **2.** BX

_____ **3.** ACL

_____ **4.** Bonding wire

_____ **5.** Anti-short bushing

A. Lead-covered armored cable

B. Serves same purpose as grounding conductor

C. Trade name for armored cable

D. Protects conductor's insulation

E. Two or more insulated wires protected by a flexible metal jacket

Installing Armored Cable

_____   **1.** Octagonal box

_____   **2.** Support required if over 12″

_____   **3.** Inner radius of bends is not less than 5 times the diameter

_____   **4.** Bar hanger

_____   **5.** Cable secured within 12″ of box

_____   **6.** Top plate

_____   **7.** Bridging

_____   **8.** Stud

_____   **9.** Receptacle box

_____   **10.** Staple

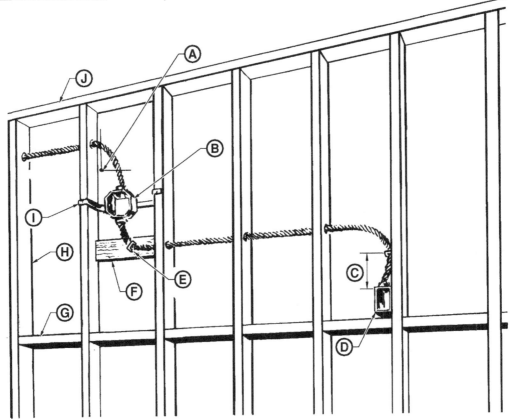

Wiring with Conduit 9

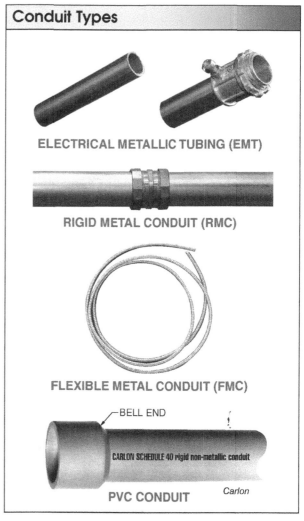

Conduit Types

ELECTRICAL METALLIC TUBING (EMT)

RIGID METAL CONDUIT (RMC)

FLEXIBLE METAL CONDUIT (FMC)

BELL END

CARLON SCHEDULE 40 rigid non-metallic conduit

PVC CONDUIT   Carlon

Chapter 9 provides an overview of how to install electrical metallic conduit. A conduit (metal pipe) wiring system is probably the most challenging of residential electrical systems because of the extra planning, special tools, and basic skill required to create a professional-looking system. Chapter 9 provides several tips and techniques for additions and repairs to existing conduit systems. Any design layout, however, must be performed in accordance with the National Electrical Code® and local building inspector.

TYPES OF CONDUIT

Conduit is a rugged protective tube (typically metal) through which wires are pulled. Although several types of conduit are used in the electrical industry, only four types are commonly found in residential installations. The four most common types of conduit used in residential installations are electrical metallic tubing (EMT), rigid metal conduit (RMC), flexible metal conduit (FMC), and polyvinyl chloride (PVC) conduit. PVC conduit is the most common type of nonmetallic (plastic) conduit and is sometimes referred to as rigid nonmetallic conduit. PVC is typically installed in environments that have a high level of moisture, such as underground from a garage to a dwelling. **See Figure 9-1.**

Figure 9-1. *Electrical metallic tubing (EMT) is the most common conduit used in residential wiring.*

Electrical Metallic Tubing (EMT)

Electrical metallic tubing (EMT) is a light-gauge electrical pipe often referred to as thin-wall conduit. EMT has a wall thickness that is about 40% the thickness of rigid conduit. Because EMT is lighter, easier to bend, and requires no threading, EMT is typically used for a complete residential layout. Additional information on the installation of EMT can be found in Article 358 of the NEC®.

Rigid Metal Conduit (RMC)

Rigid metal conduit (RMC) is a heavy-duty pipe that is threaded on the ends much like standard plumbing pipe. For residential use, rigid conduit is limited mainly to risers for service entrances. Rigid conduit can be cut with a standard hacksaw and must be threaded with a standard cutting die (plumbing die) providing ¾" per foot taper. Many electrical supply houses carry a number of conduit sizes in various lengths that are prethreaded and ready for use. Additional information on the installation of rigid metal conduit can be found in Article 344 of the NEC®.

Flexible Metal Conduit (FMC)

Flexible metal conduit (FMC) is a conduit that has no wires and can be bent by hand. The conductors (wires) are not installed until the system is complete. Generally, flexible metal conduit is used where some type of movement or vibration is present. For example, flexible conduit is suited for wiring to a motor. Flexible conduit (Greenfield) is also used where other conduits might be difficult to bend. Additional information on the installation of flexible metal conduit can be found in Article 348 of the NEC®.

Conduit benders are available in hydraulically powered models to help reduce fatigue of the electrician. Hydraulic-powered conduit benders can bend conduit with diameters of 1" through 4" and are available with a hand-powered hydraulic pump or a ½ HP electric motor driven pump.

INSTALLING EMT

To install EMT conduit, an individual must purchase, cut, bend, and secure the conduit.

Purchasing EMT

EMT is typically purchased in bundles of 100'. The bundles contain ten lengths of conduit, each 10' long. **See Figure 9-2.** Most electrical supply houses warehouse conduit; therefore, individual lengths of 10' can typically be purchased. The trade sizes for conduit range from ½" to 4". **See Figure 9-3.** All sizes of conduit have limitations on the size and number of conductors that are allowed in the conduit for an application such as residential wiring. Further information on the allowable number of conductors per size of conduit can be obtained from Annex C of the NEC®. **See Figure 9-4.**

Standard EMT Bundling

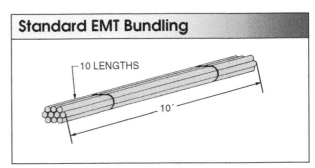

Figure 9-2. *EMT and rigid conduit are typically sold in 100' bundles.*

EMT Conduit Specifications

Inside Diameter Nearest ¹⁄₁₆"	Inside Diameter	Outside Diameter	Trade Size in Inches
⅝	0.622	0.706	½
¹³⁄₁₆	0.824	0.922	¾
1¹⁄₁₆	1.049	1.163	1
1⅜	1.380	1.508	1¼
1⅝	1.610	1.738	1½
2¹⁄₁₆	2.067	2.195	2
2¾	2.731	2.875	2½
3⅜	3.356	3.500	3
3¹³⁄₁₆	3.834	4.000	3½
4⁵⁄₁₆	4.334	4.500	4

Figure 9-3. *EMT conduit is typically specified by nominal size.*

Maximum Number of Conductors Allowed

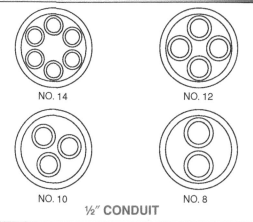

NO. 14   NO. 12

NO. 10   NO. 8

½″ CONDUIT

EMT and RIGID CONDUIT		
SIZE THHN AWG WIRE	½″ CONDUIT	¾″ CONDUIT
14	6*	11*
12	4*	8*
10	3*	5*
8	2*	3*

*Half the wire amount indicated by the NEC® for some applications due to typical residential amperage ratings

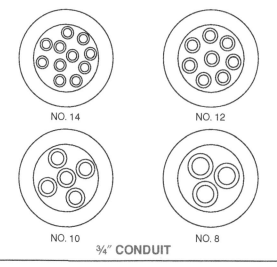

NO. 14   NO. 12

NO. 10   NO. 8

¾″ CONDUIT

Figure 9-4. *Each conduit size has a maximum number of wires allowed.*

Cutting EMT

Conduit is typically cut to length using a hacksaw. **See Figure 9-5.** However, all the rough edges from cutting must be removed before wire can be pulled through conduit. Removing the rough edges is called "deburring" and is accomplished with a conduit deburring tool designed for this purpose. **See Figure 9-6.** Conduit can also be deburred with a file or a special deburring tool called a reamer.

Cutting EMT

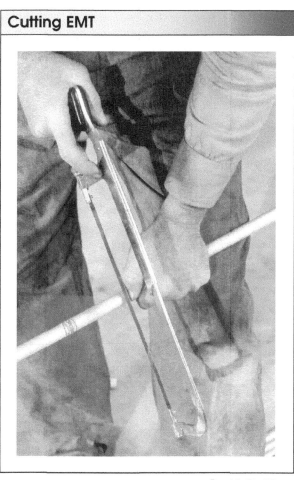

Republic Steel Corp.

Figure 9-5. *EMT conduit is typically cut with a hacksaw.*

Conduit Deburring Tool

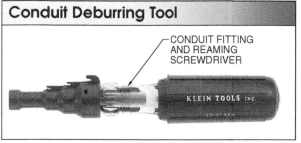

CONDUIT FITTING AND REAMING SCREWDRIVER

KLEIN TOOLS INC.

Klein Tools, Inc.

Figure 9-6. *When conduit has been cut, the conduit must be deburred to remove rough edges.*

Bending EMT

EMT conduit bends and offsets require a certain amount of planning and practice. Using proper procedures, techniques, and tools, bends and offsets can be successfully created.

Conduit Benders. Bending EMT conduit properly requires the aid of a hand conduit bender. **See Figure 9-7.** A hand rigid or heavywall conduit bender is sometimes referred to in the trade as a "hickey." A hand conduit bender has high supporting sidewalls to prevent flattening or kinking of the EMT conduit. **See Figure 9-8.**

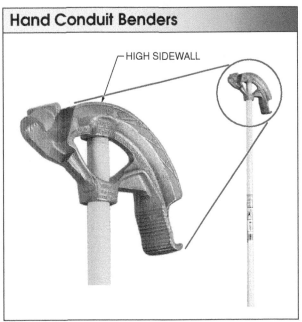

Hand Conduit Benders

HIGH SIDEWALL

Klein Tools, Inc.

Figure 9-7. *Hand conduit benders are specifically designed to bend EMT conduit.*

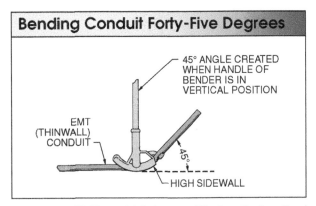

Bending Conduit Forty-Five Degrees

45° ANGLE CREATED WHEN HANDLE OF BENDER IS IN VERTICAL POSITION

EMT (THINWALL) CONDUIT

45°

HIGH SIDEWALL

Figure 9-8. *EMT conduit is bent to 45° when the handle of the hand conduit bender is exactly vertical.*

Forty-Five Degree Bends. To create a 45° angle (bend), a hand bender is placed on EMT conduit. The handle of the bender is raised until the handle is in the vertical position. The handle is pulled until the angle required (45°) is completed. The hand bender is removed, releasing the EMT conduit with a smooth-flowing 45° bend. Hand benders also have a long arc that permits making 90° bends in a single sweep without moving the bender to a new position.

Ninety Degree Bends. Ninety degree bends are used to turn corners or to form a predetermined length for use as a stub up through a floor that will be added onto later. **See Figure 9-9.** The typical method for laying out a 90° bend requires that the conduit be marked to a predetermined measurement.

Note: The distance marked on the tubing must be visually lined up with the hand bender by the individual.

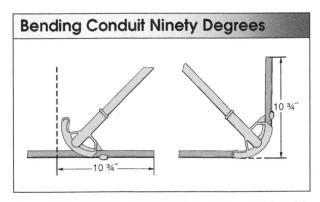

Bending Conduit Ninety Degrees

10 ¾"

10 ¾"

Figure 9-9. *A typical method of creating a 90° bend in EMT conduit is to bend the conduit to a predetermined measurement.*

The accuracy to which an individual lines up the marks on the bender determines how accurate the conduit bend will be. For the novice, it is recommended that the accuracy of bends be checked using a level. **See Figure 9-10.**

Using proper technique to secure conduit during bending assures conduit bends are made to the needed requirements. **See Figure 9-11.**

Checking Ninety Degree Bends

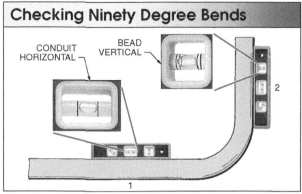

CONDUIT HORIZONTAL

BEAD VERTICAL

2

1

Klein Tools, Inc.

Figure 9-10. *A level is typically used to check conduit for a properly created 90° bend.*

Technique for Bending Conduit Ninety Degrees

Figure 9-11. *Proper technique for securing conduit while making a bend with a hand conduit bender is critical.*

Back-to-Back Bends. When running conduit from one outlet box to another box, the run can require that two 90° bends be made to one piece of conduit. The use of two 90° bends is called back-to-back bending. To make back-to-back bends, the first bend is made as a regular 90° bend. **See Figure 9-12.** Then the distance between the boxes is laid out on the conduit as distance *D*. The direction of the hand bender is reversed and point A of the bender is placed over the mark at distance *D*. Raising the hand bender to form a normal 90° bend finishes the second right angle. When creating back-to-back bends, individuals must take care to ensure that both bends line up with each other (are on the same plane or flat).

Creating Back-to-Back Ninety Degree Bends

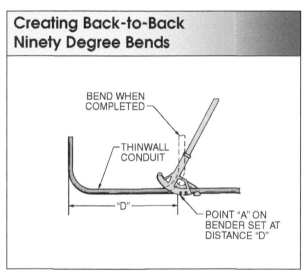

BEND WHEN COMPLETED

THINWALL CONDUIT

"D"

POINT "A" ON BENDER SET AT DISTANCE "D"

Figure 9-12. *When creating back-to-back bends, electricians must ensure that both bends line up with each other (are on the same plane or flat).*

Offsets. If all conduit bends were as simple as 45° and 90° bends, conduit forming would be a relatively simple art. Unfortunately, conduit runs are often interrupted by obstructions requiring changes in direction that cannot be accomplished by using just 45° and 90° bends. An *offset* is a compound bend in conduit used to bypass many types of obstructions. **See Figure 9-13.** An offset is laid out with the aid of parallel lines drawn on a floor or other smooth surface. Always double-check all dimensions for an offset bend on the floor layout. The hand bender must be repositioned on the conduit to create both offset bends. **See Figure 9-14.** A simple method for checking the accuracy of an offset bend uses a pocket level. **See Figure 9-15.**

Offset Bends

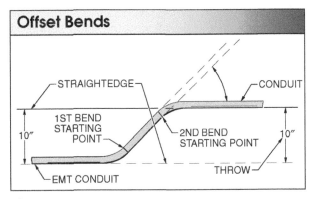

Figure 9-13. *The second bend of an offset is readily determined by using parallel lines drawn on the floor.*

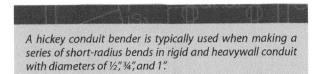

A hickey conduit bender is typically used when making a series of short-radius bends in rigid and heavywall conduit with diameters of ½", ¾", and 1".

Checking Offset Bends

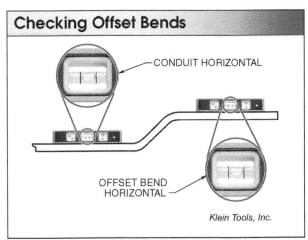

Klein Tools, Inc.

Figure 9-15. *Pocket levels are used to check the accuracy of offset bends.*

Double Offsets (Saddles). A *double offset*, also known as a saddle, is a common complex bend made in conduit to bypass obstructions. **See Figure 9-16.** Standard procedures are used in creating a double offset. **See Figure 9-17.**

Note: Use parallel lines to provide a frame of reference for a double offset. Parallel lines on each side of the conduit are used to help eliminate any deviation from center.

Creating Offset Bends

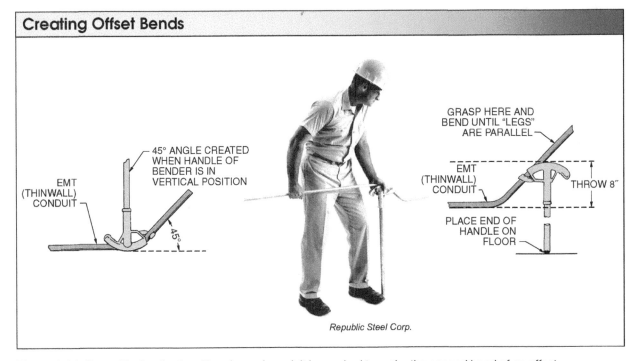

Republic Steel Corp.

Figure 9-14. *Repositioning the hand bender and conduit is required to make the second bend of an offset.*

Double Offsets (Saddles)

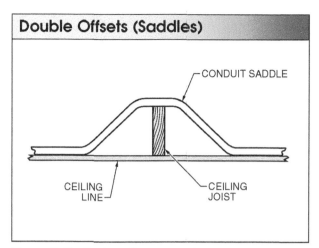

Figure 9-16. *Double offsets or saddles are used to clear obstructions.*

Standard hand conduit benders are made of aluminum or iron and include angle markings of 22½°, 25°, 45°, 60°, and 90° on both sides of the bender.

Creating Double Offsets (Saddles)

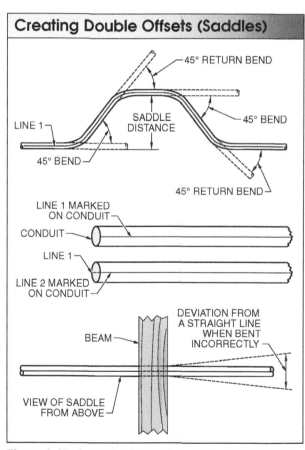

Figure 9-17. *A standard procedure is used to create double offsets (saddles).*

An alternative method can occasionally be used to create a saddle. Typically, the alternative method of creating a saddle is used where the bend will not be seen. **See Figure 9-18.**

Alternating Method for Creating a Saddle

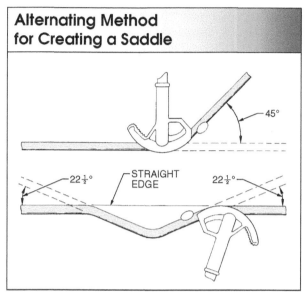

Figure 9-18. *The alternative method of creating a saddle is used where the saddle will not be seen, such as in a drop ceiling.*

Straightening Conduit. When electricians are first learning to bend conduit, some bends will miss the mark. To avoid scrapping the conduit, the conduit can sometimes be rebent to straightness. A common technique is used to straighten incorrect bends. **See Figure 9-19.** A piece of rigid conduit, the handle of a conduit bender or plumbing pipe, is typically used as a straightening tube.

Straightening Conduit

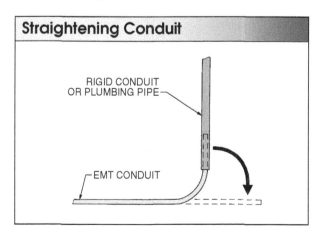

Figure 9-19. *When straightening bent conduit, place the conduit on a hard surface.*

Rules Concerning Conduit Bends. Per Article 358.24 of the NEC®, *Bends – How Made,* bends in tubing shall be made so that the tubing will not be damaged and that the internal diameter of the tubing will not be effectively reduced.

In addition, per Article 358.26 of the NEC®, *Bends – Number in One Run,* a run between pull points (between outlet and outlet, between fitting and fitting, or between outlet and fitting) shall not contain more than the equivalent of four 90° bends (360° total), including bends located immediately at the outlet box or fitting. The NEC® table Annex –

Table 2, *Radius of Conduit and Tubing Bends,* indicates the minimum acceptable radius for various sizes of conduit.

Securing EMT

The NEC® requires that EMT be installed as a complete system and be securely fastened in place at least every 10′ and within 3′ of each outlet box, junction box, cabinet, or fitting. **See Figure 9-20.** Conduit must be installed into smooth-flowing pathways through which wires can be easily pulled. **See Figure 9-21.**

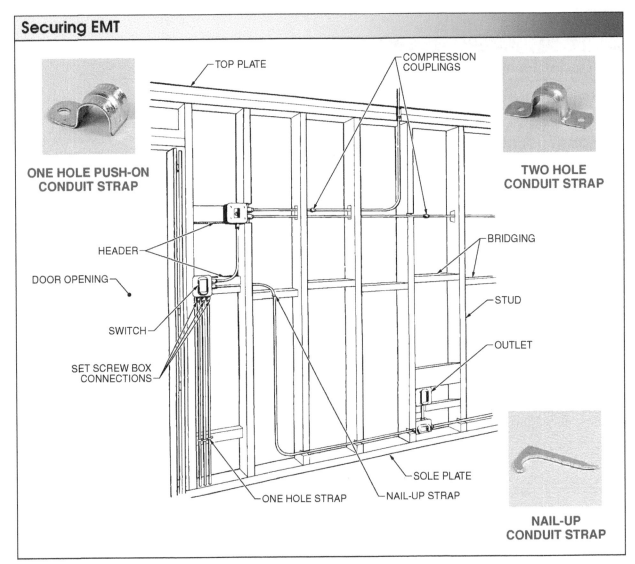

Securing EMT

ONE HOLE PUSH-ON CONDUIT STRAP

TWO HOLE CONDUIT STRAP

TOP PLATE

COMPRESSION COUPLINGS

HEADER

DOOR OPENING

SWITCH

SET SCREW BOX CONNECTIONS

BRIDGING

STUD

OUTLET

ONE HOLE STRAP

NAIL-UP STRAP

SOLE PLATE

NAIL-UP CONDUIT STRAP

Figure 9-20. *Conduit straps secure conduit in position. Straps must be used for every 10′ of conduit run and within 3′ of a box.*

Conduit is secured in place and attached to electrical boxes and other conduit using straps, connectors, and couplings.

Conduit Connectors

EMT is firmly secured to electrical boxes using compression, indenter, and set screw connectors. **See Figure 9-22.** Each type of connector has a distinctly different technique for holding the conduit securely. Various types of connectors are accepted by the NEC® for residential use.

Compression Connectors. A *compression connector* is a type of box fitting that firmly secures conduit

to a box by utilizing a nut that compresses a tapered metal ring (ferrule) into the conduit. **See Figure 9-23.** As the compression nut is tightened, the nut forces the ferrule into the conduit, locking the conduit in position. Compression connectors can be loosened to remove the conduit and reused multiple times to attach the same piece of conduit.

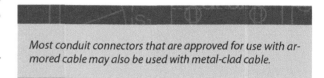

Most conduit connectors that are approved for use with armored cable may also be used with metal-clad cable.

Smooth-Flowing Pathways

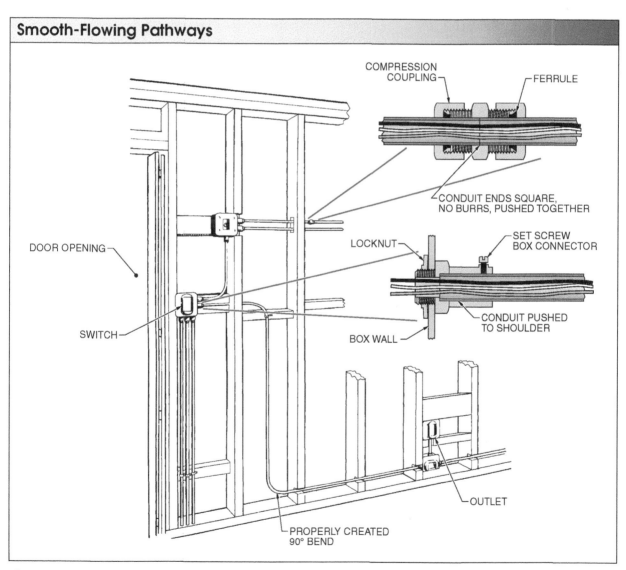

Figure 9-21. *Conduit and fittings that are properly installed form a smooth-flowing pathway through which wires can be run.*

Conduit Connectors

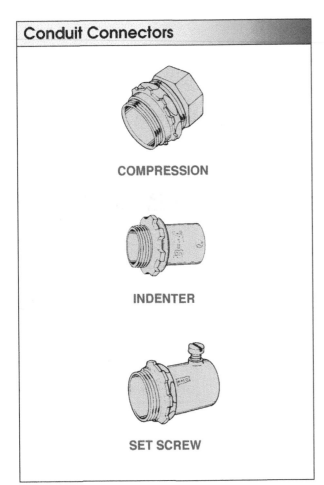

COMPRESSION

INDENTER

SET SCREW

Figure 9-22. *Conduit connectors hold conduit securely to boxes.*

Compression Connectors

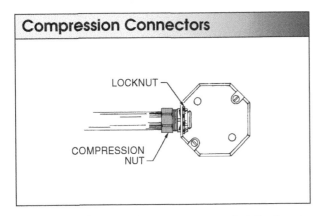

LOCKNUT

COMPRESSION NUT

Figure 9-23. *Compression connectors are held in place by a locknut, and secure conduit in place with a compression nut and ferrule.*

Indenter Connectors. An *indenter connector* is a type of box fitting that secures conduit to a box with the use of a special indenting tool. **See Figure 9-24.** Indenter coupling connectors are not reusable and must be cut off with a hacksaw to be removed.

Indenter Connectors

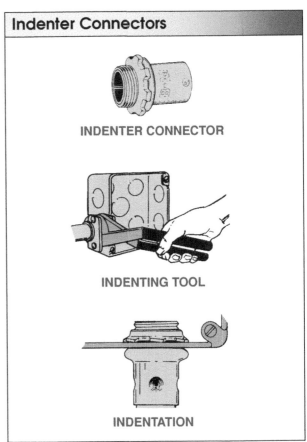

INDENTER CONNECTOR

INDENTING TOOL

INDENTATION

Raco Inc.

Figure 9-24. *Indenter connectors require a special indenting tool for installation.*

Set Screw Connectors. A *set screw connector* is a type of box fitting that relies on the pressure of a screw against conduit to hold the conduit in place. **See Figure 9-25.** Set screw connectors are also reusable.

Conduit Couplings. Unlike metallic and nonmetallic cable, conduit must be joined together using conduit couplings when a run exceeds 10′. A *conduit coupling* is a type of fitting used to join one length of conduit to another length of conduit and still maintain a smooth inner surface. **See Figure 9-26.** Conduit couplings are similar in design to back-to-back conduit connectors and operate on the same principles of compression, indentation, and pressure.

Set Screw Connectors

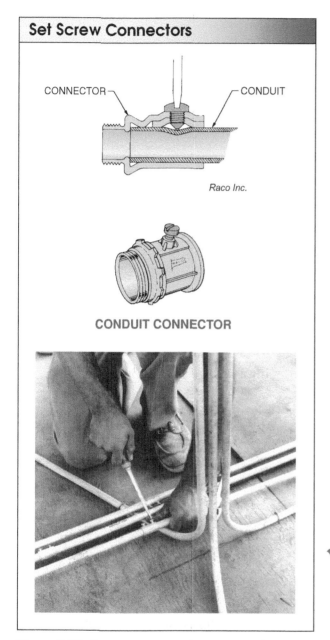

CONNECTOR — — CONDUIT

Raco Inc.

CONDUIT CONNECTOR

Figure 9-25. *Set screw connectors are quickly and securely installed with a screwdriver.*

Compression and set-screw connectors and couplings are designed for use with all rigid conduit and compresion fittings are impervious to rain, moisture, and concrete.

Conduit Couplings

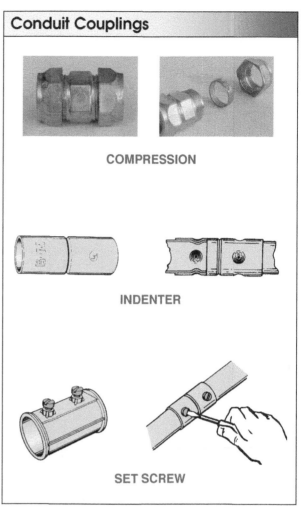

COMPRESSION

INDENTER

SET SCREW

Figure 9-26. *Conduit couplings use the same techniques as conduit connectors (compression, indenting, and set screw) to join two lengths of conduit.*

Tech Tip:
RMC Installation

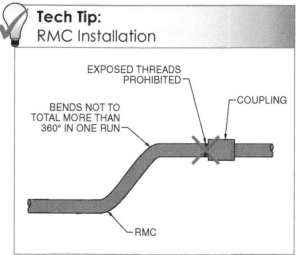

EXPOSED THREADS PROHIBITED

BENDS NOT TO TOTAL MORE THAN 360° IN ONE RUN

COUPLING

RMC

Applications that require conduit to be run where water is present require that special liquid-tight conduit and watertight conduit couplings be used. **See Figure 9-27.**

Article 350 of the NEC® provides guidelines for liquid-tight flexible metal conduit (LFMC) which is used to connect equipment that vibrates during normal operation and is located outdoors or near wet locations.

Liquid-Tight Conduit Connector

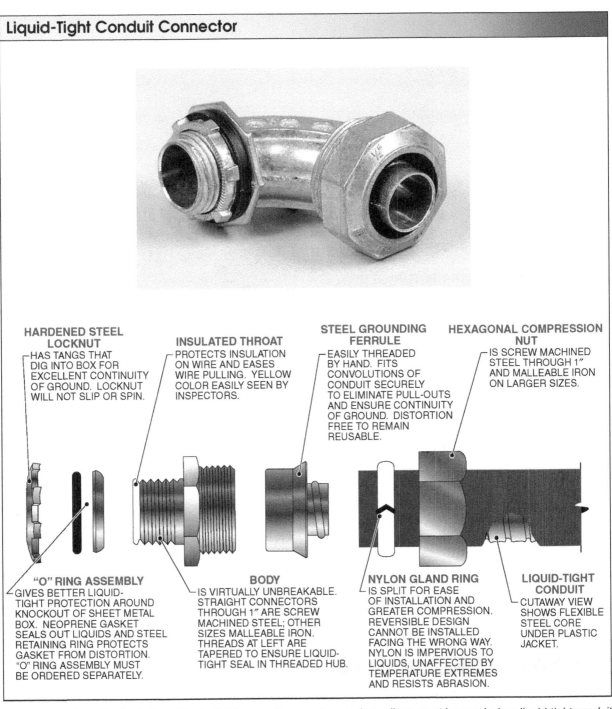

HARDENED STEEL LOCKNUT
—HAS TANGS THAT DIG INTO BOX FOR EXCELLENT CONTINUITY OF GROUND. LOCKNUT WILL NOT SLIP OR SPIN.

INSULATED THROAT
—PROTECTS INSULATION ON WIRE AND EASES WIRE PULLING. YELLOW COLOR EASILY SEEN BY INSPECTORS.

STEEL GROUNDING FERRULE
—EASILY THREADED BY HAND. FITS CONVOLUTIONS OF CONDUIT SECURELY TO ELIMINATE PULL-OUTS AND ENSURE CONTINUITY OF GROUND. DISTORTION FREE TO REMAIN REUSABLE.

HEXAGONAL COMPRESSION NUT
—IS SCREW MACHINED STEEL THROUGH 1" AND MALLEABLE IRON ON LARGER SIZES.

"O" RING ASSEMBLY
—GIVES BETTER LIQUID-TIGHT PROTECTION AROUND KNOCKOUT OF SHEET METAL BOX. NEOPRENE GASKET SEALS OUT LIQUIDS AND STEEL RETAINING RING PROTECTS GASKET FROM DISTORTION. "O" RING ASSEMBLY MUST BE ORDERED SEPARATELY.

BODY
—IS VIRTUALLY UNBREAKABLE. STRAIGHT CONNECTORS THROUGH 1" ARE SCREW MACHINED STEEL; OTHER SIZES MALLEABLE IRON. THREADS AT LEFT ARE TAPERED TO ENSURE LIQUID-TIGHT SEAL IN THREADED HUB.

NYLON GLAND RING
—IS SPLIT FOR EASE OF INSTALLATION AND GREATER COMPRESSION. REVERSIBLE DESIGN CANNOT BE INSTALLED FACING THE WRONG WAY. NYLON IS IMPERVIOUS TO LIQUIDS, UNAFFECTED BY TEMPERATURE EXTREMES AND RESISTS ABRASION.

LIQUID-TIGHT CONDUIT
—CUTAWAY VIEW SHOWS FLEXIBLE STEEL CORE UNDER PLASTIC JACKET.

Figure 9-27. *Specifically designed (watertight) conduit connectors and couplings must be used when liquid-tight conduit is being used in a wet environment.*

INSTALLATION OF RIGID METAL CONDUIT

Bending rigid metal conduit requires many of the skills used to bend EMT. The primary difference is that rigid metal conduit is heavier and much more difficult to bend, requiring use of heavy-duty hydraulic machinery. Due to the difficulty in bending, rigid metal conduit is typically threaded. The threads allow threaded fittings such as elbows to be used instead of bending the conduit. **See Figure 9-28.**

Fortunately, very little rigid metal conduit is used in residential construction. For residences, rigid metal conduit is typically only used for overhead service risers and for underground services. For service installations, rigid metal conduit can be purchased precut and prethreaded. Electrical supply houses or hardware stores typically stock a variety of lengths and sizes of rigid metal conduit ready for use. Electrical supply houses or stores also carry stock fittings such as 45° and 90° bends. When special lengths are required, rigid metal conduit can be cut and threaded by a supply house for an additional fee.

INSTALLATION OF FLEXIBLE METAL CONDUIT

Flexible metal conduit (Greenfield) uses the same techniques for installation as metallic sheathed cable. Flexible metallic conduit is a metal conduit, with no wires, that can be bent by hand. Flexible metal conduit differs from metallic sheathed cable in that wires must be pulled through flexible metal conduit after installation. **See Figure 9-29.**

Pulling Wires through Conduit

Fishing is a term used for the process of pulling wires through conduit. **See Figure 9-30.** A *fish tape* is a device used to pull wires through conduit. The fish tape is extended and pushed through the conduit until the tape reaches an opening. At the opening the wires being installed are firmly secured to the fish tape. The fish tape is retrieved by pulling the tape out of the conduit with the wires attached. The fish tape must be pulled evenly and smoothly to allow the wires to move easily through the conduit. Typically, one electrician operates the fish tape while another electrician feeds the wires into the conduit.

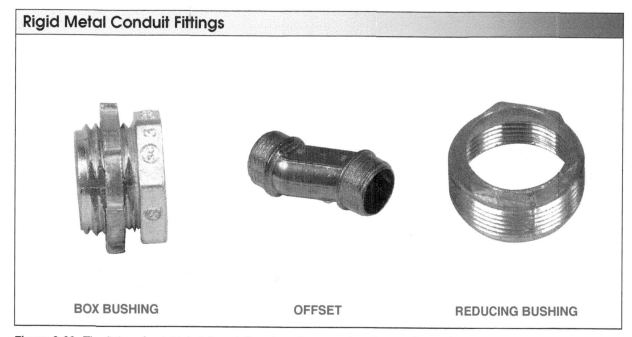

Rigid Metal Conduit Fittings

BOX BUSHING   OFFSET   REDUCING BUSHING

Figure 9-28. *The fittings for rigid metal conduit perform the same functions as fittings for other conduits, except that rigid metal conduit has threads for connections.*

Flexible Metal Conduit Installation

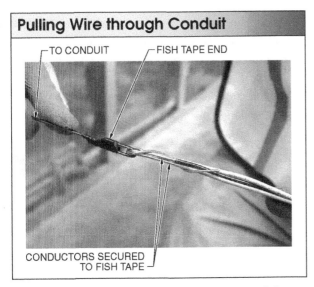

Figure 9-29. *Flexible metal conduit requires that wires be pulled through the conduit after installation.*

Pulling Wire through Conduit

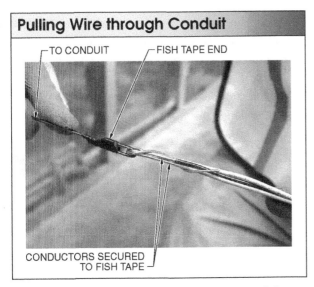

Figure 9-30. *Always securely fasten wires to a fish tape before pulling the wires through conduit.*

Two types of fish tapes are typically used by electricians. **See Figure 9-31.** One type of fish tape is made of a rigid steel ribbon while the other type is a more flexible polyethylene tape. A *rigid fish tape* is a wire-pulling device used for pulling wires through conduit and pulling wires through walls and ceilings during remodeling. A *polyethylene fish tape* is a wire-pulling device typically used to pull wire within conduit systems.

Occasionally, wires being pulled through conduit can become difficult to pull. Two aids are used by electricians when pulling wires through conduit, pulling grips and lubricants. **See Figure 9-32.** A *pulling grip* is a device that is attached to a fish tape to allow more leverage. A *lubricant* is a wet or dry compound that is applied to the exterior of wires to allow the wires to slide better. If pulling the wires is still difficult with a hand grip and lubricant, compare the number and size of wires in the conduit to NEC® standards.

Fish Tapes

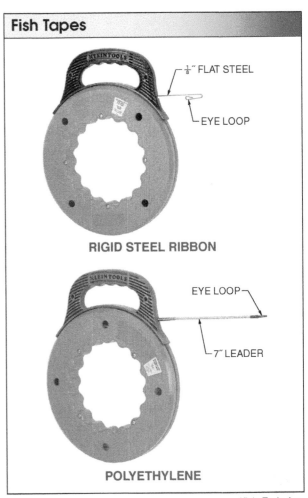

Klein Tools, Inc.

Figure 9-31. *Steel ribbon fish tapes are the more common tape used by electricians because of their suitability for general use.*

Underground conduit frequently fills with moisture, and even interior conduit saturated with wire pulling lubricants can rust a steel fish tape. A nylon-coated tape is impervious to moisture and will last longer under these conditions.

Wire Pulling Aids

FLEXIBLE - EYE
PULLING GRIP

SQUEEZE BOTTLE

LUBRICANT

Klein Tools, Inc.

Figure 9-32. *Wire pulling aids are helpful when pulling wires through conduit.*

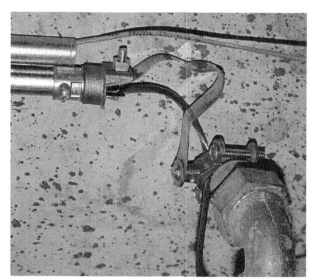

Rigid metal conduit can be grounded to the water system piping of a dwelling.

Trimming Out a Conduit System

For the most part, installing switches and receptacles in a conduit system is the same as in any other system. In a properly installed metal conduit system, switches and receptacles are grounded through the conduit. However, an additional green wire pulled through the system and attached to the green grounding screw provides additional grounding security. **See Figure 9-33.** Conduit is also used to add on to or extend existing systems. **See Figure 9-34.**

Trimming Out Conduit System

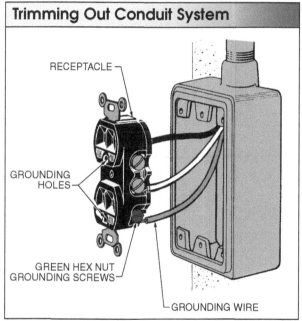

RECEPTACLE

GROUNDING HOLES

GREEN HEX NUT GROUNDING SCREWS

GROUNDING WIRE

Harvey Hubbell, Inc.

Figure 9-33. *Pulling and connecting a separate ground wire can establish additional switch and receptacle grounding security.*

A power cable puller is used to pull large cables and wires into place. Power cable pullers are similar to fish tapes but use a gas- or electric-powered motor to pull wire. This reduces the strain and fatigue of the operator. Power cable pullers can pull cables up to distances of 200 and use up to 2000 lb of force.

Extending Existing Conduit Systems

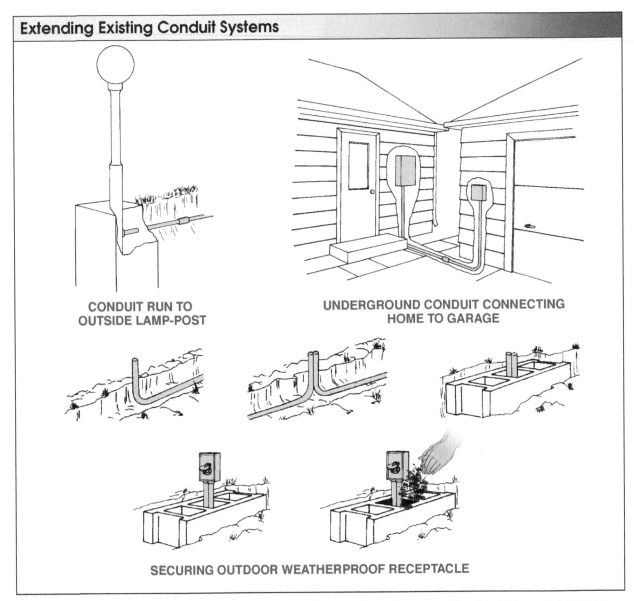

CONDUIT RUN TO
OUTSIDE LAMP-POST

UNDERGROUND CONDUIT CONNECTING
HOME TO GARAGE

SECURING OUTDOOR WEATHERPROOF RECEPTACLE

Figure 9-34. *Conduit with weatherproof connectors, couplings, and boxes is required for all outside conduit systems.*

Polyvinyl Chloride Conduit

Polyvinyl chloride (PVC), or nonmetallic, conduit is used for both outdoor and underground installations since it does not rust from outdoor exposure or moisture. PVC conduit is flexible, provides good insulation, and has high impact resistance and tensile strength. PVC conduit is available in varying diameters from ½″ to 6″. It is rated as Schedule 40 (thin wall) and Schedule 80 (heavy wall). Schedule 40 PVC does not have the strength of similar sized metal conduit and cannot withstand severe physi-

cal damage. Schedule 80 PVC is durable and has strength properties similar to rigid metal conduit (RMC). **See Figure 9-35.**

Solvent-Cementing PVC. Solvent cementing is the process of fusing plastic conduit and fittings together by softening the adjoining surfaces through a chemical reaction. A chemical is applied to soften both of the surfaces to be joined together. The surfaces are then forced together to form a solid joint that is as strong as the conduit walls. The three considerations that must be taken into account to achieve a watertight

and airtight seal include the following:
• application of appropriate primer for the PVC
• verification that the sections to be joined together have a good interference fit
• application of the proper installation techniques

PVC Conduit

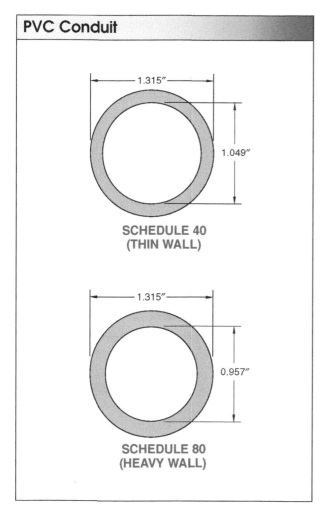

Figure 9-35. *PVC conduit is rated as Schedule 40 (thin wall) and Schedule 80 (heavy wall).*

The appropriate primer and solvent cement must be used for PVC. A *primer* is a chemical agent that cleans and softens a surface and allows solvent cement to penetrate more effectively into the surface. *Solvent cement* is a chemical agent that penetrates and softens the surface of plastic pipe and fittings. Precautions for the safe handling of solvent cements and primers include the following:

• Avoid breathing solvent cement vapors and keep work area properly ventilated.
• Keep cleaners, primers, and solvent cements away from ignition sources, sparks, heat, and open flames.
• Keep storage containers tightly closed when product is not in use.
• Properly dispose of all cloths and rags used for cleaning excess material.
• Wear proper PPE such as goggles and neoprene gloves.
• Use proper application tools.

Proper preparation and installation techniques must be used to ensure watertight and airtight joints. **See Figure 9-36.** PVC conduit is prepared for installation by applying the following procedure:
1. Cut PVC in a square manner with a chop saw, universal saw, or plastic pipe cutter.
2. Smooth the pipe ends with a deburring tool or hand file.
3. Ensure that pipe ends are clean and dry and test fit together to ensure proper interference fit. The pipe should be able to be inserted only about halfway into the socket.
4. Apply primer to plastic pipe fittings by first applying primer to inside diameter of fitting and then to the outside of the pipe to the depth that will be seated in the fitting. *Note:* Application of too much primer results in a puddle, which can cause flow restrictions.
5. Wait 10 sec to 15 sec, and apply solvent cement with a brush or roller in the same manner as primer application. *Note:* Application of too much solvent results in a puddle, which can cause flow restrictions.
6. Fit and position the pipe and fitting together before solvent cement evaporates, ensuring that pipe is properly seated in the fitting socket. Once properly seated, turn pipe ¼ turn and hold pipe and fitting in place for 10 sec to 20 sec to allow solvent cement to bond the two surfaces together.
7. Apply a bead of solvent cement around entire diameter of pipe and fitting.
8. Remove any excess solvent cement with a clean, dry cloth.

Schedule 80 PVC conduit has a wall thickness ⅓ greater than that of standard Schedule 40 PVC and is dark gray. Because it has a heavier wall, Schedule 80 is used where conduit is subject to physical damage such as outdoors or in a concrete encasement.

Solvent Cementing

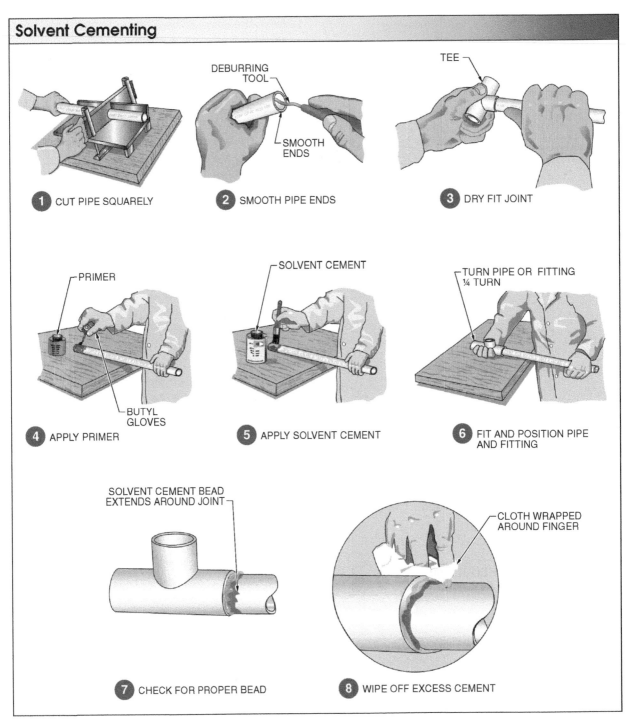

1 CUT PIPE SQUARELY

2 SMOOTH PIPE ENDS

DEBURRING TOOL

SMOOTH ENDS

3 DRY FIT JOINT

TEE

4 APPLY PRIMER

PRIMER

BUTYL GLOVES

5 APPLY SOLVENT CEMENT

SOLVENT CEMENT

6 FIT AND POSITION PIPE AND FITTING

TURN PIPE OR FITTING ¼ TURN

7 CHECK FOR PROPER BEAD

SOLVENT CEMENT BEAD EXTENDS AROUND JOINT

8 WIPE OFF EXCESS CEMENT

CLOTH WRAPPED AROUND FINGER

Figure 9-36. *The proper preparation and installation techniques must be used to ensure watertight and airtight solvent-cemented joints.*

Wiring with Conduit

TEST

9

Name_____   Date_____

Wiring with Conduit

_____ 1. Electrical metallic tubing is also known as ___ conduit.

_____ 2. ___ metal conduit is threaded on the ends.

_____ 3. ___ metal conduit is similar to armored cable but without the wires.

_____ 4. EMT is typically sold in ___′ bundles.

_____ 5. EMT is ___.

        A. lighter than rigid conduit   **C.** threaded conduit
        B. 10′ in length per piece   **D.** both A and B

_____ 6. EMT shall be securely fastened ___.

        A. within 36″ of boxes   **C.** at every bend
        B. at least every 10′   **D.** both A and B

_____ 7. A run of EMT shall not contain more than four quarter bends totaling ___°.

_____ 8. Pulling wire through conduit is also known as ___.

    T   F   9. Lubricants may be used to facilitate wire pulls through EMT.

    T   F   10. Receptacles are typically grounded by metal piping when using EMT.

    T   F   11. Compression connectors are firmly secured through tightening by hand.

    T   F   12. EMT shall be securely fastened in place within 5′ of a junction box.

_____ 13. A pocket ___ may be used to check the accuracy of an offset.

_____ 14. A hand conduit bender is also known as a(n) ___.

_____ 15. The trade sizes for EMT range from ½″ to ___″ in diameter.

_____ 16. Conduit must be ___ after being cut to length.

_____ 17. Conduit ___ are used to join pieces of EMT together.

217

_____ **18.** Bends in tubing must not reduce the ___ diameter.

T   F   **19.** A compression connector may be used to secure EMT to an electrical box.

T   F   **20.** A compression connector can only be used one time.

T   F   **21.** Set screw connectors can be reused.

T   F   **22.** An indenter connector can only be used one time.

_____ **23.** A total of ___ bends are required to produce a saddle bend.

_____ **24.** A total of ___ bends are required to produce an offset bend.

_____ **25.** A(n) ___° angle is produced when the bender handle is raised to the vertical position.

Conduit Straps

_____ **1.** One hole push-on

_____ **2.** One hole

_____ **3.** Two hole

_____ **4.** Nail-up

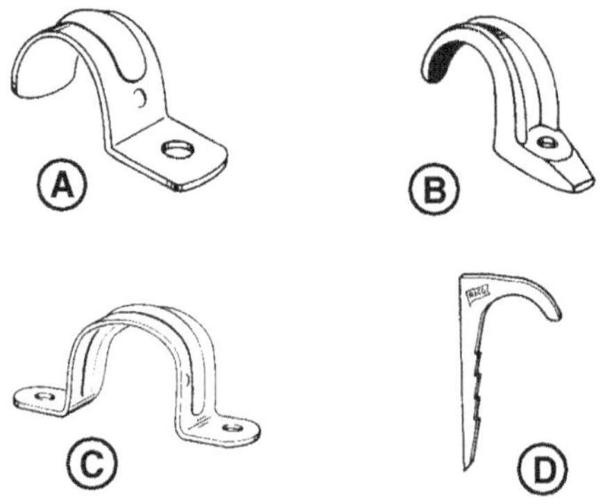

Allowable Number of Conductors

_____ **1.** ½″ conduit; No. 14 conductors

_____ **2.** ½″ conduit; No. 12 conductors

_____ **3.** ¾″ conduit; No. 10 conductors

_____ **4.** ¾″ conduit; No. 12 conductors

_____ **5.** ¾″ conduit; No. 14 conductors

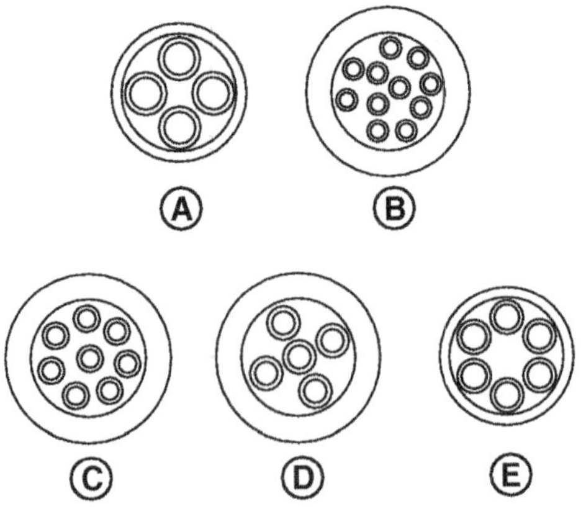

Installing Conduit

_____ 1. Outlet

_____ 2. Sole plate

_____ 3. Stud

_____ 4. Headers

_____ 5. Top plate

_____ 6. Bridging

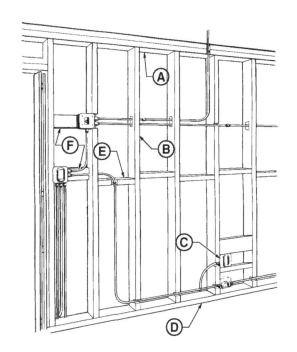

Conduit Bends

_____ 1. Double offset bend

_____ 2. 45° bend

_____ 3. Back-to-back bend

_____ 4. Offset bend

_____ 5. Alternate double offset bend

_____ 6. 90° bend

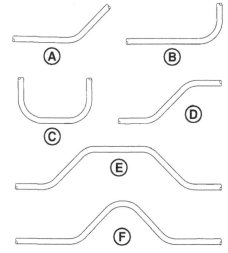

Conduit Types

_____ 1. Flexible conduit

_____ 2. Electrical metallic tubing

_____ 3. Rigid conduit

_____ 4. PVC conduit

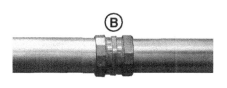

CARLON SCHEDULE 40 rigid non-metallic conduit

Conduit Connectors and Couplings

_____ 1. Indenter connector

_____ 2. Compression connector

_____ 3. Set screw connector

_____ 4. Compression coupling

_____ 5. Set screw coupling

_____ 6. Indenter coupling

Ⓐ

Ⓑ

Ⓒ

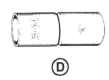

Ⓓ

Ⓔ

Ⓕ

Liquid-Tight Conduit Connector

_____ 1. O-ring assembly

_____ 2. Steel grounding ferrule

_____ 3. Hexagonal compression nut

_____ 4. Hardened steel locknut

_____ 5. Nylon gland ring

_____ 6. Flexible steel core

_____ 7. Insulated throat

_____ 8. Body

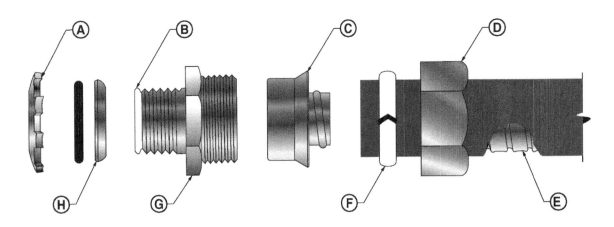

Service Entrances 10

Chapter 10 provides an overview of how service entrance equipment is installed. The local building inspector and power company must be consulted before installing service equipment. Properly installed service equipment will provide years of maintenance-free service.

Installers of electrical service equipment should verify that the equipment is suitable for use as service equipment. There are many types of identification used to mark the suitability of service equipment. Equipment should be marked "Suitable only for use as service equipment." Use of the wrong equipment violates the NEC® requirements and could introduce the possibility of a fire and/or shock hazard.

RESIDENTIAL SERVICE ENTRANCES

Installing a service entrance is typically the first priority when wiring a new home. A *service entrance* is the connecting link between a residence and the power company. Service entrances provide all the equipment necessary to obtain electricity and distribute it throughout a residence.

Electrical service to a residence is provided by wires that run overhead from a utility pole to the residence, or by wires to the residence that are buried in the ground. The two methods of electrical service to a residence are the service drop and underground lateral. The *service drop* method of electrical service runs wires from a utility pole to a service head on or above the residence. The *underground (service lateral)* method of electrical service buries the wires to a residence in the ground. **See Figure 10-1.**

Clearance requirements are provided to protect the service-drop conductors from physical damage and to protect personnel from contact with conductors.

Service Entrances

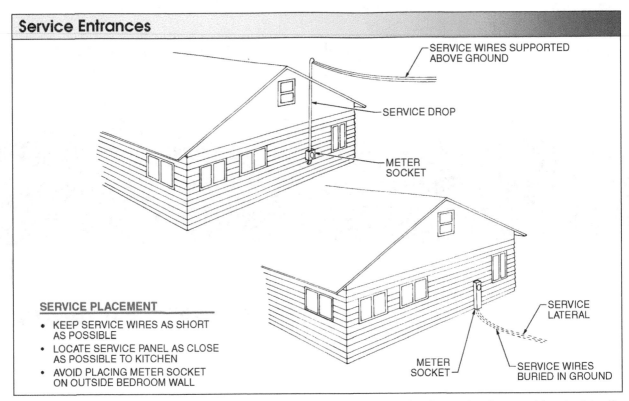

SERVICE WIRES SUPPORTED ABOVE GROUND

SERVICE DROP

METER SOCKET

SERVICE PLACEMENT

• KEEP SERVICE WIRES AS SHORT AS POSSIBLE
• LOCATE SERVICE PANEL AS CLOSE AS POSSIBLE TO KITCHEN
• AVOID PLACING METER SOCKET ON OUTSIDE BEDROOM WALL

SERVICE LATERAL

METER SOCKET

SERVICE WIRES BURIED IN GROUND

Figure 10-1. *The two service entrance methods used for residences are the service drop and underground (service lateral).*

Placement of Service Entrances

Placement of a service entrance must take into consideration both cost and convenience. Electricians must try to make the service panel of the residence as accessible as possible using the minimum of materials. The three tips that are useful in determining the placement of a service entrance are the following:
• Keep the service wires as short as possible.
• Locate the service panel as close to the kitchen as possible to avoid costly wire runs to major appliances.
• Avoid having the meter socket placed on the outside of a bedroom wall.

Sizing Service Entrances

Service entrances for single family dwellings require a minimum 100 A capability. For larger residences and structures using electric heat, a 200 A or larger service entrance can be required. Residential service entrances are typically rated in amperes because voltage is typically a standard (115 V/230 V).

Further information concerning service entrances can be obtained from Article 230 of the NEC® or the local building inspector.

ROUGHING-IN SERVICE ENTRANCES

Electricians need not be apprehensive about installing an electrical service. To make service installation to a residence easier, manufacturers provide service entrance installation kits. **See Figure 10-2.** Depending on local codes and the specific power company, service entrances can be installed by anyone safely, with a professional look and at minimal cost.

Types of Service Entrances

Not all electrical services are installed in the same manner. Services differ because of the application or because of the place in which the service is located. Typically four methods are used to bring electrical service to a residence.

Service Entrance Parts

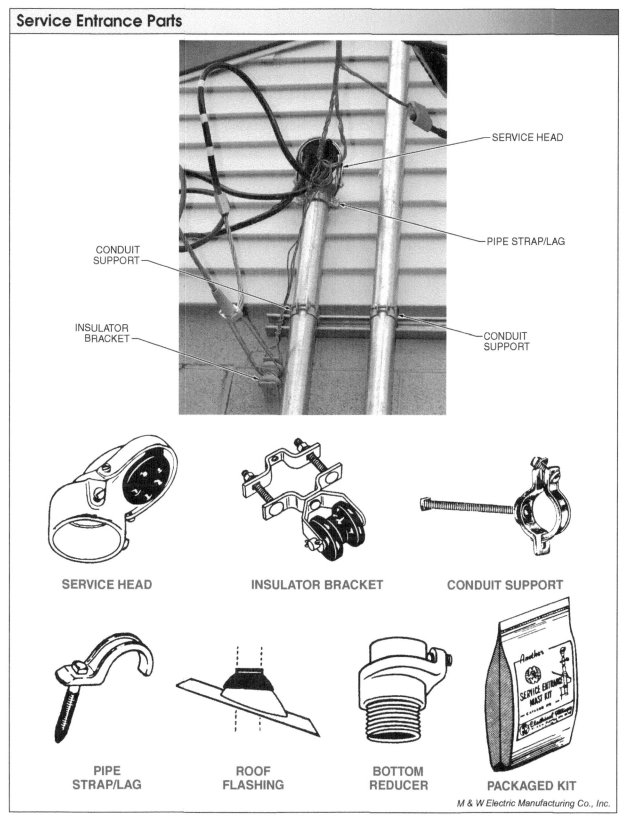

M & W Electric Manufacturing Co., Inc.

Figure 10-2. *To make service installation to a residence easier, manufacturers provide service entrance installation kits for electricians.*

Overhead Riser Service Entrances

Many established neighborhoods and urban areas use the overhead riser method of electrical service. **See Figure 10-3.** The *overhead riser service entrance* has wires running from a utility pole to a service head, where the meter socket and riser are firmly secured to the exterior of a dwelling, while the service panel is placed inside. An appropriate-size conduit typically connects the meter socket and meter to the service panel. Several methods are used to connect the meter and socket to the service panel with conduit.

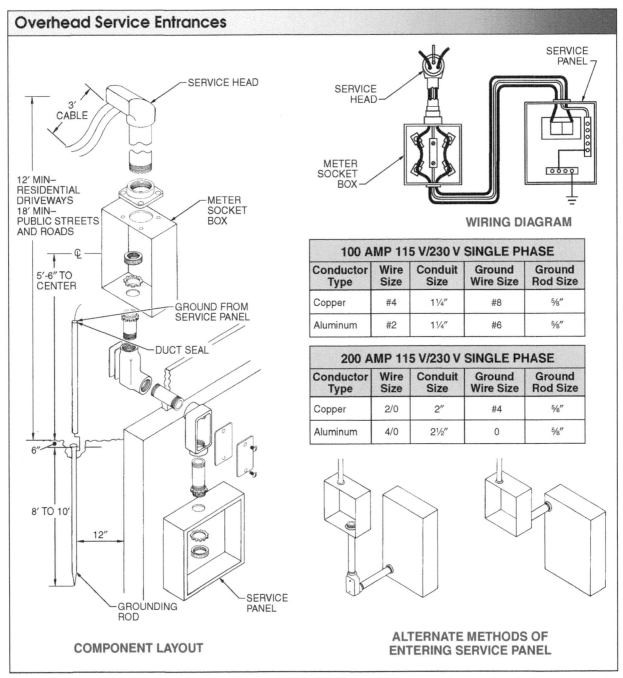

Overhead Service Entrances

SERVICE HEAD

3' CABLE

12' MIN– RESIDENTIAL DRIVEWAYS
18' MIN– PUBLIC STREETS AND ROADS

5'-6" TO CENTER

METER SOCKET BOX

GROUND FROM SERVICE PANEL

DUCT SEAL

6"

8' TO 10'

12"

GROUNDING ROD

SERVICE PANEL

COMPONENT LAYOUT

SERVICE HEAD

SERVICE PANEL

METER SOCKET BOX

WIRING DIAGRAM

100 AMP 115 V/230 V SINGLE PHASE				
Conductor Type	Wire Size	Conduit Size	Ground Wire Size	Ground Rod Size
Copper	#4	1¼″	#8	⅝″
Aluminum	#2	1¼″	#6	⅝″

200 AMP 115 V/230 V SINGLE PHASE				
Conductor Type	Wire Size	Conduit Size	Ground Wire Size	Ground Rod Size
Copper	2/0	2″	#4	⅝″
Aluminum	4/0	2½″	0	⅝″

ALTERNATE METHODS OF ENTERING SERVICE PANEL

Figure 10-3. *The overhead riser method of electrical service wiring has the meter socket and riser firmly secured to the exterior of a dwelling, while the service panel is placed inside.*

Lateral Service Entrances (Conduit)

In most new developments it is customary to bury all service wires underground. **See Figure 10-4.** A *lateral service entrance (conduit)* is a service entrance where all service wires are buried underground, creating a condition where the wires are subjected to less damage from the environment (weather) and allowing unsightly poles and service wires to be removed from streets and alleys. Typically, underground services are built using conduit.

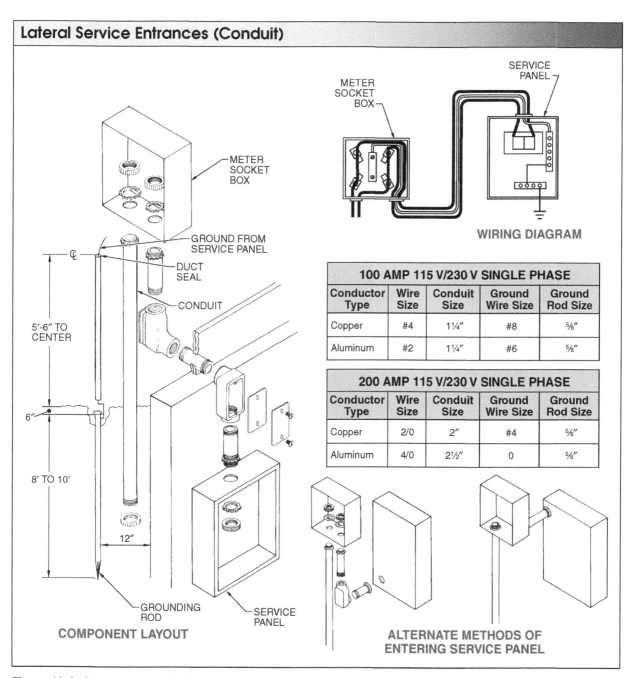

Lateral Service Entrances (Conduit)

METER SOCKET BOX

GROUND FROM SERVICE PANEL

DUCT SEAL

CONDUIT

℄

5'-6" TO CENTER

6"

8' TO 10'

12"

GROUNDING ROD

SERVICE PANEL

COMPONENT LAYOUT

SERVICE PANEL

METER SOCKET BOX

WIRING DIAGRAM

100 AMP 115 V/230 V SINGLE PHASE				
Conductor Type	Wire Size	Conduit Size	Ground Wire Size	Ground Rod Size
Copper	#4	1¼"	#8	⅝"
Aluminum	#2	1¼"	#6	⅝"

200 AMP 115 V/230 V SINGLE PHASE				
Conductor Type	Wire Size	Conduit Size	Ground Wire Size	Ground Rod Size
Copper	2/0	2"	#4	⅝"
Aluminum	4/0	2½"	0	⅝"

ALTERNATE METHODS OF ENTERING SERVICE PANEL

Figure 10-4. *A lateral service entrance is a service entrance where all service wires are buried underground, creating a condition where the wires are subjected to less damage from the environment (weather) and allowing unsightly poles and service wires to be removed from streets and alleys.*

Lateral Service Entrances (Cabinet)

The cabinet type of underground service is another method of installing an underground service to a residence. **See Figure 10-5.** A *lateral service entrance (cabinet)* utilizes a cabinet that encloses the service wires, meter socket, and meter. The entire enclosure is then covered with a metal cover that is typically locked. A cabinet type of underground service is very neat and makes the routing of wires much easier.

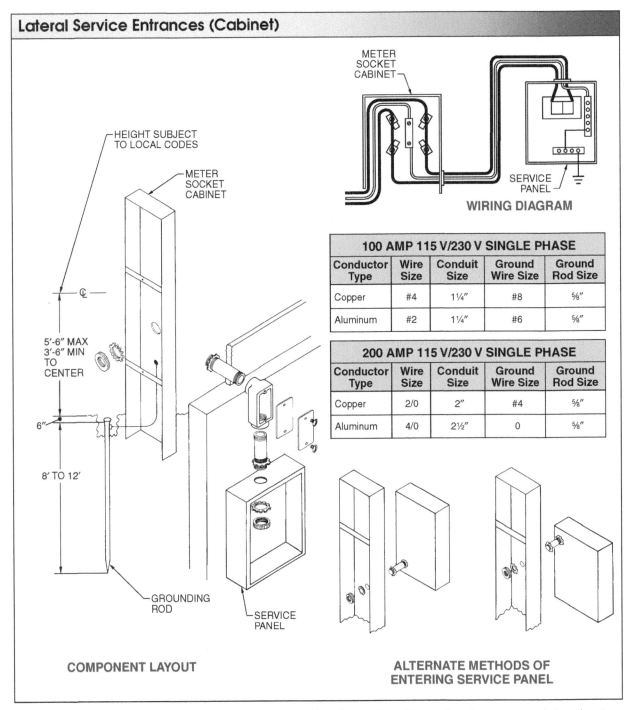

Lateral Service Entrances (Cabinet)

100 AMP 115 V/230 V SINGLE PHASE

Conductor Type	Wire Size	Conduit Size	Ground Wire Size	Ground Rod Size
Copper	#4	1¼″	#8	⅝″
Aluminum	#2	1¼″	#6	⅝″

200 AMP 115 V/230 V SINGLE PHASE

Conductor Type	Wire Size	Conduit Size	Ground Wire Size	Ground Rod Size
Copper	2/0	2″	#4	⅝″
Aluminum	4/0	2½″	0	⅝″

Figure 10-5. *A lateral service entrance (cabinet) utilizes a cabinet that encloses the service wires, meter socket, and meter.*

Mobile Home Service Entrances

Mobile homes present a unique problem for service entrances because the service is not typically attached directly to the mobile home. Instead, the service is permanently mounted near the mobile home and a cable is run to the mobile home.

See Figure 10-6. Overhead and lateral methods are used to connect electrical service to a mobile home. A cable from the meter runs to a disconnect, then a cable from the disconnect runs to the service panel of the mobile home. The service panel then distributes electricity throughout the mobile home.

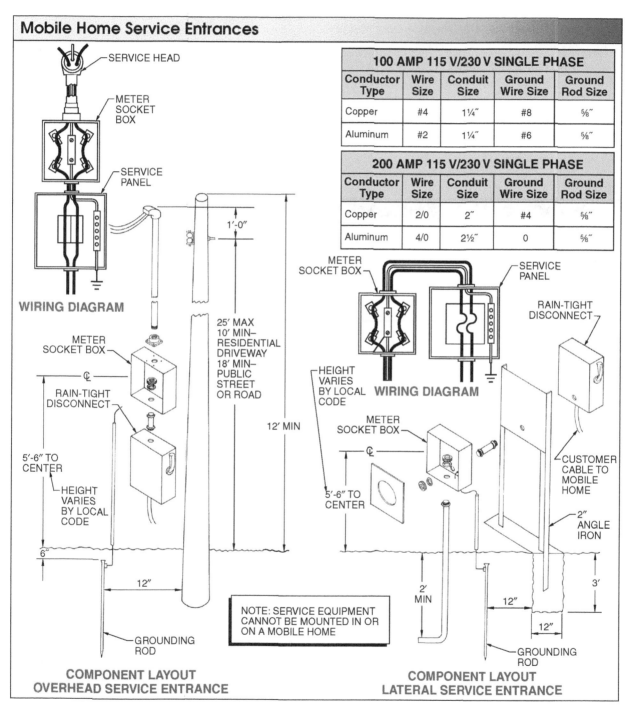

Figure 10-6. *Overhead and lateral methods are used to connect electrical service to a mobile home.*

Securing Service Equipment

Typically, the service wires are not connected to the power company until the service and inside of the residence are properly wired. All boxes, panels, conduit, and wires must be securely held in place to avoid problems from wind, rain, snow, hail, and other environmental concerns.

Knockouts. During the installation of a service entrance, large and small box knockouts must be removed from the service panel. Whenever possible, large knockouts should be removed before the service panel is mounted. Removing the knockouts before mounting reduces any chance of loosening the panel. **See Figure 10-7.** Knockouts are removed using standard procedures.

An enclosure is a device that protects electrical control devices. Enclosures are categorized by the protection they provide. An enclosure is selected based on the location of the equipment and the NEC® requirements. For example, a Type 3 enclosure is used outdoors for applications subjected to windblown dust, rain, sleet, and ice. Type 3 enclosures must pass rain, external icing, windblown dust, and rust resistance tests. Type 3 enclosures, however, do not protect against internal condensation or internal icing.

Tech Tip:
Single-Ring Knockout

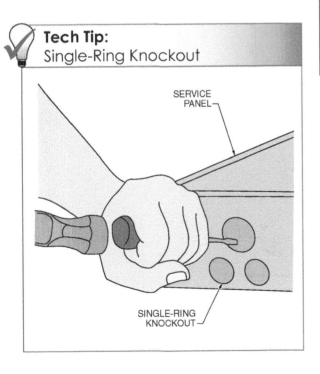

Concentric Knockout Removal

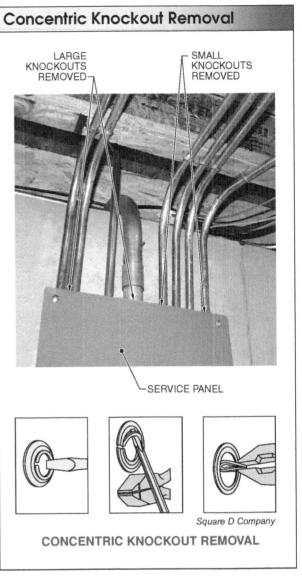

CONCENTRIC KNOCKOUT REMOVAL

Figure 10-7. *Removing knockouts before mounting a panel reduces any chance of loosening the panel.*

Fasteners. Meter sockets and service panels must be firmly secured before any wires (cables) are pulled or conduit attached. When service equipment is mounted on wood, wood screws or lag bolts are used to hold the equipment in place. When service equipment is mounted to masonry, drills and special bolts with anchors are used. **See Figure 10-8.** Plastic anchors support a considerable amount of weight when properly installed.

Mounting Service Equipment to Masonry

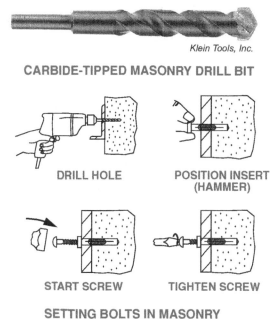

Klein Tools, Inc.

CARBIDE-TIPPED MASONRY DRILL BIT

DRILL HOLE

POSITION INSERT (HAMMER)

START SCREW

TIGHTEN SCREW

SETTING BOLTS IN MASONRY

Figure 10-8. *When service equipment is mounted to masonry, drills and special bolts with anchors are used.*

SERVICE PANELS

The electric power brought to a residence is typically a three-wire 115 V/230 V, 1φ system. Red, black, and white wires for the service panel enter the residence through a service head. **See Figure 10-9.** The red and black wires represent the hot or live wires, and the white wire represents the neutral. As with most residential services, 230 V can be obtained between the red and black hot wires. One hundred and fifteen volts can also be obtained between either of the hot wires and neutral. The 115 V and 230 V voltages are constant throughout the meter socket and into the service panel.

Inside the service panel, where the wires are terminated, metal conductors called busbars continue the system. Busbars distribute the power to the circuit breakers in an organized manner. **See Figure 10-10.** Service panel busbars convey power to single-pole and double-pole circuit breakers to form 115 V and 230 V circuits.

Service Equipment Voltages

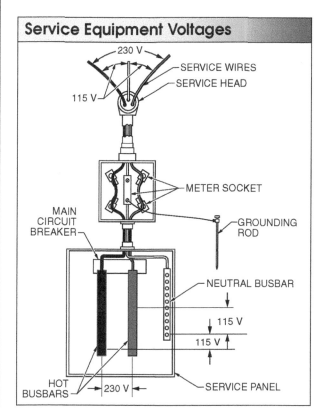

Figure 10-9. *A standard 115 V/230 V system consists of two incoming hot lines (red and black) and a neutral (white).*

Busbar Electrical Distribution

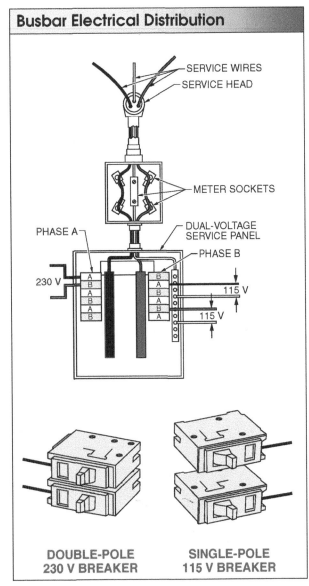

Figure 10-10. *Busbars alternately supply circuit breakers with phase 1 and phase 2 to form 115 V and 230 V circuits.*

Any single circuit breaker can only furnish power to one 115 V circuit. Two consecutive circuit breakers are used to provide power for a 230 V circuit. The circuit breakers must be consecutive so that the breaker handles can be fastened together as one unit. Fastening the handles of two circuit breakers together ensures that, in the event of a 230 V overload, both hot lines are disconnected simultaneously. **See Figure 10-11.** Circuit breakers are snapped into position when being installed to the busbars of a service panel.

230 V Circuit Breakers

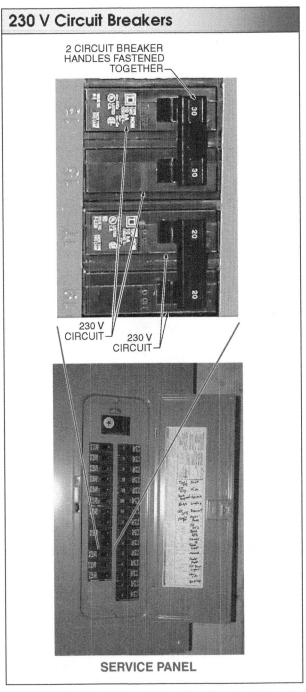

SERVICE PANEL

Figure 10-11. *Fastening the handles of a circuit breaker together ensures that, in the event of a 230 V overload, both hot lines are disconnected simultaneously.*

Service Entrance Cables

Instead of individual wires being pulled through conduit, quite often service entrance cables are used. **See Figure 10-12.** Service entrance cables have a bare conductor that is wound around the insulated conductors. The spiral winding prevents tampering with the conductors ahead of the meter and also serves as the neutral conductor.

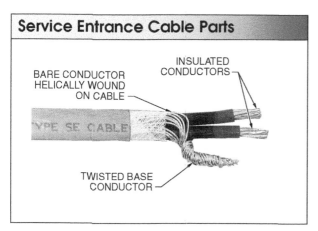

Figure 10-12. *The twisted bare conductor in a service cable protects the current-carrying wires and serves as the neutral wire.*

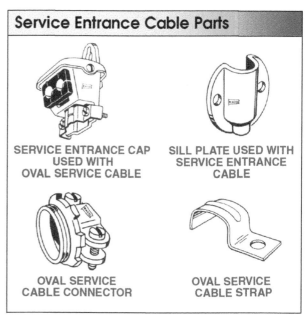

Figure 10-13. *Service entrance cables are accepted in many areas and require no special tools to install.*

Service entrance cables are accepted in many areas and require no special tools to install. Service entrance caps are shaped to accept oval service cables. **See Figure 10-13.** The sill plate covers the cable and the cable opening where the service enters the dwelling. Sealing compound or caulking is applied to make the sill plate watertight.

Service cable connectors are available as standard box connectors or as watertight connectors when used on top of the meter socket. Mounting straps are used to secure service cables in place. No matter who performs the work, the service entrance cable and devices must be installed as if an electrician performed the work. **See Figure 10-14.**

A service head or gooseneck should be installed at a point that is above the point of attachment for the service-drop conductors. This ensures that water does not enter the service raceway or service-entrance cable. There are installations where it is impractical to locate the service head or gooseneck above the point of attachment. NEC® Article 230.54(C) Ex. permits the point of attachment not to be located above the service head or gooseneck, but the service head or gooseneck must be located less than 24" from the point of attachment.

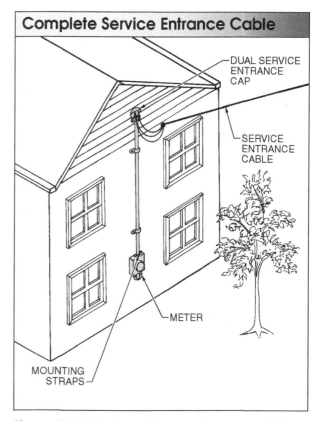

Figure 10-14. *Service cable connectors are available as standard box connectors or as watertight connectors when used on top of the meter socket.*

SERVICE DROPS

A service drop consists of overhead wires and devices that connect the power company power lines to a residence. Service drops are owned and maintained by the power company. Homeowners and electricians must be aware of the important rules governing the placement and elevation of service drops for new installations. The National Electrical Code® (NEC®) provides guidelines for the installation of service drops in relation to platforms, windows, service heads, and structure elevations.

Platforms

To comply with NEC® standards, a service head must be at least 10′ above and 3′ to the side of any platform. **See Figure 10-15.**

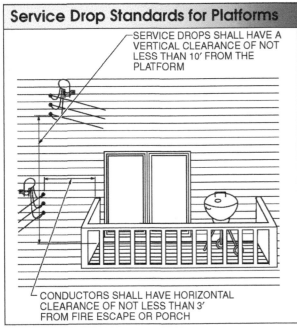

Service Drop Standards for Platforms

SERVICE DROPS SHALL HAVE A VERTICAL CLEARANCE OF NOT LESS THAN 10′ FROM THE PLATFORM

CONDUCTORS SHALL HAVE HORIZONTAL CLEARANCE OF NOT LESS THAN 3′ FROM FIRE ESCAPE OR PORCH

Figure 10-15. *NEC® standards govern service drops in relation to platforms (porches).*

Windows

The NEC® has standards governing service drops in relation to windows. **See Figure 10-16.** Service conductors must be at least 3′ from the bottom and sides of a window. Conductors that run above a window are considered out of reach from the window.

Service Drop Standards for Windows

CONDUCTORS RUN ABOVE THE WINDOW ARE CONSIDERED OUT OF REACH FROM THE WINDOW

CONDUCTORS SHALL HAVE CLEARANCE OF NOT LESS THAN 3′ FROM A WINDOW

Figure 10-16. *NEC® standards govern service drops in relation to windows.*

Elevations

NEC® standards governing the elevation of service conductors specify a 3′ clearance over adjacent buildings, and various elevations are required for sidewalks, private drives, alleys, and streets. **See Figure 10-17.**

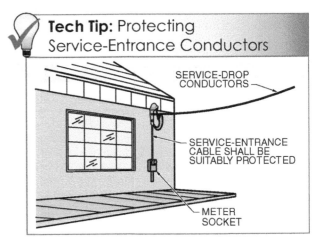

Tech Tip: Protecting Service-Entrance Conductors

SERVICE-DROP CONDUCTORS

SERVICE-ENTRANCE CABLE SHALL BE SUITABLY PROTECTED

METER SOCKET

Service Drop Standards for Elevations

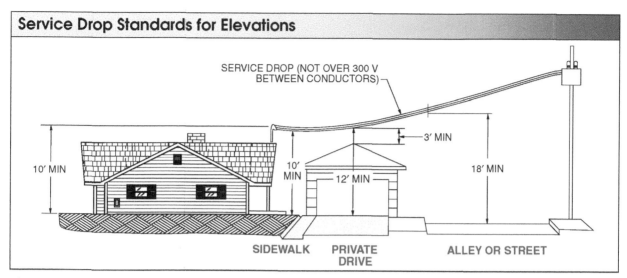

Figure 10-17. *Depending on the type of structure, service drops are required to maintain specific heights above grade.*

Service Heads

The NEC® has standards relating to the use of service heads. **See Figure 10-18.** Very explicit specifications are in place to ensure the performance and safety of service head installations. Careful consideration of each NEC® standard during installation ensures a permanent weatherproof connection out of the reach of potentially destructive obstacles.

The authority having jurisdiction (AHJ) has the final word regarding electrical installations, regardless of requirements stated by the NEC®. For example, many local AHJs require that all electrical services be installed in metal conduit. This wiring method may contradict the wiring methods of the NEC®, however, it must be used. Because of the potential for differing practices, AHJ and NEC® requirements must be distinguished from one another prior to installing electrical services.

Service Drop Standards for Service Heads

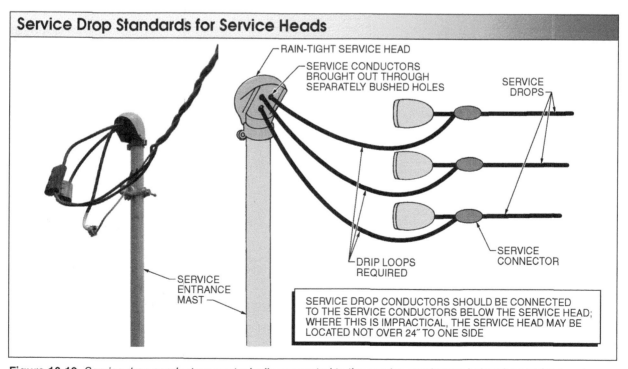

Figure 10-18. *Service drop conductors are typically connected to the service conductors below the service head.*

SERVICE LATERALS

A *service lateral* is any service to a residence that is achieved by burying the wires underground. The four types of service laterals used for residences are the following:

- from utility manhole in street to service panel disconnect
- from utility sidewalk handhole to service panel disconnect
- from pole riser to service panel disconnect
- from transformer pad to service panel disconnect

When service originates in a manhole, the work to be performed must be arranged with the utility company for joint installation of all conduit and wires. The manhole requirement is necessary because a manhole is a potentially dangerous space and is restricted to utility workers.

When service originates in a sidewalk handhole, electricians are typically permitted to run conduit and wires into the handhole. **See Figure 10-19.** The utility company will make the appropriate connections in the handhole.

When a pole riser is used, the service installation is shared by the homeowner and utility company. **See Figure 10-20.** In many instances, electricians simply install conduit to a point at least 8′ above grade and pull enough wire to reach the crossarm at the top of the pole. The utility company extends the protection for the drop cables with approved molding to the crossarm upon final connection. In some cases a short piece of conduit with a service head is used to terminate the run at the crossarm. When the utility company extends protection, the homeowner furnishes the conduit and service head. In other cases, homeowners must furnish conduit all the way with a service head at the end. For this type of service, see the utility company to avoid waste and delay.

When a service entrance originates from a transformer pad, special direct-burial cable is used to make the installation. **See Figure 10-21.** Transformer pad service entrances require the homeowner to furnish a transformer pad in addition to the regular electrical service equipment. The utility company typically creates the trench from the transformer pad to the dwelling, buries the cable, and makes all final connections. Always check with the utility company ahead of time to avoid delays and problems.

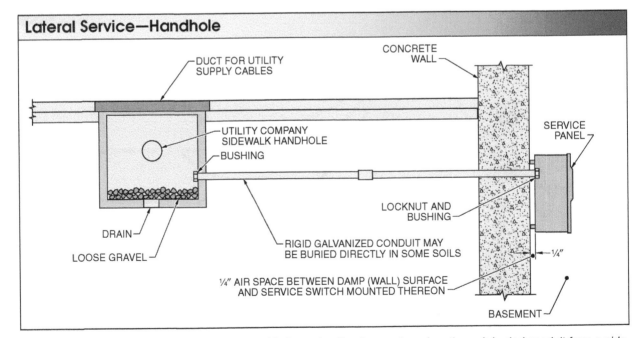

Lateral Service—Handhole

DUCT FOR UTILITY SUPPLY CABLES

CONCRETE WALL

UTILITY COMPANY SIDEWALK HANDHOLE

BUSHING

SERVICE PANEL

DRAIN

LOOSE GRAVEL

LOCKNUT AND BUSHING

RIGID GALVANIZED CONDUIT MAY BE BURIED DIRECTLY IN SOME SOILS

¼″

¼″ AIR SPACE BETWEEN DAMP (WALL) SURFACE AND SERVICE SWITCH MOUNTED THEREON

BASEMENT

Figure 10-19. *Underground service can be provided to a dwelling by running wires through buried conduit from a sidewalk handhole.*

Lateral Service—Pole Riser

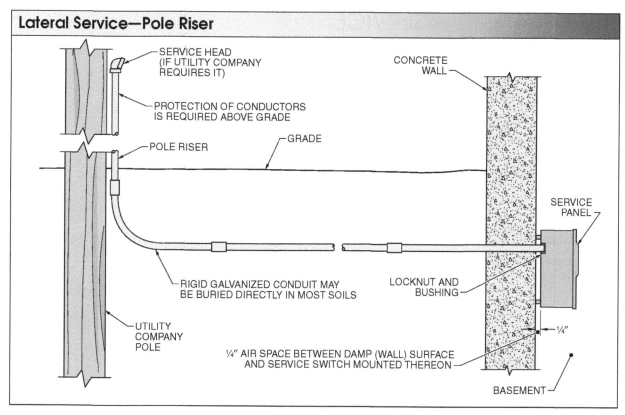

Figure 10-20. *Service laterals can originate from a pole riser.*

Lateral Service—Transformer Pad

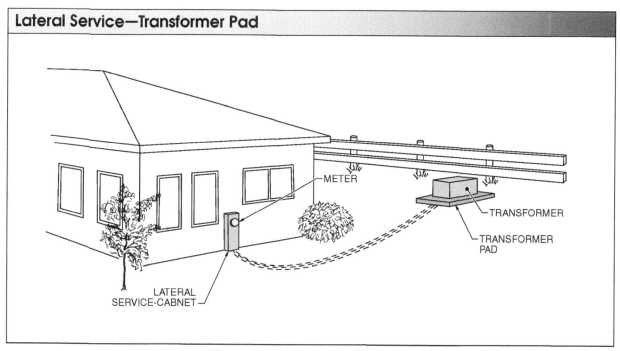

Figure 10-21. *Transformer pad service entrances require the homeowner to furnish a transformer pad and service equipment.*

TRIMMING OUT A SERVICE ENTRANCE

Trimming out or finishing a service entrance requires the installation of internal service conductors, circuit breakers, and branch circuit wiring. Wires must be bent to shape to fit some service heads. **See Figure 10-22.** Similar bends are required when wiring the meter socket and service panel.

> ⚠️ **CAUTION**
>
> *Never force a wire into a connection; always bend and shape wires to fit.*

Bending Wires for Service Caps

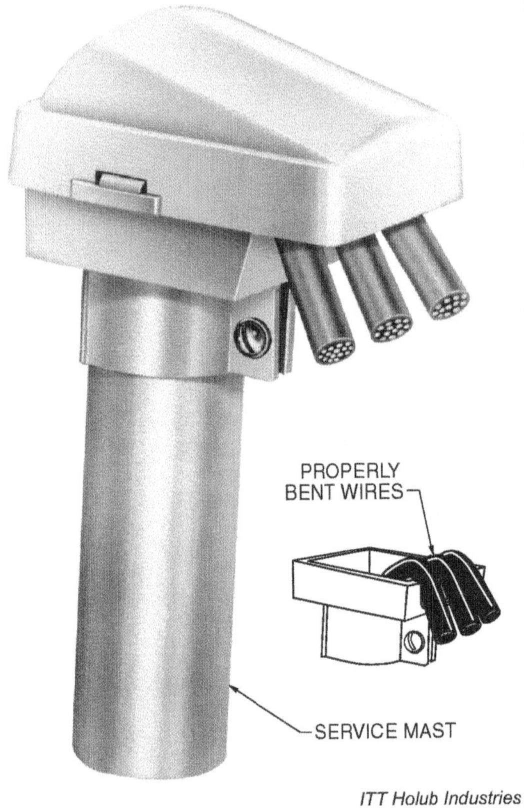

PROPERLY BENT WIRES

SERVICE MAST

ITT Holub Industries

Figure 10-22. *Service wires must be properly bent and fitted into service caps for maximum protection.*

Wire (cable) coming from branch circuits must also be bent to fit the configuration of the service panel. **See Figure 10-23.** Electricians should always take care and provide a professional appearance and organization to the wires to each circuit breaker. Although a novice may not be able to accomplish a professional look immediately, the professional look can be achieved given time and practice.

Service Panel Neatness

Figure 10-23. *Taking care to route wires in a service panel in an organized manner makes it easier to distinguish one wire from another.*

After all the circuit breakers have been properly installed and each circuit tested, electricians must prepare the trim cover to the service panel. **See Figure 10-24.** Proper technique must be used to remove the breaker openings from the service panel trim cover. Remove only the openings necessary for the circuit breakers installed.

The service panel cover is installed to finish trimming out a service panel. **See Figure 10-25.** The cover panel will have a directory on the inside of the door. The directory is filled out to indicate the location and use of each circuit breaker.

Removing Service Panel Breaker Knockouts

FIRST POSITION

SECOND POSITION

SERVICE PANEL TRIM COVER

Figure 10-24. *Circuit breaker knockouts should only be removed for the positions where circuit breakers are installed.*

Service Panel Breaker Directory

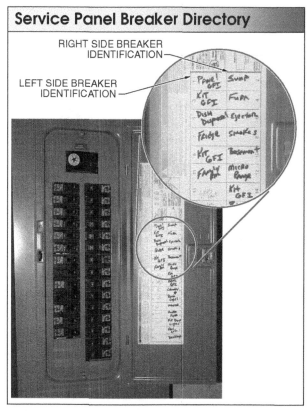

RIGHT SIDE BREAKER IDENTIFICATION

LEFT SIDE BREAKER IDENTIFICATION

Figure 10-25. *A service panel is finished when the cover is installed and each breaker is labeled according to position and operation (circuit).*

AC GENERATORS

A *generator* is an electromechanical device that converts mechanical energy into electrical energy by means of electromagnetic induction. AC generators (alternators) convert mechanical energy into AC voltage and current. AC generators consist of field windings, an armature, slip rings, and brushes. **See Figure 10-26.**

AC Generators

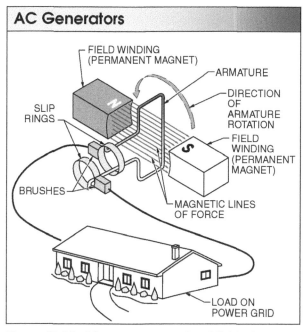

FIELD WINDING (PERMANENT MAGNET)

ARMATURE

DIRECTION OF ARMATURE ROTATION

FIELD WINDING (PERMANENT MAGNET)

SLIP RINGS

BRUSHES

MAGNETIC LINES OF FORCE

LOAD ON POWER GRID

Figure 10-26. *AC generators consist of field windings, a coil (armature), slip rings, and brushes.*

Field windings are magnets used to produce the magnetic field in a generator. The magnetic field in a generator can be produced by permanent magnets or electromagnets. Most generators use electromagnets, which must be supplied with current. An *armature* is the movable coil of wire in a generator that rotates through the magnetic field. The armature may consist of many coils. The ends of the coils are connected to slip rings. *Slip rings* are metallic rings connected to the ends of the armature and are used to connect the induced voltage to the generator brushes. When the armature is rotated in the magnetic field, a voltage is generated in each half of the armature coil. A *brush* is the sliding contact that rides against the slip rings and is used to connect the armature to the external circuit (power grid).

AC generators are similar in construction and operation to DC generators. The major difference between AC and DC generators is that DC generators contain a commutator that reverses the connections to the brushes every half cycle. The commutator maintains a constant polarity of voltage to the load. AC generators use slip rings to connect the armature to the external circuit (load). The slip rings do not reverse the polarity of the output voltage produced by the generator. The result is an alternating sine wave output. **See Figure 10-27.**

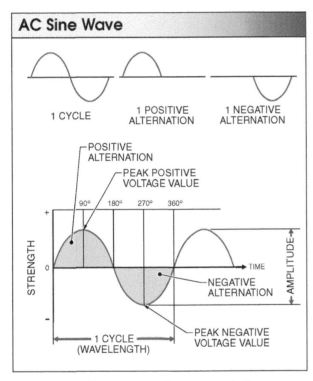

Figure 10-27. *The armature of a generator induces a varying voltage (AC) output that is positive then negative.*

As the armature is rotated, each half cuts across the magnetic lines of force at the same speed. The strength of the voltage induced in one side of the armature is always the same as the strength of the voltage induced in the other side of the armature. However, since the two halves of the coil are connected in a closed loop, the voltages add to each other. The result is that the total voltage of a full rotation of the armature is twice the voltage of each coil half. The total voltage is obtained at the brushes connected to the slip rings, and is applied to an external circuit (grid).

SINGLE-PHASE AC GENERATORS

Each complete rotation of the armature in a single-wire (1ϕ) AC generator produces one complete alternating current cycle. **See Figure 10-28.** When the armature is in the vertical position (Position A), before the armature begins to rotate in a clockwise direction, there is no voltage and no current in the external load circuit because the armature is not cutting across any magnetic lines of force (0° of rotation).

As the armature rotates from position A to position B, each half of the armature cuts across magnetic lines of force, producing current and voltage in the armature and external circuit. The voltage increases from zero to its maximum value while current travels in the same direction. The changing value of voltage is represented by the first quarter (90° of rotation) of the sine wave.

As the armature rotates from position B to position C, current continues to travel in the same direction. The voltage decreases from the maximum value to zero. The changing value of voltage is represented by the second quarter (91° to 180° of rotation) of the sine wave.

As the armature continues to rotate to position D, each half of the coil cuts across the magnetic lines of force in the opposite direction. Cutting across magnetic lines in the opposite direction changes the direction of current flow. During this time, voltage increases from zero to the maximum negative value. The constantly changing value of voltage is shown by the third quarter (181° to 270° of rotation) of the sine wave.

As the armature continues to rotate to position E (position A), voltage decreases to zero. After 360° of armature rotation, one cycle of the AC sine wave is completed.

Single-Phase AC Generators

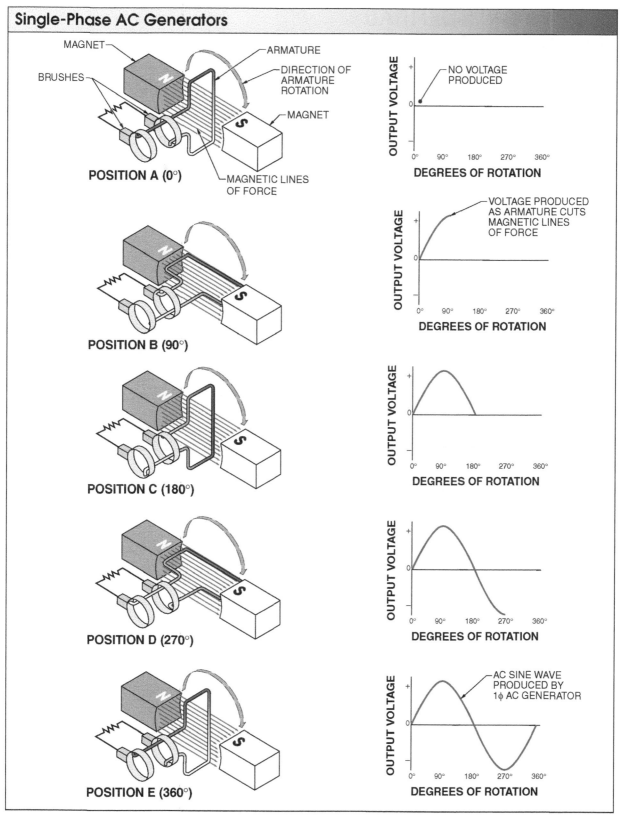

Figure 10-28. *In a 1φ AC generator, as the armature rotates through 360° of motion, the voltage generated is a continuously changing AC sine wave.*

PORTABLE RESIDENTIAL EMERGENCY STANDBY GENERATORS

Portable residential emergency standby generators are installed during new construction or retrofitted into existing residences and businesses. Residential generators are installed outside similarly to the condenser of a home air conditioning unit. As with air conditioning condensers, standby generators are typically placed on a solid surface such as packed gravel or a concrete pad. **See Figure 10-29.** Portable standby generators are powered by gasoline, natural gas, or LPG. Portable generators have a limited capacity. Typical residential emergency standby generators range in capacity from 6500 W to 40,000 W. The size of a generator is determined by the number of preselected lighting and appliance circuits that must be powered during a power outage.

Emergency Standby Generators

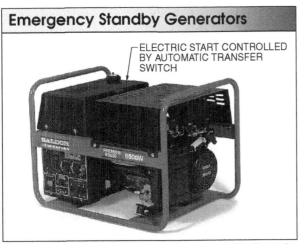

ELECTRIC START CONTROLLED BY AUTOMATIC TRANSFER SWITCH

Baldor Electric Co.

Figure 10-29. *Residential emergency standby generators typically range in capacity from 6500 W to 40,000 W.*

Portable Residential Emergency Standby Generator Connections

Portable residential emergency standby generators should be located as close as possible to the main electrical panel of a residence. The longer the extension cord, the more the voltage drop that will occur, which reduces the power available to the residence. For ease of connection, an extension cord with a twist-lock plug and grounding conductor should be run from the generator through a window, wall, or doorway into the residence. **See Figure 10-30.** To prevent exhaust gases from entering the residence, the generator should be placed in a well-ventilated area away from any doors and windows.

Twist-Lock Extension Cords

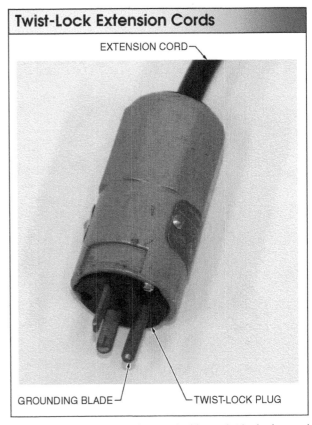

EXTENSION CORD

GROUNDING BLADE

TWIST-LOCK PLUG

Figure 10-30. *An extension cord with a twist-lock plug and grounding conductor should be run from a generator to a residence.*

Portable generators have a standard 15 A, 120 V ground-fault receptacle on the housings for the connection of appliances. There is an additional 120/240 V receptacle that resembles a standard receptacle with two horizontal, rather than vertical, slots. A 120/240 V receptacle is used to connect multiple appliances or to connect appliances that require much power. Also commonly used is a three-slot twist-lock receptacle that supplies 120 V. This receptacle has amperage ratings of 20 A or 30 A. If the generator must supply an entire residence with power, it must have a female, round twist-lock receptacle with four slots. This receptacle has a rating of 125/250 V. **See Figure 10-31.**

Portable Generator Receptacle Configurations

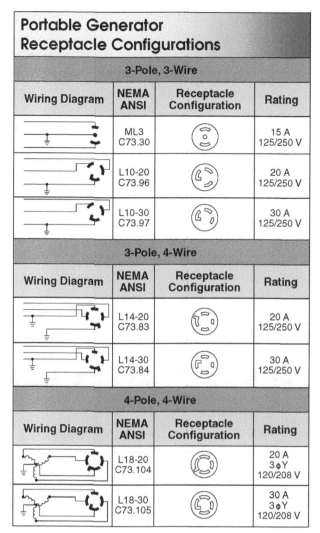

3-Pole, 3-Wire			
Wiring Diagram	NEMA ANSI	Receptacle Configuration	Rating
	ML3 C73.30		15 A 125/250 V
	L10-20 C73.96		20 A 125/250 V
	L10-30 C73.97		30 A 125/250 V

3-Pole, 4-Wire			
Wiring Diagram	NEMA ANSI	Receptacle Configuration	Rating
	L14-20 C73.83		20 A 125/250 V
	L14-30 C73.84		30 A 125/250 V

4-Pole, 4-Wire			
Wiring Diagram	NEMA ANSI	Receptacle Configuration	Rating
	L18-20 C73.104		20 A 3øY 120/208 V
	L18-30 C73.105		30 A 3øY 120/208 V

Figure 10-31. *In addition to standard 120 V receptacles, most portable generators have 3-wire or 4-wire receptacles rated at 125/250 V or 120/208 V.*

Transfer Switches

A *transfer switch* is a switch that isolates a generator from a power grid. A portable integrated generator must connect to residence wiring through a transfer switch that isolates the generator from the power grid. A transfer switch approved by the NEC® ensures that power from the generator will not enter the utility lines and harm power company employees working on the wires. It also protects the generator, which could burn out trying to serve the load of the entire power network. The transfer switch also reconnects the utility system when the power resumes.

PERMANENT WHOLE-HOUSE GENERATORS

A *permanent whole-house generator* is a permanently installed standby generator system that can supply temporary power to an entire residence. During a power outage, it allows the continued use of essential appliances, such as electric range ovens, air conditioners, and refrigerators, and specialty equipment, such as computers and televisions. **See Figure 10-32.** A permanent whole-house generator can start automatically if it is already connected to the service entrance panel and a fuel source such as liquid petroleum or natural gas.

Permanent Whole-House Generators

20,000 kW STANDBY GENERATOR

CONCRETE PAD

Briggs & Stratton Corporation

Figure 10-32. *Permanent whole-house generators are permanently installed standby generator systems that can supply temporary power to entire residences.*

An automatic generator system has an automatic transfer switch that senses the power outage, isolates the dedicated electrical circuits from the utility grid, and starts the generator. When power is restored, the system reconnects the dedicated circuits back to the utility grid and turns itself OFF. Automatic transfers can take from 5 sec to 20 sec to complete.

Permanent Whole-House Generator Connections

It is recommended that a licensed electrician make all electrical connections for a permanent whole-house generator. Depending on the local municipal regulations, approval for installation may be required by the electric utility company. If a large fuel tank is to be part of the system, a permit also may be required.

Automatic Transfer Switches

An *automatic transfer switch* is an electrical device that transfers the load of a residence from public utility circuits to the output of a standby generator during a power failure. An automatic transfer switch (ATS) has electronic circuitry that can sense an interruption in power. **See Figure 10-33.** When a power failure is detected, an ATS turns on a standby generator. The transfer switch also disconnects all the dedicated circuits in the service panel from the public utility source and transfers the dedicated circuits to the output of the emergency generator. When power is transferred, an ATS prevents electricity from flowing into the public service lines and accidentally injuring electric company workers who are working on the system to restore power.

Automatic transfer switches come in various sizes depending upon the power rating required. An ATS is installed in close proximity to the electrical distribution panel and should be installed by a licensed electrician.

Voltage, Amperage, and Wattage

Because generators are chosen based on capacity in watts, it is important to understand the relationship between volts (V), amperage (A), and watts (W). The basic formula for calculating wattage when the voltage (V) and amperage (A) are known is watts equal volts times amps, or $W = V \times A$. When an appliance is plugged into a 120 V outlet and draws 5 A, the watts are 600 W (600 W = 120 V × 5 A). In most cases an appliance has a nameplate stating the start and running wattages. **See Figure 10-34.**

A standby generator can also be connected to a combination transfer switch. A combination transfer switch is a switch that can function as either a manual transfer switch or an automatic transfer switch. An essential electrical system uses an automatic or combination transfer switch to transfer power between normal and alternate sources.

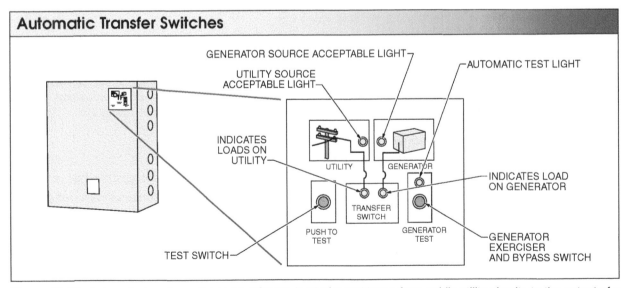

Automatic Transfer Switches

Figure 10-33. *Automatic transfer switches transfer the load of a residence from public utility circuits to the output of a standby generator during a power failure.*

Nameplate Information

CLOTHES DRYER

SUMP PUMP

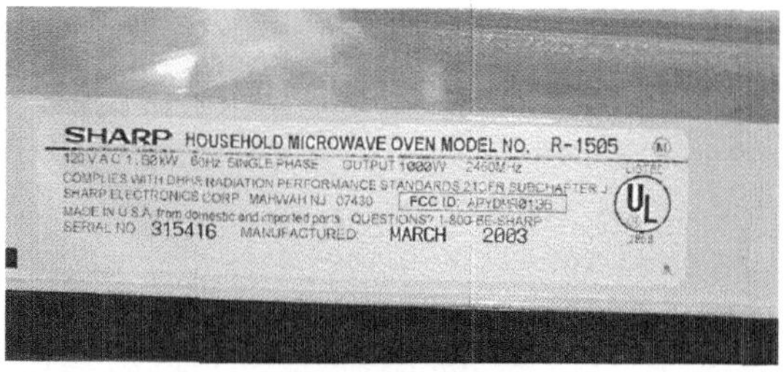

MICROWAVE

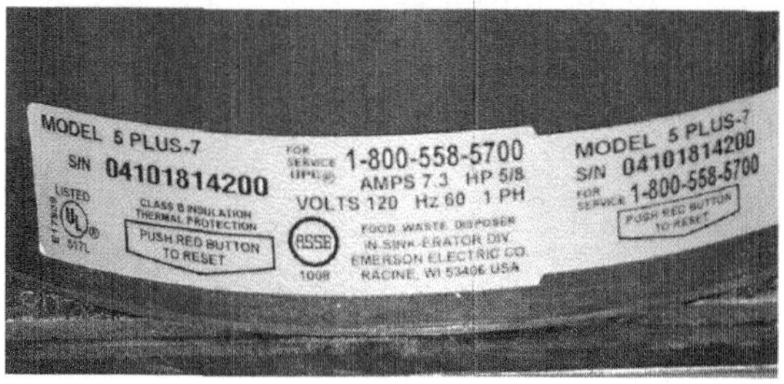

GARBAGE DISPOSAL

Figure 10-34. *Ratings are typically found on the name-plates of electrical devices. When a nameplate cannot be easily located, refer to the operation or installation manual from the manufacturer for the information.*

When computing the total wattage required from a generator, the calculation must account for some appliances that require more power (watts) to start than to run. **See Figure 10-35.** It is also extremely important to understand which circuits (appliances) are considered critical circuits and must have power supplied during a power outage.

For example, in winter it is critical that the furnace of a residence operate, but in summer it is more important to have air conditioning and a freezer operating. **See Figure 10-36.** All appliances, HVAC systems, lighting, and receptacles of residences have wattage ratings for determining the total load created.

Measuring and Calculating Power

Although adding the wattage provided in a table can produce an estimate of the total power required, the best method to determine power is to measure current draw and calculate the wattages ($P = V \times A$). A clamp-on ammeter can also be used to calculate total power required by taking measurements of the starting and running current.

Intelligent Controls

Automatic transfer switches with AC power control modules are available. A power control module allows loads to be prioritized. The module monitors all load current in the circuits and temporarily turns off lower priority lines to allow major appliances, such as an air conditioner, to turn on. This option can increase the cost of a generator.

Generator Sound Levels

Sound intensity is expressed in decibels. A *decibel (dB)* is a unit used to express the relative intensity of sound. Sound intensity ranges from 0 dB (least perceptible) to 140 dB (deafening). The higher the decibel, the louder the sound. For air-cooled standby generators, the sound level is about 65 dB at 20′. For example, a normal conversation can be held next to an air-cooled standby generator running at a sound level of 65 dB. **See Figure 10-37.**

Appliance Starting and Running Wattages

Appliance	Wattage (Starting)*	Wattage (Running)*
Refrigerator	2800	700
Freezer	2500	500
Well Pump-2 HP	6000	2000
Well Pump-3 HP	9000	3000
Gas Furnace Fan	500 to 2350	300 to 875

* in watts

Figure 10-35. *Appliances have typical wattage ratings. It is extremely important that each residence log the specific wattage ratings of appliances being used.*

Typical Appliance Wattages

Appliance	Typical Wattage*
Lights	4500
Electric Water Heater	10,000
Electric Cooktop Top-8″	2100
Washing Machine	1200
Electric Clothes Dryer	6000
Electric Fry Pan	1000
Electric Oven	6500
Cofeemaker	1100
Microwave	600 to 1200
Dishwasher	700
Fan	275
Toaster	200
Hair Dryer	1500
Computer	700 to 1000
Computer Monitor	200 to 800
Printer	400 to 800
Radio	40 to 225
Television	300 to 800

* in watts

Figure 10-36. *Some appliances draw more power when starting up than when running. Ensure startup wattages are known for use in circuit load calculations.*

Maintenance

A properly maintained generator should provide over 20 years of operation when operated as an emergency standby generator. Air-cooled generators are not intended to provide continuous power. The liquid cooled units are designed for longer operation but not for continuous duty year round. The expected life for liquid cooled generators is 10,000 hr to 12,000 hr, with proper maintenance.

Sound Levels

Decibel (dB)	Loudness	Examples
140	Deafening	Jet airplane taking off, air raid siren, locomotive horn
130	Pain threshold	
120	Feeling threshold	
110	Uncomfortable	
100	Very loud	Chainsaw
90	Noisy	Shouting, auto horn
80	Moderately loud	Vacuum cleaner
70	Loud	Telephone ringing, loud talking
65	Moderate	Air-cooled standby generator
60	Moderate	Normal Conversation
50	Quiet	Hair dryer
40	Moderately quiet	Refrigerator running
30	Very quiet	Quiet conversation, broadcast studio
20	Faint	Whispering
10	Barely audible	Rustling leaves, soundproof room, human breathing
0	Hearing threshold	Intolerably quiet

Figure 10-37. *A decibel is a unit used to express the relative intensity of sound.*

Generator Selection

Some appliances require extra power and current to start. For example, most HVAC compressors and motors need two to three times their run wattage for starting. A 2 HP pump needs 6000 W to start pumping water into the residence but needs only 2000 W to maintain the pump. The best generators have built-in surge current capabilities for when a pump motor or other motor goes on-line.

Generator Safety Labels

A *safety label* is a label that indicates areas or tasks that can pose a hazard to personnel and/or equipment. Safety labels can appear in several ways on equipment and in equipment manuals. Safety labels use signal words to communicate the severity of a potential problem. The three most common signal words are danger, warning, and caution. **See Figure 10-38.**

Safety Labels

Safety Label	Box Color	Symbol	Significance
⚠ DANGER **HAZARDOUS VOLTAGE** • Ground equipment using screw provided. • Do not use metallic conduits as a ground conductor.	red	⚠	**DANGER** – Indicates an imminently hazardous situation which, if not avoided, will result in death or serious injury
⚠ WARNING **MEASUREMENT HAZARD** When taking measurements inside the electric panel, make sure that only the test lead tips touch internal metal parts.	orange	⚠	**WARNING** – Indicates a potentially hazardous situation which, if not avoided, could result in death or serious injury
⚠ CAUTION **MOTOR OVERHEATING** Use of a thermal sensor in the motor may be required for protection at all speeds and loading conditions. Consult motor manufacturer for thermal capability of motor when operated over desired speed range.	yellow	⚠	**CAUTION** – Indicates a potentially hazardous situation which, if not avoided, may result in minor or moderate injury, or damage to equipment; may also be used to alert against unsafe work practices
WARNING Disconnect electrical supply before working on this equipment.	orange	⚡	**ELECTRICAL WARNING** – Indicates a high voltage location and conditions that could result in death or serious injury from an electrical shock
WARNING Do not operate the meter around explosive gas, vapor, or dust.	orange		**EXPLOSION WARNING** – Indicates location and conditions where exploding electrical parts may cause death or serious injury

Figure 10-38. *Safety labels are used to indicate situations with different degrees of likelihood that injury or death could occur.*

Other signal words may also appear with danger, warning, and caution signal words used by manufacturers. ANSI Standard Z535.4, *Product Safety Signs and Labels,* provides additional information concerning safety labels. Additional signal words may be used alone or in combination on safety labels.

Danger Signal Word. A *danger* is a word used to indicate an imminently hazardous situation, which could result in death or serious injury. The information indicated by a danger signal word indicates the most extreme type of potential situation, and must be followed. The danger symbol is an exclamation mark enclosed in a triangle followed by the word "danger" in bold letters in a red box.

Warning Signal Word. A *warning* is a word used to indicate a potentially hazardous situation, which could result in death or serious injury. The information indicated by a warning signal word indicates a potentially hazardous situation and must be followed. The warning symbol is an exclamation mark enclosed in a triangle followed by the word "warning" in bold letters in an orange box.

Caution Signal Word. A *caution* is a word used to indicate a potentially hazardous situation, which could result in minor or moderate injury. The information indicated by a caution signal word indicates a potential situation that may cause a problem to people and/or equipment. A caution signal word also warns of problems due to unsafe work practices. The caution symbol is an exclamation mark enclosed in a triangle followed by the word "caution" in bold letters in a yellow box.

Briggs & Stratton Corporation
Transfer switches are installed inside a residence and resemble load-center breaker boxes.

Electrical Warning Signal Word. An *electrical warning* is a word used to indicate a high-voltage location and conditions that could result in death or serious injury from an electrical shock if proper precautions are not taken. An electrical warning safety label is usually placed where there is potential for coming in contact with energized electrical wires, terminals, or parts. The electrical warning symbol is a lightning bolt enclosed in a triangle. The safety label may be shown without words or may be preceded by the word "warning" in bold letters.

Explosion Warning Signal Word. An *explosion warning* is a word used to indicate a location and conditions where exploding parts may cause death or serious personal injury if proper precautions and procedures are not followed. The explosion warning symbol is an explosion enclosed in a triangle. The safety label may be shown with no words or may be preceded by the word "warning" in bold letters.

Generator Hazards

Despite the safe design of generators, operating equipment improperly, neglecting maintenance, or being careless can cause possible injury or death. Only responsible, qualified persons should be permitted to install, operate, and maintain equipment. Parts of the generator are rotating and/or hot during operation. Caution must be used near running generators. Installation must always comply with applicable codes, standards, laws, and regulations. A running generator emits carbon monoxide, which is an odorless, colorless, and poisonous gas. Inhaling carbon monoxide can cause headaches, fatigue, dizziness, nausea, vomiting, confusion, fainting, seizures, and death.

Grounding Hazards. The National Electrical Code® (NEC®) may require the frame and external electrically conductive parts of the generator to be connected to an approved earth ground, depending on the conditions of the installation. Local electrical codes also may require proper grounding of the generator electrical system. Grounding requirements should be verified with the local authority having jurisdiction (AHJ) before installing or using a generator.

Explosion Hazards. Flammable fluids, such as natural gas and liquid propane, are extremely explosive. To avoid the possibility of an explosion, the fuel supply system must be installed according to applicable fuel-gas codes. Before placing a residential standby electric system into service, fuel system lines must be properly purged and leak tested according to applicable codes. After installation, the fuel system should be inspected periodically for leaks. Leakage is not permitted.

Furthermore, smoking and flames should not be permitted in the immediate vicinity of a generator. Fuel and oil spills should be cleaned up immediately. It must be ensured that combustible materials are not left in/on the generator housing or near the generator, as a fire or explosion may result. The area surrounding the generator must be kept clean and free from debris.

Transfer Switch Safety. A transfer switch should be installed indoors on a firm, sturdy supporting structure.

To prevent switch distortion, the switch should be level. To verify that the switch is level, washers may be placed between the switch enclosure and mounting surface. A switch should never be installed where water or any corrosive substance might drip onto the enclosure. Switches must be protected at all times against excessive moisture, dust, dirt, and lint.

Transfer Switch Battery Installation

Prior to installing a battery for a generator transfer switch, the battery must be filled with the proper electrolyte fluid, if necessary, and be fully charged. Before installing and connecting the battery, apply the following procedure:

1. Set the AUTO/OFF/MANUAL switch of the generator to the OFF position.
2. Turn off utility power supply to the transfer switch.
3. Remove the fuse from the generator control panel. *Note:* Battery cables were factory connected at the generator.
4. Connect the red battery cable (from the starter contactor) to the battery post indicated by a positive identification mark ("POS" or "+").
5. Connect the black battery cable (from frame ground) to the battery post indicated by a negative identification mark ("NEG" or "–").

Cable leads and connectors must be inspected for fraying, looseness, and burning of the contact areas. Cable leads and connectors should be repaired or replaced as needed. **See Figure 10-39.** Batteries can fail or malfunction due to any of the following conditions:

- corrosion
- frayed or broken cables
- sealing compound defects
- excessive dirt
- overfilled with electrolyte
- missing vent cap
- cracked cell cover
- cracked case
- low electrolyte level (accounts for over 50% of failures)
- cell connector corrosion
- excessive vibration from loose hold-down device
- misapplication of battery
- use of an undersized battery
- undercharging
- freezing

Battery Safety. If the AUTO/OFF/MANUAL switch is not set to the OFF position, the generator can crank and start as soon as the battery cables are connected. If the utility power supply is not turned off, sparking can occur at the battery posts and cause an explosion. To prevent an explosion, the following safety precautions must be followed:

- Do not smoke near the battery.
- Do not have open flames or sparks in battery area.
- Be sure the AUTO/OFF/MANUAL switch is set to the OFF position before connecting the battery cables. If the switch is set to AUTO or MANUAL, the generator can crank and start as soon as the battery cables are connected.
- Do not dispose of the battery in a fire. The battery is capable of exploding.

A battery presents the risk of electrical shock from high short-circuit current. To prevent electrical shock when handling a battery, the following safety precautions must be followed:

- Remove the fuse from the generator control panel.
- Remove jewelry or other metal objects from the body before working on or near the generator.
- Only use tools with insulated handles.
- Wear rubber gloves with leather insulators and shoes with electrical hazard (EH) rated soles.
- Do not lay tools or metal parts on top of the battery.
- Disconnect the charging source prior to connecting or disconnecting battery terminals.

Electrolyte is a solution of sulfuric acid and water that is harmful to the skin and eyes. It is electrically conductive and corrosive. Full eye protection and protective clothing must be worn when handling electrolyte. If electrolyte contacts the skin, the affected area should be flushed immediately with water for 10 min to 15 min. If electrolyte contacts the eyes, the eyes should be flushed thoroughly and immediately with water for 10 min to 15 min, and medical attention should be sought.

Many modern transfer switches are available with maintenance-free, marine-grade, sealed batteries. These batteries keep their strength for up to 36 months, do not require periodic maintenance, and are guaranteed to operate during occasional power outages.

Battery Failure Causes

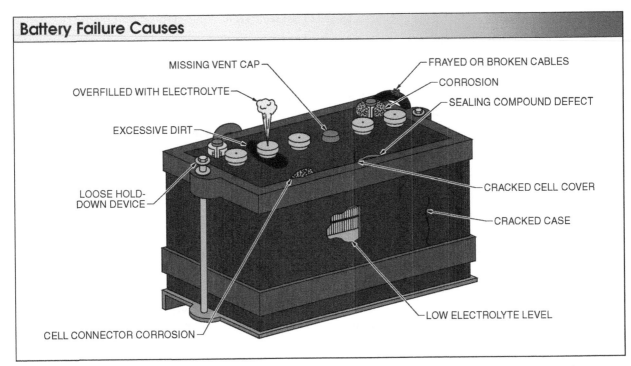

Figure 10-39. *A poorly maintained battery could develop several problems that will eventually cause battery failure.*

Spilled electrolyte is to be washed down with an acid-neutralizing agent. To neutralize electrolyte, a solution of 1 lb (500 g) of bicarbonate of soda to 1 gal (4 l) of water is normally used. The bicarbonate of soda solution is to be added until foaming (evidence of the reaction) has ceased. The resulting liquid is to be flushed with water and the area dried.

READING AN ELECTRIC METER

An electric meter has four dials that look like clocks. Each dial has a hand. **See Figure 10-40.** *Note:* Electric meters have two dials that read clockwise and two dials that read counterclockwise.

Always use the number that the hand has just passed (CW or CCW) on each of the dials. For example, the reading of the dials at the beginning of the month is 3456 kilowatt-hours (kWh). At the end of the month the readings of the dials is 3592 kWh. The difference is as follows:

End of month reading	3592
Beginning of month reading	− 3456
Kilowatt-hours used in month	136

Calculating Appliance Operating Costs

To calculate the cost to operate an appliance for a month, the nameplate on the appliance must be checked for amperage and wattage specifications. When amperage is provided, the power formula converts amperage and voltage into watts. The formula states that $1 \text{ A} \times 115 \text{ V} = 115 \text{ W}$. Therefore, a 10 A iron consumes 1150 W ($10 \text{ A} \times 115 \text{ V} = 1150 \text{ W}$). When the wattage of an appliance is known, operating cost per hour can be determined. **See Figure 10-41.**

A checklist to use when installing a home emergency generator is as follows:
- Generator should be Underwriters Laboratory listed (UL listed).
- Generator should be properly sized for the calculated load.
- Generator should be properly positioned so the noise generated is acceptable.
- Generator should be properly housed for outdoor environmental exposure, such as to temperature, snow, and rain.
- Generator should operate with the available fuel, whether gasoline, natural gas, or LPG.

Reading Electric Meters

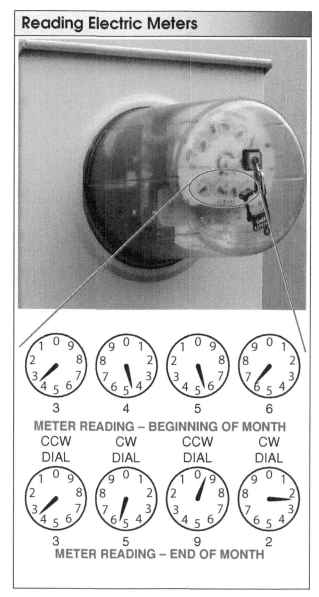

METER READING – BEGINNING OF MONTH

3   4   5   6

| CCW DIAL | CW DIAL | CCW DIAL | CW DIAL |

METER READING – END OF MONTH

3   5   9   2

Figure 10-40. *Electric meters have two dials that read clockwise (CW) and two dials that read counter-clockwise (CCW).*

- Follow any special considerations due to the type of fuel.
- Generator battery should be charged automatically.
- Generator should have automatic safety shut downs to protect engine and generator in event of low oil pressure, high temperature, or overspeed.
- Generator should have a seven-day exerciser programmed to run the system for several minutes each week to maintain top running condition.

Voltage Changes

AC generators are designed to produce a specific amount of output current at a rated voltage. In addition, all electrical and electronic equipment is rated for operation at a specific voltage. *Rated voltage* is a voltage range that is typically within ±10% of ideal voltage. Today, however, with many components derated to save energy and operating cost, the range is typically +5% to −10%. A voltage range is used because an overvoltage is more damaging than an undervoltage. Equipment manufacturers, utility companies, and regulating agencies must routinely compensate for changes in system voltage.

Back-up (emergency standby) generators are used to compensate for voltage changes and are powered by diesel, gasoline, natural gas, or propane engines connected to the generator. When there is time between the loss of main utility power and when the generator starts providing power, the generator is classified as a standby (emergency) power supply. Voltage changes in an electrical system are categorized as momentary, temporary, or sustained. **See Figure 10-42.**

Small Appliance Hourly Operating Costs

Wattage Consumed by Appliance	0.03 kWh	0.04 kWh	0.05 kWh	0.06 kWh
100	10 hr for 3¢	7.5 hr for 3¢	6 hr for 3¢	5 hr for 3¢
300	3 hr for 2.7¢	2,5 hr for 3¢	2 hr for 3¢	1.66 hr for 3¢
500	2 hr for 3¢	1 hr for 2¢	1.2 hr for 3¢	1 hr for 3¢
700	1.4 hr for 3¢	1.1 hr for 3¢	1 hr for 3.5¢	1 hr for 4.2¢
900	1.1 hr for 3¢	1 hr for 3.6¢	1 hr for 4.5¢	1 hr for 5.4¢
1000	1 hr for 3¢	1 hr for 4¢	1 hr for 5¢	1 hr for 6¢

Figure 10-41. *To calculate the cost of operating an appliance for a month, the nameplate on the appliance must be checked for amperage and wattage specifications.*

Voltage Changes

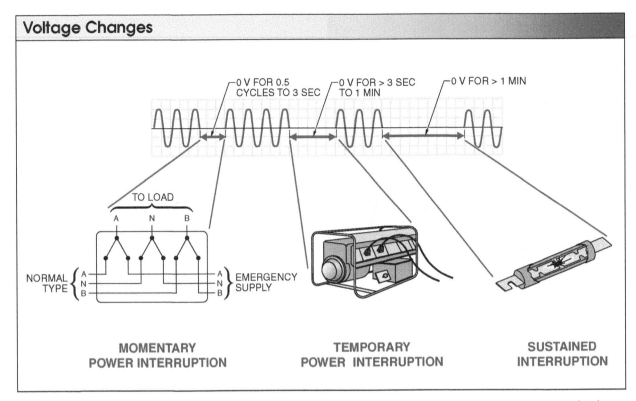

Figure 10-42. *Voltage changes in an electrical system can be categorized as momentary, temporary, or sustained.*

Momentary Power Interruptions. A *momentary power interruption* is a decrease to 0 V on one or more power lines lasting from 0.5 cycles up to 3 sec. All power distribution systems have momentary power interruptions during normal operation. Momentary power interruptions can be caused when lightning strikes nearby, by utility grid switching during a problem situation (a short on one line), or during open circuit transition switching. *Open circuit transition switching* is a process in which power is momentarily disconnected when switching a circuit from one voltage supply or level to another.

Temporary Power Interruptions. A *temporary power interruption* is a decrease to 0 V on one or more power lines lasting between 3 sec and 1 min. Automatic circuit breakers and other circuit protection equipment protect all power distribution systems. Circuit protection equipment is designed to remove faults and restore power. An automatic circuit breaker takes from 20 cycles to about 5 sec to close. When power is restored, the power interruption is only temporary. If power is not restored, a temporary power interruption becomes a sustained power interruption. A temporary power interruption can also be caused by a time gap between a power interruption and when a back-up power supply (generator) takes over, or if someone accidentally opens the circuit by switching the wrong circuit breaker switch.

Sustained Power Interruptions. A *sustained power interruption* is a decrease to 0 V on all power lines for a period of more than 1 min. All power distribution systems have a complete loss of power at some time. Sustained power interruptions (outages) are commonly the result of storms, tripped circuit breakers, blown fuses, and/or damaged equipment.

The effect of a power interruption on a load depends on the load and the application. When a power interruption causes equipment, production, and/or security problems that are not acceptable, an uninterruptible power system is used. An *uninterruptible power system (UPS)* is a power supply that provides constant on-line power when the primary power supply is interrupted. For long-term power interruption protection, a generator/UPS is used. For short-term power interruptions, a static UPS is used.

Transients

A *transient* is a temporary, unwanted voltage in an electrical circuit. Transient voltages are typically large erratic voltages, or spikes that have a short duration and a short rise time. Computers, electronic circuits, and specialized electrical equipment require protection against transient voltages. Protection methods commonly include proper wiring, grounding, shielding of the power lines, and use of surge suppressors. A *surge suppressor* is a receptacle that provides protection from high-level transients by limiting the level of voltage allowed downstream from the surge suppressor. Surge suppressors are installed at service entrance panels and at individual loads. **See Figure 10-43.**

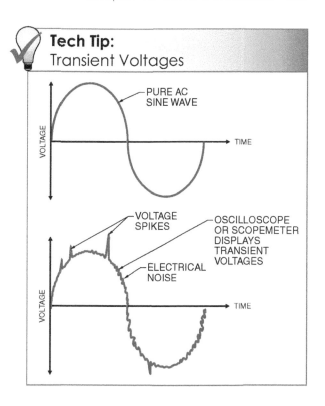

Tech Tip:
Transient Voltages

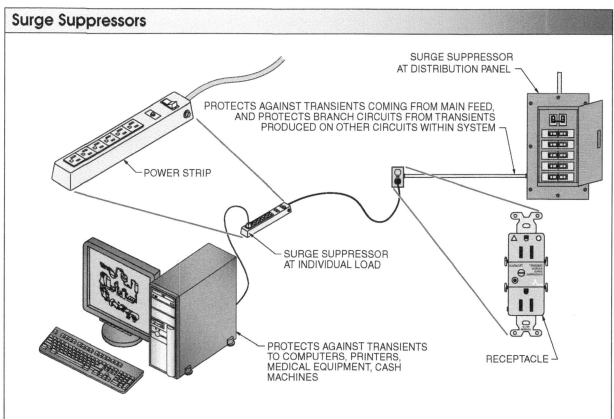

Surge Suppressors

Figure 10-43. *A surge suppressor is a receptacle that provides protection from high-level transients by limiting the level of voltage allowed downstream from the surge suppressor.*

Name_____ Date _____

Service Entrances

_____ **1.** A service ___ has conductors run overhead.

_____ **2.** A service ___ has conductors run underground.

T   F   **3.** Service conductors should be as long as possible.

T   F   **4.** A service panel should be close to the kitchen to avoid costly runs to major appliances.

T   F   **5.** The preferred position for a meter socket is on an outside bedroom wall.

T   F   **6.** A service entrance for a one-family dwelling should have a minimum ampere rating of 100 A.

_____ **7.** Residential services have a standard voltage of ___ V, 1φ.

_____ **8.** The minimum size Cu conductor used for service entrances is #___.

_____ **9.** The minimum size Al conductor used for service entrances is #___.

_____ **10.** The minimum conduit size for service entrance conductors is ___″.

_____ **11.** The minimum ground wire size for copper service entrances is #___.

_____ **12.** The minimum ground wire size for aluminum service entrances is #___.

_____ **13.** The minimum ground rod size for service entrances is ___″ in diameter.

T   F   **14.** The minimum requirements of conductor sizes for a mobile home are the same as for a one-family dwelling.

T   F   **15.** Large knockouts should be removed after the panel is mounted.

T   F   **16.** Sealing compound or caulking may be applied to make the sill plate watertight.

_____ 17. In a typical residential electrical service, ___ V is obtained between the red and black service wires.

        A. 115   **C.** 480

        B. 230   **D.** none of the above

_____ 18. Service entrance conductors that run ___ a window are considered out of reach.

_____ 19. The service drop is owned and maintained by the ___.

_____ 20. The minimum clearance for service drop conductors passing over a private drive is ___′.

_____ 21. The minimum clearance for service drop conductors passing over an alley or street is ___′.

T   F   22. Drip loops are required between the service drop conductors and the service head.

T   F   23. A service head may be located not over 12″ to one side of the point of attachment for the service drop conductors.

_____ 24. ___ or finishing the service requires the installation of internal service conductors, circuit breakers, and branch circuit wiring.

_____ 25. Direct ___ cable is used to make the hookup when the service originates from a transformer pad.

T   F   26. Rigid galvanized conduit may be buried directly in most soils.

T   F   27. The minimum clearance for service drop conductors passing over a residential sidewalk is 10′.

T   F   28. A service head should be at least 8′ above a platform.

_____ 29. Service drop conductors shall have not less than ___′ clearance from a window.

_____ 30. The white wire is the ___.

Service Entrance Components

_____   **1.** Cable connector

_____   **2.** Sill plate

_____   **3.** Cap

_____   **4.** Strap

Ⓐ Ⓑ

Ⓒ Ⓓ

Service Entrance Parts

_____   **1.** Roof flashing   _____   **5.** Conduit
                                                                 support
_____   **2.** Bottom reducer
                                          _____   **6.** Pipe strap
_____   **3.** Service head

_____   **4.** Insulator
                          bracket

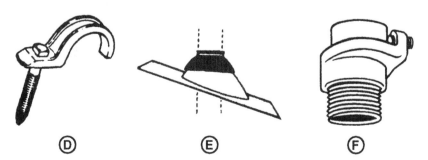

Ⓐ   Ⓑ   Ⓒ

Ⓓ   Ⓔ   Ⓕ

Service Drop

_____ **1.** 18' minimum

_____ **2.** 3' minimum

_____ **3.** Not over 300 V between conductors

_____ **4.** 10' minimum

_____ **5.** 12' minimum

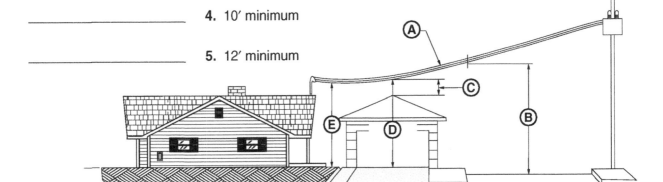

115 V – 230 V Electrical Service

_____ **1.** The voltage at A is ___ V.

_____ **2.** The voltage at B is ___ V.

_____ **3.** Service wires enter the ___ at C.

_____ **4.** Service wires are routed through a(n) ___ at D.

_____ **5.** A(n) ___ is shown at E.

_____ **6.** The ___ is shown at F.

_____ **7.** The voltage at G is ___ V.

_____ **8.** The voltage at H is ___ V.

_____ **9.** The voltage at I is ___ V.

_____ **10.** The panel has two hot ___ at J.

_____ **11.** The ___ breaker is shown at K.

_____ **12.** The voltage at L is ___ V.

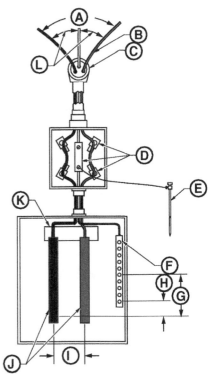

Service Drop Clearances

_____ **1.** Conductors considered out of reach.

_____ **2.** Conductors shall have 3′ minimum vertical clearance below windows.

_____ **3.** Conductors shall have 3′ minimum clearance from fire escape or porch.

_____ **4.** Conductors shall have 3′ minimum horizontal clearance from windows.

_____ **5.** Service drop shall have 10′ minimum clearance above platforms.

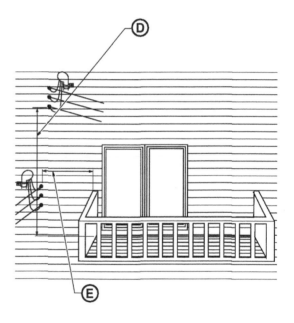

Service Lateral

_____   **1.** ¼″ air space

_____   **2.** Grade

_____   **3.** Concrete wall

_____   **4.** Rigid galvanized conduit

_____   **5.** Utility company pole

_____   **6.** Service head

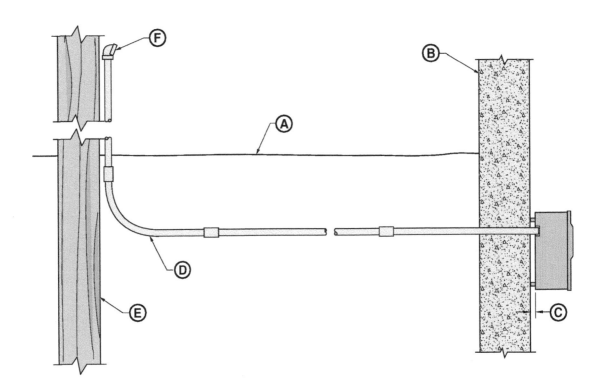

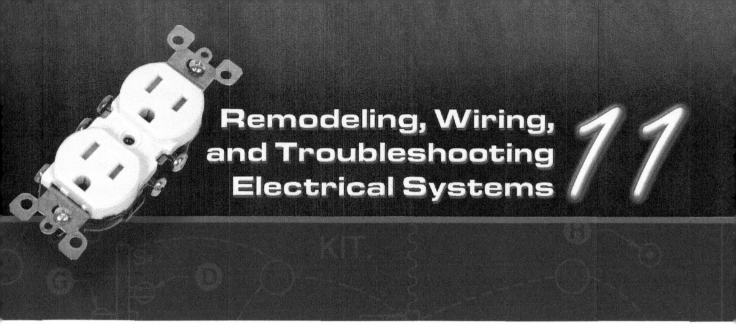

Remodeling, Wiring, and Troubleshooting Electrical Systems 11

Chapter 11 provides an overview of how to test existing residential electrical circuits for defective parts and how to add devices and components to the system. Using fish tapes to route cables and wires through ceilings and walls to install boxes and raceways is also covered. Chapter 11 provides an overview of residential wood frame construction to aid in the process of pulling wires.

Construction drawings for dwellings are not as detailed as true electrical drawings. Typically, an electrician will only have a drawing of a basic floor plan to help aid in the design of an electrical system. Typical floor plans show only the room arrangement with some information on type of lighting and arrangement. True electrical drawings show receptacle and switching locations in addition to lighting types and locations.

RESIDENTIAL WOOD FRAME CONSTRUCTION

To route conduit or cables through an existing drywalled or plaster home, an individual must understand how the home is constructed and where obstacles may exist that could prevent conduit or cables from being routed. Standard construction methods are used to construct the wood framing for one- and two-story residences. **See Figure 11-1.**

Floor plans and associated detail drawings are used to construct the framing for one- and two-story homes. **See Figure 11-2.** Careful study of the prints for a home will indicate which areas are best for installing new conduit or cable.

Glue laminated timbers are typically used for beams in residential construction.

Residential Construction

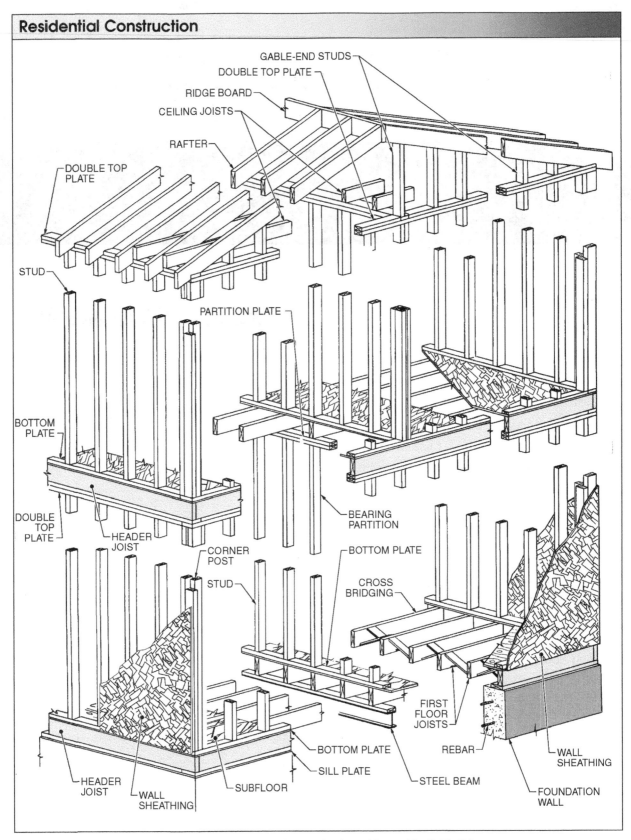

Figure 11-1. *Platform framing is the typical construction method used for residential buildings.*

Framing from Prints

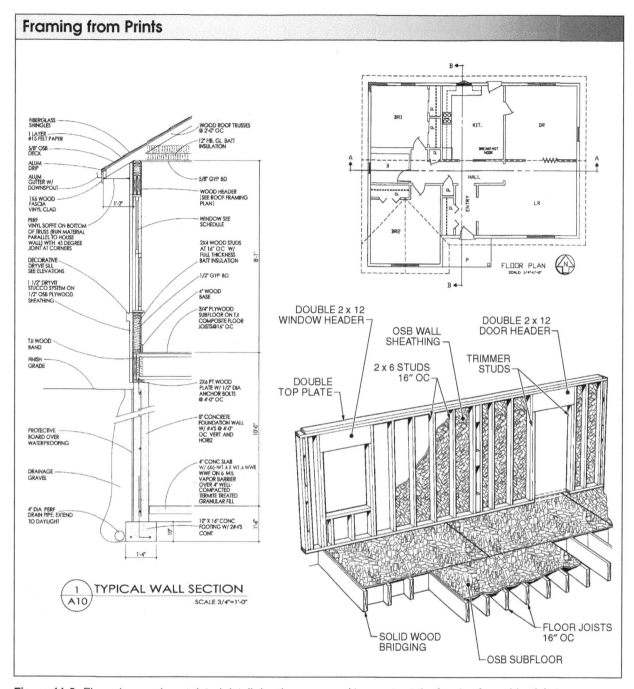

Figure 11-2. *Floor plans and associated detail drawings are used to construct the framing for residential structures.*

Residential prints include cutaway drawings that detail the construction used to attach the roof to the outside walls. **See Figure 11-3.** Routing wires through an outside wall to a ceiling fixture is one of the most difficult tasks to accomplish when remodeling.

Older homes are typically found with bridging between the studs in walls. **See Figure 11-4.** Bridging complicates the process of fishing wires through a wall. When bridging is encountered, the bridging must be notched or drilled to allow the routing of wires.

Roofing Prints and Framing

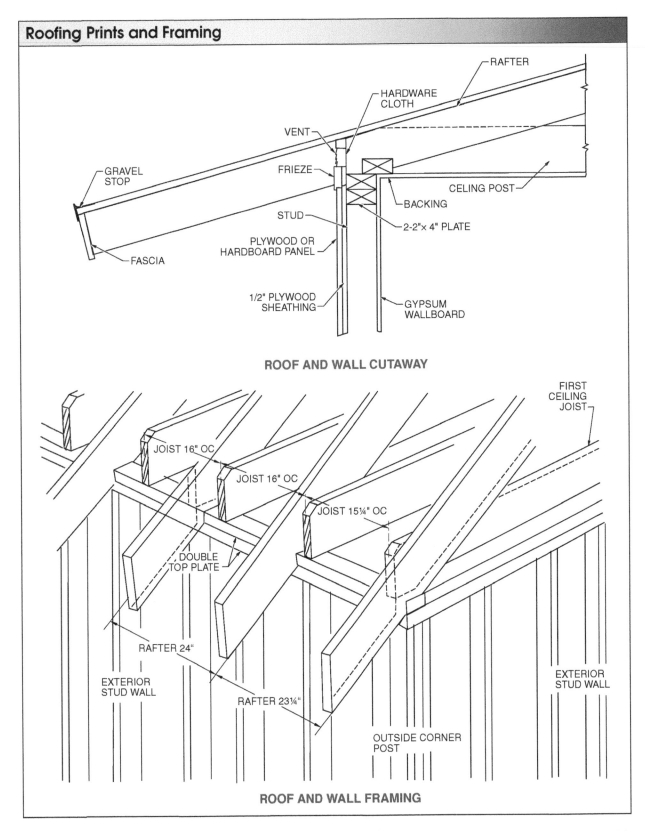

ROOF AND WALL CUTAWAY

ROOF AND WALL FRAMING

Figure 11-3. *Residential prints that include cutaway detail drawings of the construction used to attach the roof to outside wall are used to plan the routing of wires through an outside wall to a ceiling fixture.*

Bridging

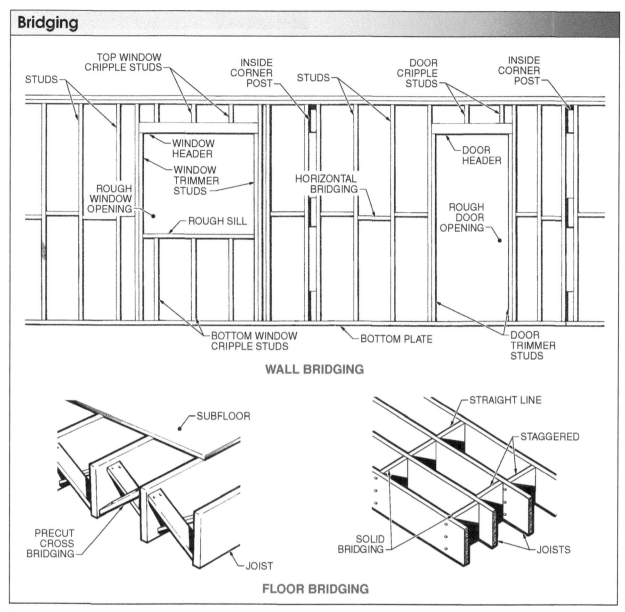

Figure 11-4. *Bridging encountered in walls, floors, and ceilings can be notched or drilled to allow the running of cables or conduit.*

INSTALLING RECEPTACLE BOXES

To install a receptacle box in an existing drywalled wall, the location for the box must first be determined. Then a hole must be cut in the plaster or drywall. The receptacle box is then mounted to a stud or to bracing attached to studs. Cable or wire must then be routed to the appropriate boxes for connection.

Determining Hole Location

When new receptacle boxes must be added to a room, holes will have to be cut in existing walls and ceilings. To avoid unnecessary holes and associated repairs, individuals must carefully select the proper location for a box before cutting any holes.

Wall studs and ceiling joists are laid out for residential construction on 16″ (sometimes 24″) centers. Once a stud or joist is located, other studs are found by measuring 16″ (or 24″) from the center of the located stud. Two methods are used to locate the first stud. The first method uses an electronic stud finder to zero in on any nails used to install the drywall. **See Figure 11-5.** The presence of a nail causes the arrow of the magnetic stud finder to point.

In addition to using a stud finder or finishing nails to locate studs, stud positions can also be located by taking simple measurements. After locating the position of one wall stud, make 16″ measurements along the bottom of the wall to locate the remaining studs. A stud finder is relatively inexpensive and powered by a 9 V battery.

Finding Wall Studs

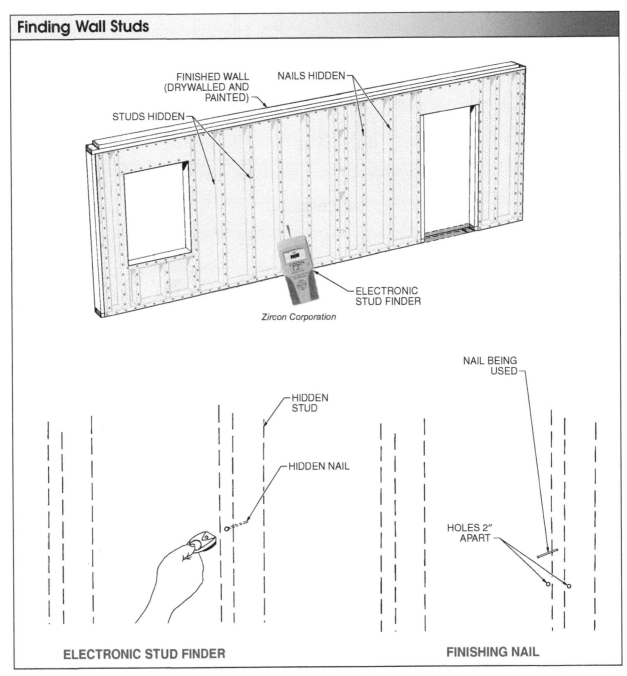

FINISHED WALL (DRYWALLED AND PAINTED)

NAILS HIDDEN

STUDS HIDDEN

ELECTRONIC STUD FINDER

Zircon Corporation

NAIL BEING USED

HIDDEN STUD

HIDDEN NAIL

HOLES 2″ APART

ELECTRONIC STUD FINDER

FINISHING NAIL

Figure 11-5. *To find a stud in a finished wall, stud finders are typically used. A finishing nail can also be used to find a stud.*

The second method used to locate a stud is to drive small finishing nails into a wall at 2″ intervals until a stud is located. The disadvantage of driving finishing nails every 2″ is that several small holes will require patching after a stud is located.

Individuals who are sure there is enough room between studs and supports must also check the area above and below the planned opening area to locate any unforeseen obstacles. **See Figure 11-6.** To check the area around a planned opening, a coat hanger with a bend larger than the planned opening is rotated 360° through a small hole. A hanger that rotates freely indicates a clear area, allowing the cutting of the hole for a box.

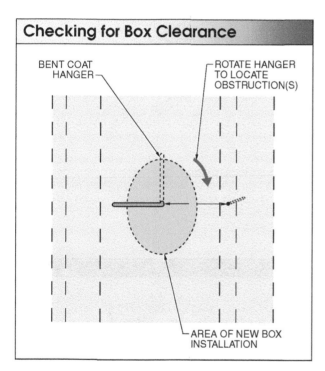

Figure 11-6. *To check the area around a planned box opening, a coat hanger with a bend is rotated 360° through a small hole to determine if the area is clear.*

Cutting Box Holes

Once an area is checked for box installation, electricians typically trace a pattern of the planned box opening on the wall. Drilling each corner of the pattern establishes a place to begin or end cutting and ensures that cutting does not overshoot the pattern. **See Figure 11-7.**

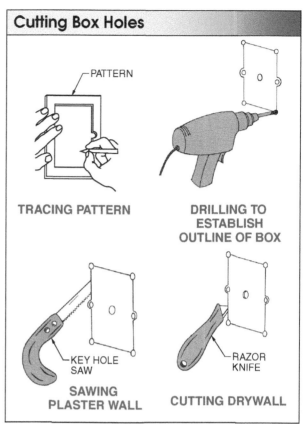

Figure 11-7. *Cutting the opening in finished walls for a new box requires defining the outer perimeter of the box in position, drilling the corners of the outer perimeter, and cutting or sawing the plaster walls or drywall.*

Modern wallboard or drywall is typically cut with a razor knife or small saw. When cutting into plaster walls, keyhole saws are the tools typically used. A faster method of cutting a box opening is to use a zip router.

When a large hole is required, the pattern is secured to the studs with thin strips of wood around the perimeter to avoid damaging the surrounding wallboard or drywall. **See Figure 11-8.**

Mounting Receptacle Boxes

Manufacturers are aware of the need of homeowners to update electrical systems in older homes. To aid homeowners, manufacturers produce products (boxes) that deal with the problems encountered when mounting boxes in existing ceilings and walls. Adding wooden braces for box mounting is not practical for most remodeling work.

A cut-in receptacle box (shallow box) is typically used for remodeling projects and attach to the face of the wall through drywall ears on the front of the box. Clamps hold the box to the back of the finished wall.

Tech Tip:
Nail Spacing for Drywall

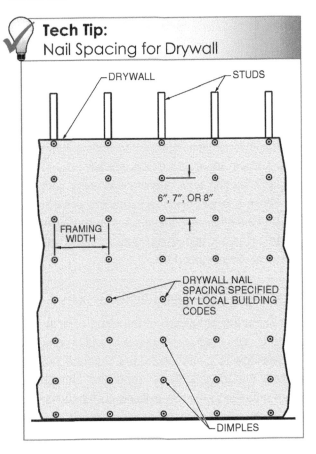

Cutting Large Access Holes

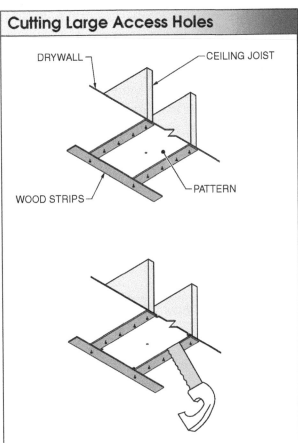

Figure 11-8. When cutting large openings in drywall or plaster, wood strips that are secured in place protect the edges of the cut opening.

A drywall cutout tool is a hand-held electric cutout tool used to make irregular cuts and holes in panels, gypsum board, and drywall for electrical boxes and duct openings.

Box manufacturers use mounting systems such as tabs, cleats, or nail plates to secure boxes to existing structures. **See Figure 11-9.** One common feature that box manufacturers use is a folding cleat on each side of the box that allows the box to be firmly secured from the inside of the wall opening. L clamps on the front of a rectangular retrofit box are used to keep the box from pulling through the drywall hole as the clamps are tightened. Rectangular retrofit boxes are useful for adding switches and receptacles to finished walls.

Installing Boxes in Existing Wall

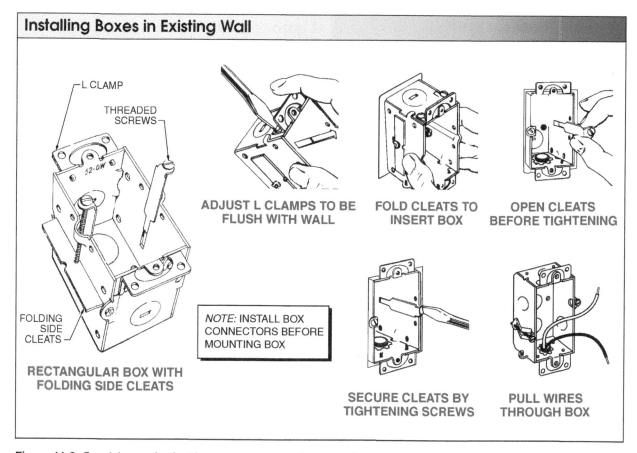

L CLAMP

THREADED SCREWS

52-0W

FOLDING SIDE CLEATS

RECTANGULAR BOX WITH FOLDING SIDE CLEATS

ADJUST L CLAMPS TO BE FLUSH WITH WALL

FOLD CLEATS TO INSERT BOX

OPEN CLEATS BEFORE TIGHTENING

NOTE: INSTALL BOX CONNECTORS BEFORE MOUNTING BOX

SECURE CLEATS BY TIGHTENING SCREWS

PULL WIRES THROUGH BOX

Figure 11-9. *Special mounting brackets are provided by box manufacturers to install boxes in existing walls and ceilings.*

Another method of securing rectangular boxes is the use of metal stampings. **See Figure 11-10.** Individual metal stampings are positioned in the wall and the tabs are bent over the box. The L clamps keep the box from pulling through the opening.

Mounting a box in a finished ceiling is a similar process to mounting a box in a finished wall. **See Figure 11-11.** Once a hole is cut, a strap with a threaded bolt is laid across the opening. Tightening the locknut through the box secures the box in place. Care must be taken to limit the size and weight of the fixture attached to a strapped box installation.

When it is not possible to have a fixture box mounted to the ceiling studs, bracing and toggle bolts can be used. **See Figure 11-12.** Holes are drilled that are large enough to accommodate large expandable toggle bolts. Tightening the toggle bolts pulls the box firmly into place. Fixture boxes are typically shallow and are totally covered by lighting fixtures.

Box Installation Accessories

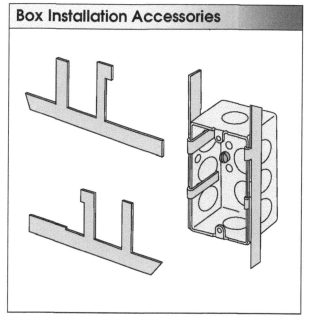

Figure 11-10. *Metal stampings can be used to secure electrical boxes in existing walls.*

Installing Boxes in Existing Ceiling

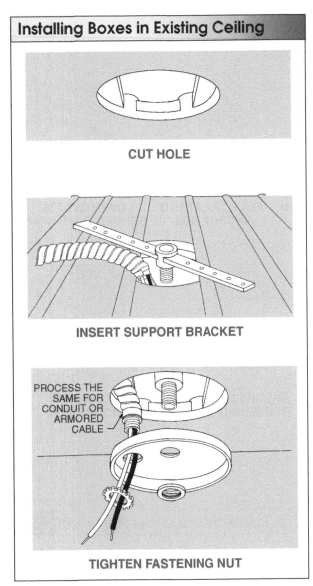

CUT HOLE

INSERT SUPPORT BRACKET

PROCESS THE
SAME FOR
CONDUIT OR
ARMORED
CABLE

TIGHTEN FASTENING NUT

Figure 11-11. *Typically a three-part sequence is used to mount a box in an existing ceiling.*

Installing Fixture Box with Toggle Bolts

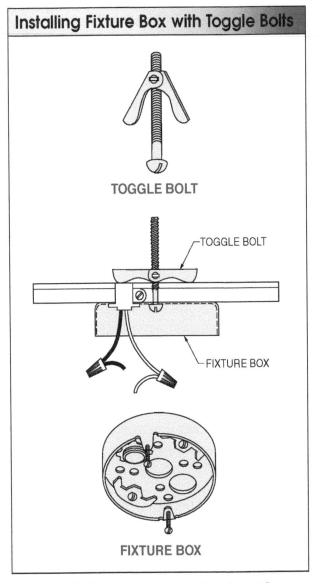

TOGGLE BOLT

TOGGLE BOLT

FIXTURE BOX

FIXTURE BOX

Figure 11-12. *Toggle bolts require that a larger than normal clearance hole be made in drywall to allow room for the wings of the toggle bolts.*

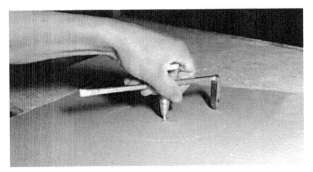

A circle cutter is a hand tool used to cut holes in drywall or thin wood sheets.

Routing Cables

Fish tapes are essential when routing cables through ceilings and walls. The proper use of fish tapes allows electricians to run wires that otherwise would be impossible to route.

Fish Tapes. Fish tapes require a five-part sequence to properly pull a wire or cable. **See Figure 11-13.** Fish tapes sometimes are required to be hooked together. Always take care when securing cable or wire to the ends of a fish tape. A little extra time taken to secure the wire to the fish tape ensures that the wire will not fall off halfway through a pull.

Back-to-back outlets can require two fish tapes to get the pull started for a cable. **See Figure 11-14.** Fish tapes are used to pull cable or wire through conduit along a wall from receptacle to receptacle. **See Figure 11-15.** A fish tape is also used to pull cable along the baseboard of a wall from Point A to Point B. **See Figure 11-16.**

Push a fish tape in to conduit until it stops. Move to the opposite end of the conduit and place another fish tape, or a piece of 10 AWG solid wire, though the end of the conduit and push it in until you feel it contact the first fish tape. Once it contacts the first fish tape, push additional 3' to 5'. Take the second fish tape or wire and twist it inside the conduit with a circular motion. Make at least 15 to 20 circular motions. Have enough slack in the first fish tape before pulling the second fish tape. Carefully pull the second fish tape until the hooked end is through the conduit.

Pulling Cables with Fish Tapes

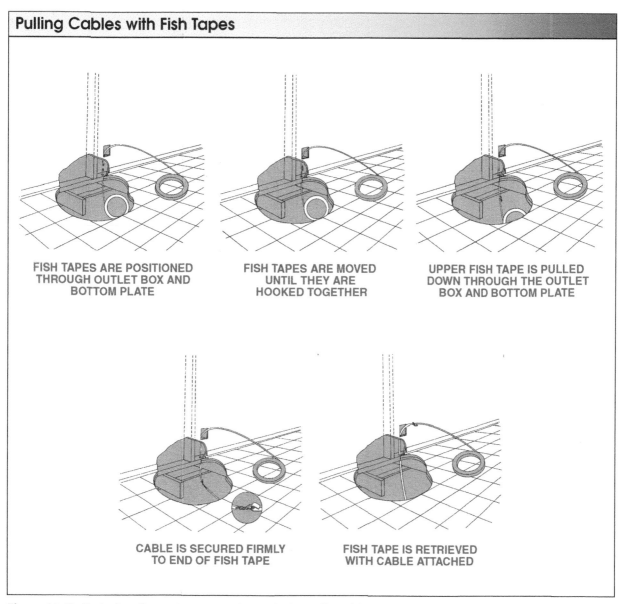

FISH TAPES ARE POSITIONED THROUGH OUTLET BOX AND BOTTOM PLATE

FISH TAPES ARE MOVED UNTIL THEY ARE HOOKED TOGETHER

UPPER FISH TAPE IS PULLED DOWN THROUGH THE OUTLET BOX AND BOTTOM PLATE

CABLE IS SECURED FIRMLY TO END OF FISH TAPE

FISH TAPE IS RETRIEVED WITH CABLE ATTACHED

Figure 11-13. *Typically a five-part sequence is required to pull a cable through the bottom plate of an existing wall.*

Pulling Wires or Cables through Back-to-Back Boxes

Figure 11-14. *Two fish tapes can be required to pull wires or cables through back-to-back boxes.*

Pulling Cable along Wall Baseboards

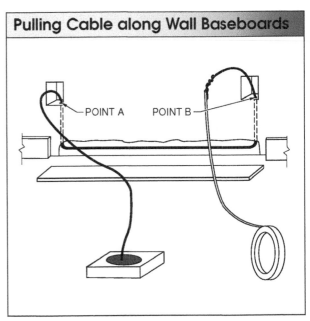

POINT A   POINT B

Figure 11-16. *Fish tapes are used to pull cables along wall baseboards to add additional outlet boxes.*

Pulling Wires or Cables from Box to Box

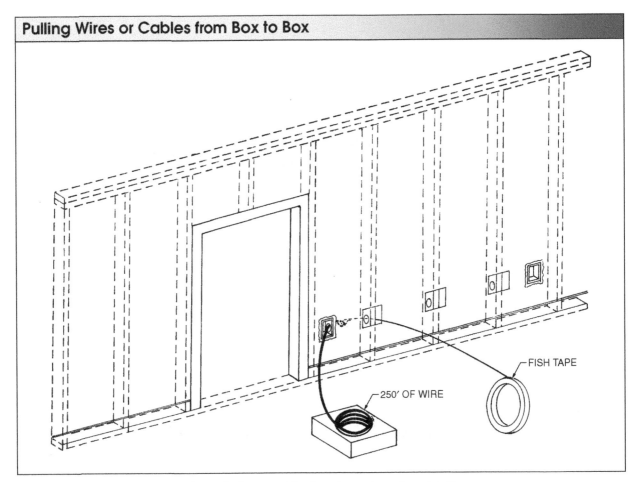

FISH TAPE

250' OF WIRE

Figure 11-15. *Fish tapes are used to pull wires or cables from box to box along a wall.*

When doorways are encountered, the doorway trim can be removed and the cable routed around the opening. **See Figure 11-17.** After securing the cable, the doorway trim is reinstalled, hiding the cable.

When it is necessary to replace a ceiling box in an older two-story building, second-floor floorboards may have to be removed. **See Figure 11-18.** Each floorboard is carefully removed so that floor joists can

be notched to accommodate the new cable or conduit. With the cable or conduit secured, the floorboards are carefully reinstalled.

Improperly connecting communication cables to wall terminals is the main cause of residential communication (VDV) problems.

Routing Cables around Doorways

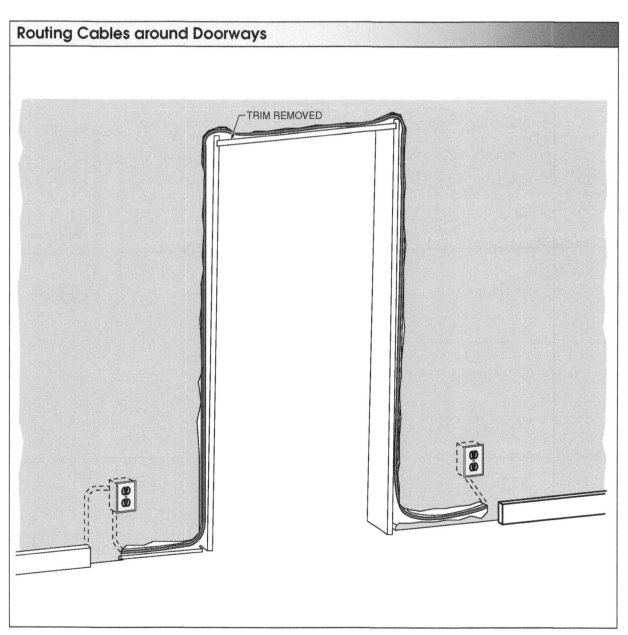

TRIM REMOVED

Figure 11-17. *Cables can be routed around doorways after the trim has been removed and the drywall grooved.*

Replacing First-Floor Ceiling Box in Two-Story Dwelling

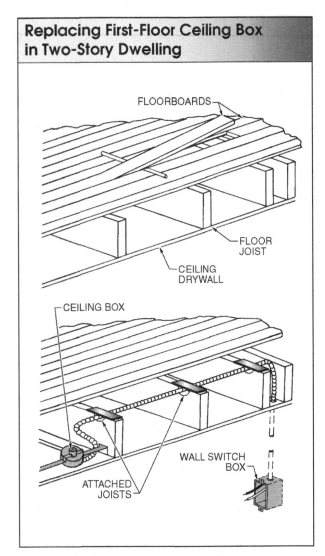

Figure 11-18. *In two-story residences, it may be necessary to remove floorboards or plywood to install a ceiling box.*

Extended Length Drill Bits

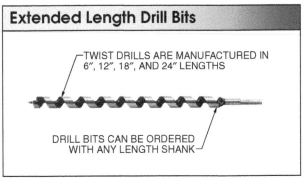

Figure 11-19. *Drill bits can be purchased in various lengths or modified to be any length required.*

Routing Cable from Basement to First-Floor Wall

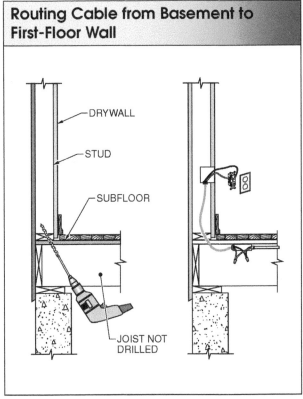

Figure 11-20. *Fish tapes aid dramatically when routing cables from the basement to first-floor walls.*

Cables or conduit can be required to run from the attic through interior walls of a home. A drill bit is used to penetrate the top plate of a wall. Several drill extensions or an extended-length drill bit may be required to penetrate obstacles inside a wall. **See Figure 11-19.** Cables or conduit can also be required to run from the basement to a first-floor receptacle box. A drill bit is used to penetrate subflooring and the wall bottom plate. **See Figure 11-20.** By pushing a fish tape up from the basement and attaching the cable, the cable can easily be pulled from the receptacle box to the junction box in the basement.

Surface Raceways

In some situations, such as with masonry walls, it is impossible to fish cables through the wall. When cables cannot be run through walls, cables must be run externally to walls. Cables being run externally to a wall must be run through surface raceways. **See Figure 11-21, Figure 11-22,** and **Figure 11-23.**

Surface Raceway – Feed

1 **Bring Feed into 2000B Base.**
Back-feed connection shown. See back
of sheet for alternate methods of feeding.

2 **Install 2000B Base on Surface.**
Starting with feed section, mount entire run of 2000B Base with no. 8
flat head screws, through screw piercings and knockouts. Cut base to
length at corners and end of run, as required.

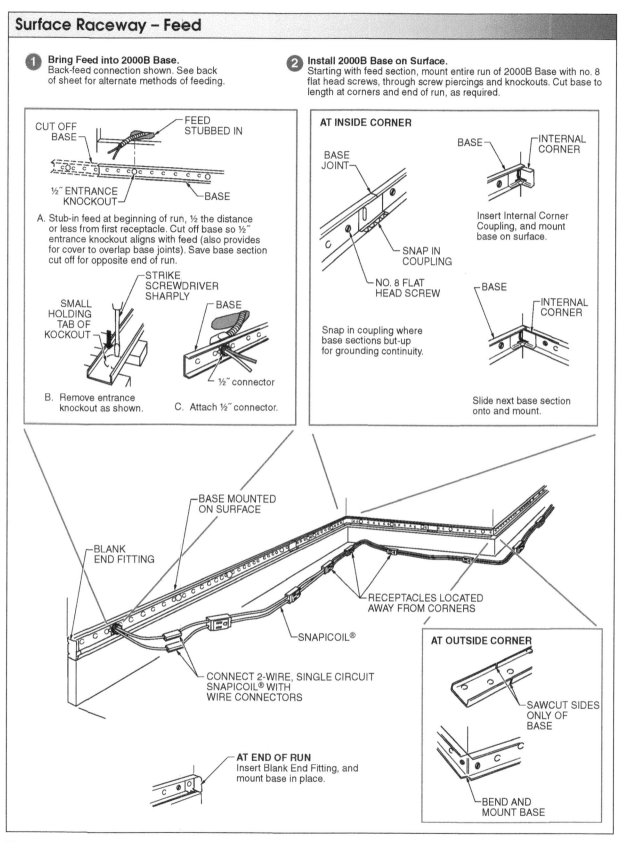

CUT OFF BASE

FEED STUBBED IN

½˝ ENTRANCE KNOCKOUT

BASE

A. Stub-in feed at beginning of run, ½ the distance
or less from first receptacle. Cut off base so ½˝
entrance knockout aligns with feed (also provides
for cover to overlap base joints). Save base section
cut off for opposite end of run.

STRIKE SCREWDRIVER SHARPLY

SMALL HOLDING TAB OF KOCKOUT

BASE

½˝ connector

B. Remove entrance
knockout as shown.

C. Attach ½˝ connector.

AT INSIDE CORNER

BASE JOINT

BASE

INTERNAL CORNER

SNAP IN COUPLING

NO. 8 FLAT HEAD SCREW

Insert Internal Corner
Coupling, and mount
base on surface.

Snap in coupling where
base sections but-up
for grounding continuity.

BASE

INTERNAL CORNER

Slide next base section
onto and mount.

BASE MOUNTED ON SURFACE

BLANK END FITTING

RECEPTACLES LOCATED AWAY FROM CORNERS

SNAPICOIL®

CONNECT 2-WIRE, SINGLE CIRCUIT
SNAPICOIL® WITH
WIRE CONNECTORS

AT OUTSIDE CORNER

SAWCUT SIDES ONLY OF BASE

BEND AND MOUNT BASE

AT END OF RUN
Insert Blank End Fitting, and
mount base in place.

Figure 11-21. *Raceways use knockouts (sometimes piercings) to connect feed (power supply) cables to raceway.*

Surface Raceway – Snapicoil®

3 **Connect Snapicoil® to Feed.**
Lay out Snapicoil® along entire run of base so that receptacles are not located over feed or in corners. Connect to feed wires with pressure type wire connectors (for common connection of 2, 3, or 4 No. 12 or No. 14 solid conductors: NOT TO BE USED to connect equipment grounding conductors). Insert only conductors of same color in a connector.

WIRE CONNECTORS

Connecting 3-wire, two-circuit Snapicoil®

HOW TO USE WIRE CONNECTORS

A. Strip wire ends to width of strip gauge (½″).

B. Insert wire ends the full stripped distance.

SNAPICOIL® GROUND WIRE

GROUND CLAMP

WIRE CONNECTORS

Connecting 3-wire Snapicoil® with insulated ground wire

SNAPICOIL® INSTALLATION TIPS

LOCKNUT RUN ONTO CONNECTOR AS SPACER

ATTACHING LOCKNUT

To shorten ½″ connector inside base for feed wire clearance.

FOLD BACK WIRES

RECEPTACLE

CUT OFF COVER TO MATCH WIRE FOLD-BACK

If receptacle falls at corner
Fold back wires to bring receptacle at least 6″ away from corner. Cut off cover to compensate for receptacle relocation before cutting it to length for corner.

Figure 11-22. *Raceway receptacles cannot be located over the raceway feed cable connection.*

Tech Tip:
Surface Raceway NEC® Requirements

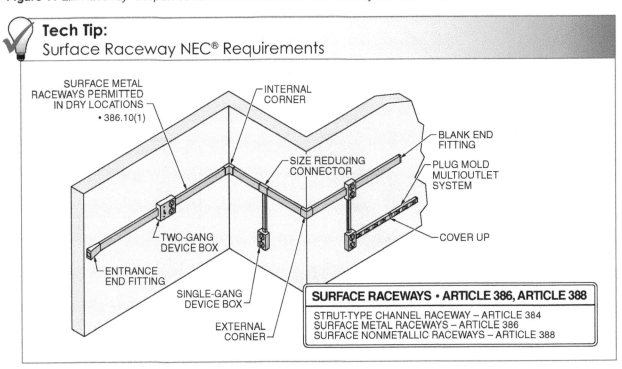

SURFACE METAL RACEWAYS PERMITTED IN DRY LOCATIONS
• 386.10(1)

INTERNAL CORNER

BLANK END FITTING

PLUG MOLD MULTIOUTLET SYSTEM

SIZE REDUCING CONNECTOR

COVER UP

TWO-GANG DEVICE BOX

ENTRANCE END FITTING

SINGLE-GANG DEVICE BOX

EXTERNAL CORNER

SURFACE RACEWAYS • ARTICLE 386, ARTICLE 388
STRUT-TYPE CHANNEL RACEWAY – ARTICLE 384 SURFACE METAL RACEWAYS – ARTICLE 386 SURFACE NONMETALLIC RACEWAYS – ARTICLE 388

Surface Raceway – Cover

④ Assemble Snapicoil® and Cover in Base.
Cover sections should overlap base joints for rigidity and better ground continuity.

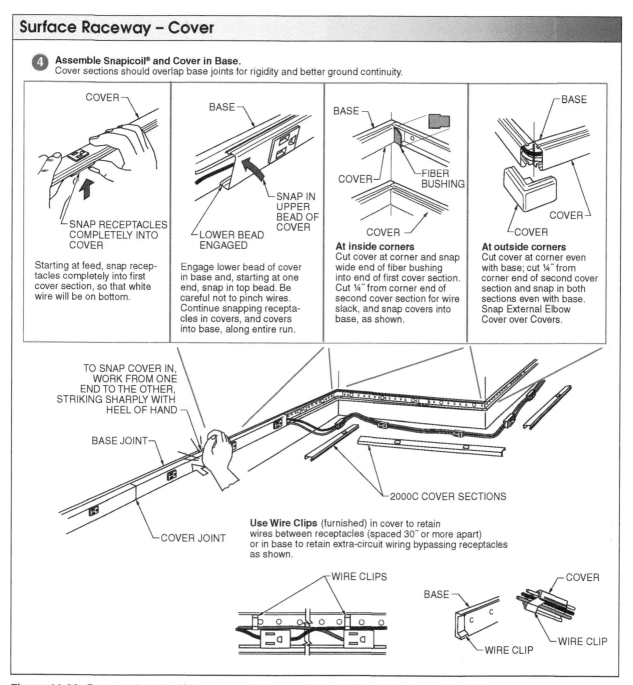

SNAP RECEPTACLES COMPLETELY INTO COVER

Starting at feed, snap receptacles completely into first cover section, so that white wire will be on bottom.

LOWER BEAD ENGAGED

Engage lower bead of cover in base and, starting at one end, snap in top bead. Be careful not to pinch wires. Continue snapping receptacles in covers, and covers into base, along entire run.

At inside corners
Cut cover at corner and snap wide end of fiber bushing into end of first cover section. Cut ¼″ from corner end of second cover section for wire slack, and snap covers into base, as shown.

At outside corners
Cut cover at corner even with base; cut ¼″ from corner end of second cover section and snap in both sections even with base. Snap External Elbow Cover over Covers.

TO SNAP COVER IN, WORK FROM ONE END TO THE OTHER, STRIKING SHARPLY WITH HEEL OF HAND

BASE JOINT

COVER JOINT

2000C COVER SECTIONS

Use Wire Clips (furnished) in cover to retain wires between receptacles (spaced 30″ or more apart) or in base to retain extra-circuit wiring bypassing receptacles as shown.

WIRE CLIPS

BASE

COVER

WIRE CLIP

WIRE CLIP

Figure 11-23. *Receptacles should be snapped into cover first with the white wire on the bottom.*

INSTALLING RECESSED LIGHTS

To install a recessed light in a ceiling, verify that there are no obstructions where the fixture is to be placed and that there is enough room to accommodate the entire fixture. Once an area is known to be clear, a hole can be cut using a hand saw or a hole saw and drill. **See Figure 11-24.** Hole saws are ideal for making holes in plaster or when creating several openings at one time. The opening created by a hole saw must provide a snug fit for the canister of the can light.

Creating Large Holes

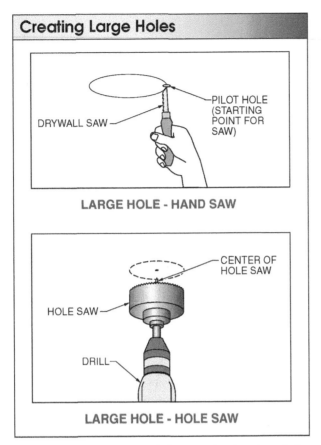

Figure 11-24. *When an area is known to be clear, large holes can be cut using a handsaw, or with a drill and hole saw.*

Typically the most difficult problem when installing a recessed (can) light is running cable or conduit inside walls and ceilings from one room to another. Nonmetallic sheathed cable or armored cable is preferred for remodeling work because cable can be easily pulled and routed between ceilings, walls, and floors with a fish tape.

Fishing Wires

When cables must be routed around wall-ceiling corners of a finished home, the corner around which the cable must be routed is typically blocked by obstructions. **See Figure 11-25.** There are two methods by which corner obstructions can be bypassed. The first method uses a hole drilled at a diagonal from an opposite room, with a second hole drilled horizontally to meet the first hole. The two drilled holes provide an opening through the wall for cable. Some patchwork is required after the cable is routed.

Routing Cable from First-Floor Wall to First-Floor Ceiling

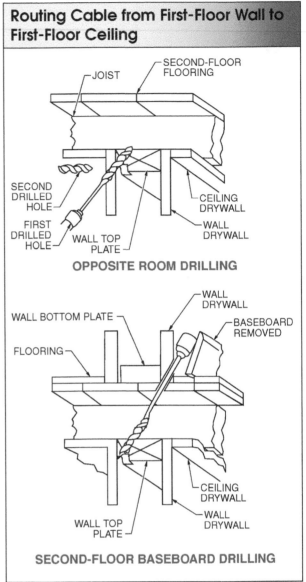

Figure 11-25. *Holes drilled from the first floor require patchwork while holes drilled from the second floor can typically be covered with the second-floor base molding.*

The second method of bypassing wall-ceiling corner obstructions is to remove the second-floor baseboard and drill a diagonal hole from one wall to the ceiling, forming the route for the cable. When the baseboard is replaced the drilled openings are not visible.

After drilling the necessary opening by either of the methods mentioned, cable is pulled through the opening using a fish tape. **See Figure 11-26.** To pull cable through the opening, push the hooked end of a fish tape through the drilled opening. Run another fish tape from the ceiling opening and hook the tape to the first fish tape. When the two hooked ends are joined, pull the first fish tape to the ceiling opening so that one fish tape is inside the wall and ceiling. Attach a continuous length of cable to the end of the fish tape and pull the cable into position.

When attic space is accessible, such as in one- or two-story houses with attics, floorboards, if present, can be lifted and a hole bored with an extended drill bit through any obstructions. **See Figure 11-27.**

Once a cable is fished, allow approximately 16″ of the cable to protrude out of the box for final wiring to a recessed light. **See Figure 11-28.**

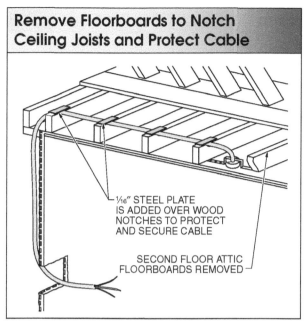

Remove Floorboards to Notch Ceiling Joists and Protect Cable

¹⁄₁₆″ STEEL PLATE IS ADDED OVER WOOD NOTCHES TO PROTECT AND SECURE CABLE

SECOND FLOOR ATTIC FLOORBOARDS REMOVED

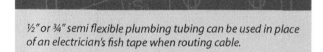

½″ or ¾″ semi flexible plumbing tubing can be used in place of an electrician's fish tape when routing cable.

Figure 11-27. *Removing second-floor or attic floorboards allows good accessibility to ceiling joists and fixture opening.*

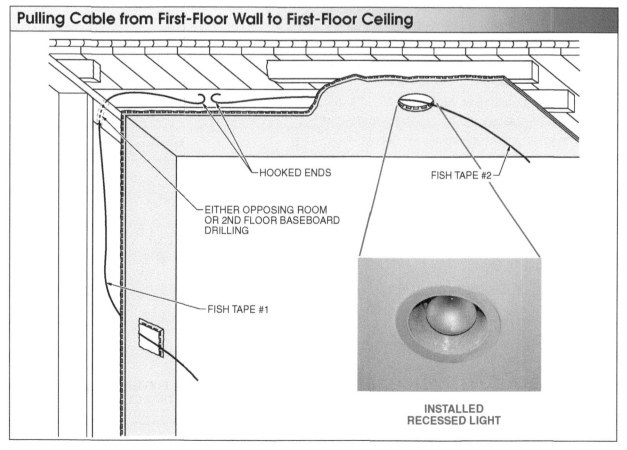

Pulling Cable from First-Floor Wall to First-Floor Ceiling

HOOKED ENDS

FISH TAPE #2

EITHER OPPOSING ROOM OR 2ND FLOOR BASEBOARD DRILLING

FISH TAPE #1

INSTALLED RECESSED LIGHT

Figure 11-26. *Routing cable from a wall to a ceiling typically requires the use of two fish tapes.*

Extra Cable for Wiring Recessed Lights

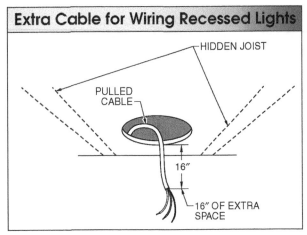

Figure 11-28. *After a cable is fished into a ceiling box, allow approximately 16" of cable to protrude out of the box for final wiring to a recessed light.*

Wiring Recessed Lights

Recessed lights are manufactured with prewired junction boxes attached to the housing (can) of the light. Once the cover plate of the junction box is removed, the cable can be secured and connected to the box. The cable must be connected so that the cable cannot be pulled out of the junction box. Use standard wiring techniques to strip and connect the wires of the cable, white to white, red to red or black, and ground to ground. **See Figure 11-29.** In some cases the wire nuts should be taped for additional security to ensure the connections do not come undone when folded back into the junction box.

Wiring Recessed Lights

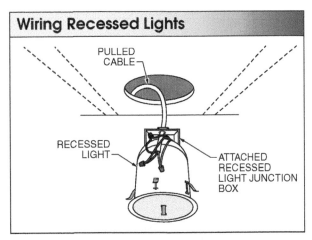

Figure 11-29. *Some circumstances require that wire nuts be taped for additional security to ensure that connections do not become undone when being folded back into the box.*

Securing Recessed Lights

Recessed lights designed for remodeling are equipped with mounting clips that clamp or secure the fixture to the ceiling. To secure the fixture, the mounting clips push down against the drywall. Once the recessed light fixture is pushed up into the hole, the clips are secured. The mounting clips provide an audible sound (short click) when pushed into clamping position. **See Figure 11-30.**

Securing Recessed Lights

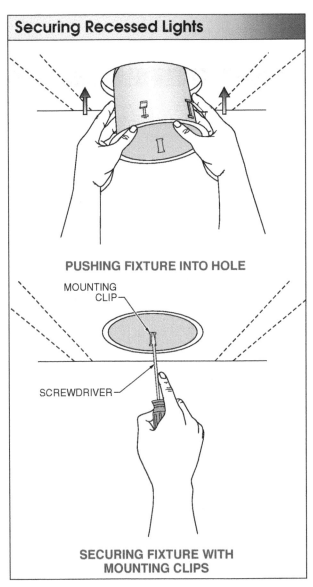

PUSHING FIXTURE INTO HOLE

SECURING FIXTURE WITH MOUNTING CLIPS

Figure 11-30. *Recessed lights designed for remodeling are equipped with mounting clips that clamp or secure the fixture to the ceiling drywall.*

Pressing a thumb against the clip on the inside of the fixture can help the process of securing a light fixture. Some fixtures require the aid of a screwdriver.

Insulating Recessed Lights

For safety reasons, choosing the correct type of recessed light avoids damage to the light and eliminates potential fire risk. Recessed light fixtures and housings are rated either noninsulated ceiling (NIC) or insulated ceiling (IC). For fire safety, NIC-rated fixtures must have at least 3″ of clearance around the fixture (including the wiring box) from thermal insulation. An IC-rated ceiling fixture can make direct contact with thermal insulation without creating a fire hazard. **See Figure 11-31.**

Recessed Light Trims

Some recessed light fixtures are manufactured with built-in pieces of trim. More commonly, the fixtures and trims are separate. **See Figure 11-32.** Ensure the trim is the same size and make as the recessed light fixture.

Recessed light trim is typically a ring that covers the gap between the ceiling material and the fixture. Trims with lenses are used for fixtures with standard light bulbs. When no lens is present, floodlight-type bulbs are typically used in the fixture.

An eyeball trim swivels so that the light emitted can be pointed in any direction to highlight a part of a room or a spot on a wall. Eyeball trim is typically used with spotlight bulbs, which produce a narrower beam of light than floodlight bulbs and can focus on an object such as a painting.

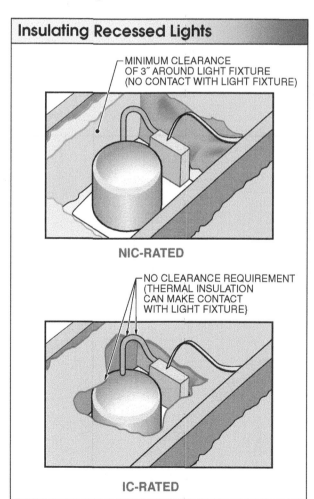

Figure 11-31. *Recessed light fixtures and housings are rated as either noninsulated ceiling (NIC) or insulated ceiling (IC).*

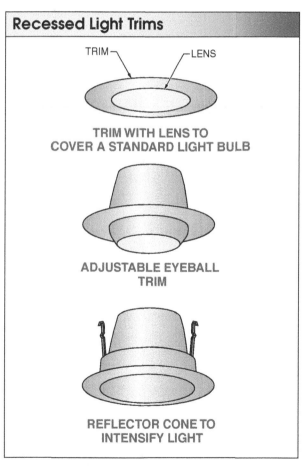

Figure 11-32. *Trims with lenses are used for recessed fixtures with standard light bulbs, and fixtures with no lenses have floodlight-type bulbs.*

Choosing Recessed Lights

Recessed lights are manufactured in a variety of sizes and wattages for locations that are both wet and dry. Suppliers should be consulted to determine the correct recessed light for an application.

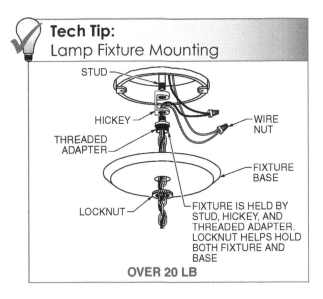

Tech Tip:
Lamp Fixture Mounting

STUD

HICKEY

WIRE NUT

THREADED ADAPTER

FIXTURE BASE

LOCKNUT

FIXTURE IS HELD BY STUD, HICKEY, AND THREADED ADAPTER. LOCKNUT HELPS HOLD BOTH FIXTURE AND BASE

OVER 20 LB

INSTALLING CEILING FANS

When installing a ceiling fan in an existing home, the fan is typically replacing a light fixture in the center of the room. If the installation is a fan only, the circuit should be able to handle the load. If the installation includes a fan and light, a new circuit may be required. Ceiling fans can be standard mount or flush mount. **See Figure 11-33.**

Mounting Ceiling Fans

When mounting a ceiling fan, the weight of the fan is the number one consideration, but the force created by the blades rotating must also be considered. All ceiling fans are required to be mounted to a NEC® listed fan box that is properly mounted. Fans heavier than 35 lb must use the structural members of the house to support the fan. **See Figure 11-34.** Various manufacturers provide accessory kits to secure fans to the structural members of a house. A common feature of ceiling fan installations is the use of substantial screws such as lag screws.

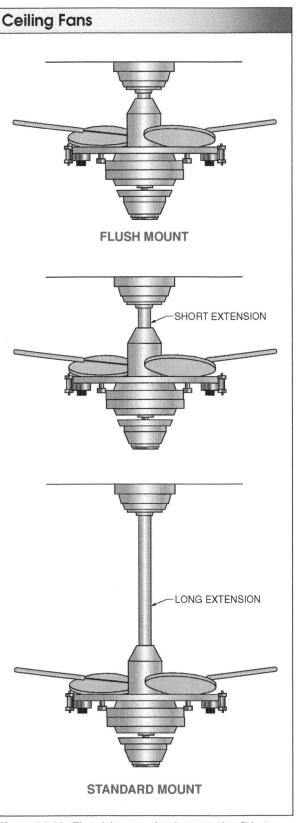

Ceiling Fans

FLUSH MOUNT

SHORT EXTENSION

LONG EXTENSION

STANDARD MOUNT

Figure 11-33. *Electricians use bar hangers that fit between ceiling joists to provide support for heavy ceiling fans.*

Ceiling Fan Mountings

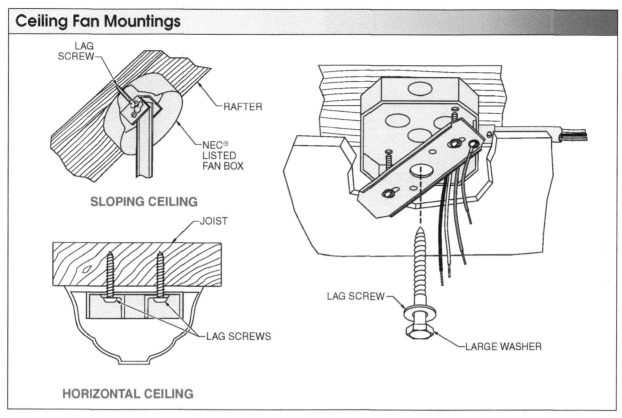

Figure 11-34. *All ceiling fans are required to be mounted to a NEC® listed fan box. Fans heavier than 35 lb must be supported by structural members of the house.*

The biggest challenge when mounting a ceiling fan arises when structural members are not directly available to secure the fan. To aid in mounting ceiling fans, manufacturers provide kits that reach out to the nearest structural member. **See Figure 11-35.** Typically electricians use bar hangers that fit between ceiling joists to provide support for heavy ceiling fans.

To install a bar hanger, position the bar in place and force the spikes at the end of the bar into the joists to hold the bar in place. Secure the bar flanges with screws. If access through the attic is not possible, there are fan brace kits that can be installed through the hole in the ceiling. **See Figure 11-36.**

Never assume a circuit is OFF even after turning the breaker OFF or removing the fuse. Equipment must be verified to be OFF by testing it with a test instrument such as a multimeter, voltmeter, DMM, voltage tester, or clamp-on ammeter.

Attic Access Bar Hangers

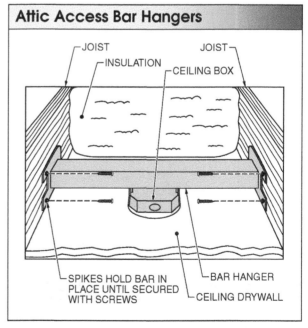

Figure 11-35. *Electricians use bar hangers that fit between ceiling joists to provide support for heavy ceiling fans.*

Hole Access Bar Hangers

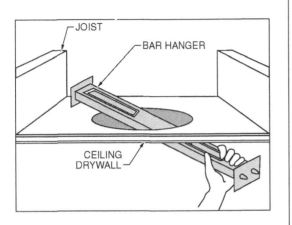

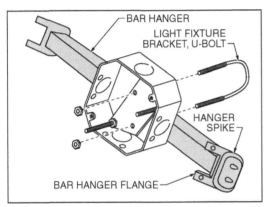

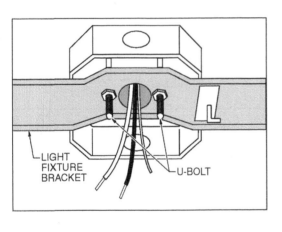

Figure 11-36. *To aid in mounting ceiling fans, manufacturers provide mounting kits that reach out to the nearest structural member.*

Most modern ceiling fans have a hook to hold the ceiling fan in place while making electrical connections, which allows for one-person installation.

Electrical ceiling boxes are typically mounted to a joist for maximum support of light fixtures and ceiling fans.

Grounding Ceiling Fans

The two types of ceiling fans are flush mount and downrod mount. **See Figure 11-37.** *Flush mount* is a type of ceiling fan mounting used to allow maximum head clearance. *Downrod mount* is a type of ceiling fan mounting used to adjust head clearance or for angle mounting to vaulted ceilings. The wiring difference between the two mountings is the type of grounding used. In flush mount the ground goes directly to the metal junction box. In the downrod mount the ground is attached to the metal junction box and the downrod.

Wiring Ceiling Fans

When wiring a ceiling fan, two-wire or three-wire cable is used, depending on the application. A two-wire cable contains a black wire (hot), white wire (neutral), and a bare wire (ground). A three-wire, metallic or nonmetallic cable contains a black wire (hot), red wire (hot), white wire (neutral), and a bare wire (ground) in one outer cover. Two-wire cables are used when a fan will only be turned on and off. **See Figure 11-38.** Three-wire cables are used when it is necessary to control a fan and a light attached to the fan. When connecting a fan with light(s), the fan and light fixture wires are green, white, black, and blue.

Ceiling Fan Grounding

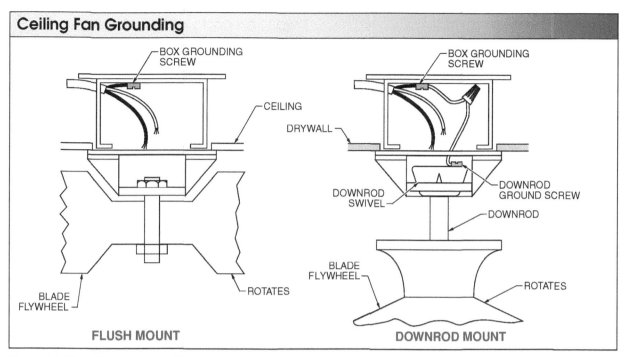

Figure 11-37. *A flush-mount ceiling fan has the ground going directly to a metal junction box. A downrod type of ceiling fan mounting has the ground attached to the metal box and the downrod.*

Wiring Ceiling Fans

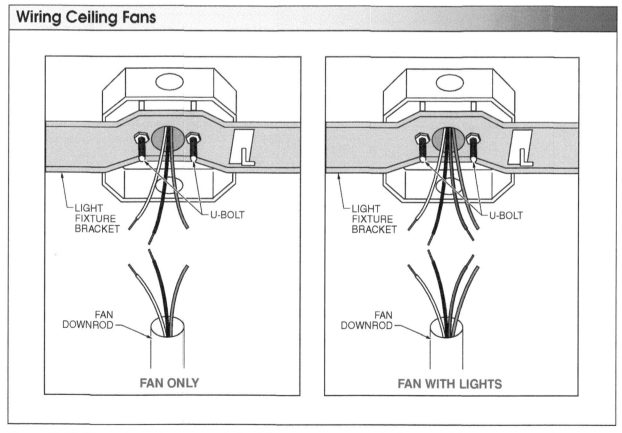

Figure 11-38. *Three-wire cables are used when it is necessary to control a fan and a light attached to the fan.*

The green grounding wire of the fixture connects to the bare ground wire of the supply cable. The white and black wires connect to the white and black wires of the cable. The red cable wire connects to the blue wire of the fan so the light(s) can be switched separately. At the switch box of a fan and light combination, there are two switches with power going to the bottom of each. **See Figure 11-39.** One switch will connect to the black wire in the cable for controlling the fan. The red wire will connect to the other switch to control the lighting circuit.

Ceiling Fan Switch Connections

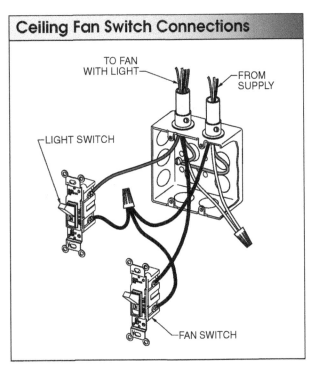

Figure 11-39. *A ceiling fan with light has two switches. One switch controls the fan and the other switch controls the light(s).*

Ceiling Fan Controls

Fans with new technology provide many additional features. A fan can be started or reversed, its speed can be controlled, and the lights dimmed, all from a remote control. Typically, fans and fans with lights are controlled by wall controls. **See Figure 11-40.** However, fans can be controlled by handheld remote controls. **See Figure 11-41.**

Wall Fan and Light Controls

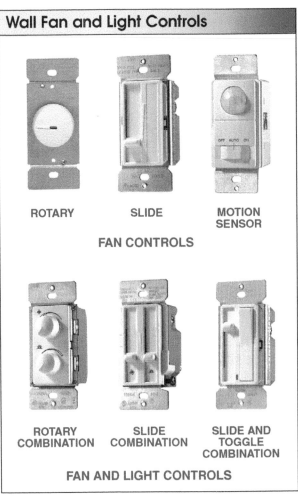

Cooper Wiring Devices

Figure 11-40. *A fan can be started or reversed, its speed can be controlled, and the lights dimmed, all from one wall box.*

Remote Fan and Light Controls

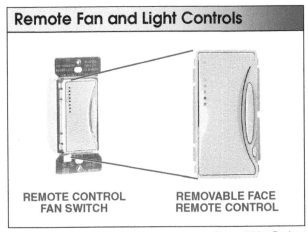

Cooper Wiring Devices

Figure 11-41. *A fan can be started or reversed, its speed can be controlled, and the lights dimmed, all from a handheld remote control.*

INSTALLING WHOLE-HOUSE EXHAUST FANS

Whole-house fans can save energy during periods when the outside air is cool enough to make the house temperature comfortable by bringing fresh air in to the home. When purchasing and installing a whole-house fan, the following considerations must be taken into account:

- There must be an adequate amount of roof vents.
- A fan with a two-speed motor should be installed, in case one speed provides too much airflow.
- The manufacturer specifications should be used to determine the fan size and amount of roof ventilation

required. A general rule is that the cubic feet per minute (CFM) of airflow that must be turned over should be approximately three times the square footage of the home.

Usually a central hallway is the best location for a whole house fan. To install a whole-house fan, use the following procedure: **See Figure 11-42.**

> ⚠ **WARNING**
>
> *Use a test instrument such as a DMM, voltmeter, or voltage tester to verify that power to the circuit is OFF before performing this procedure.*

Whole-House Fan Installation

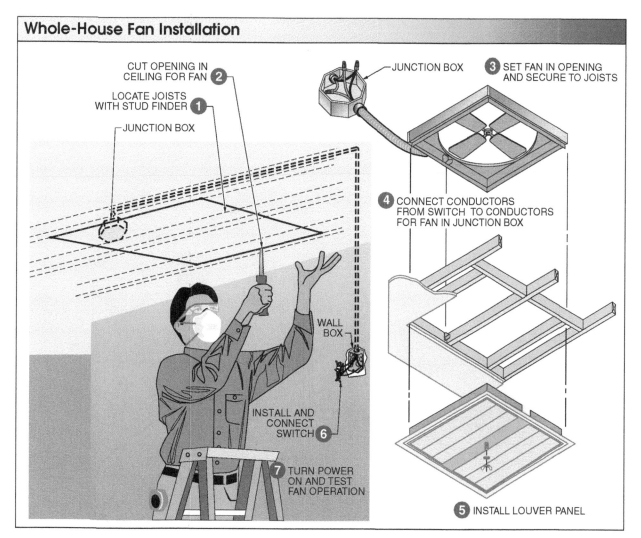

Figure 11-42. *The arched hole of a receptacle is connected through a wire or metal sheathing to the ground bar of the service panel and from there to a grounding electrode in the earth.*

1. Verify that the fan selected will fit in the opening. The size of the opening is determined by the stud locations in the ceiling. Use a stud finder to locate the opening between the joists.
2. Lay out the opening on the ceiling and use a drywall or reciprocating saw to cut out the opening. Drilling a hole in each corner will facilitate cutting and removing the drywall. Place a drop cloth under the ladder to catch drywall pieces and dust.
3. Set the fan in place in the ceiling opening and attach to studs with screws.
4. Run switch wiring and power wiring to the junction box for the fan. Be sure to use enough wire to reach the junction box.
5. Attach the louver panel to the fan with screws.
6. Install and connect the switch, making sure that it is in the OFF position.
7. Turn power ON, and test the circuit to the fan.

INSTALLING LOW-VOLTAGE OUTDOOR LIGHTING

Low-voltage outdoor lighting is used to provide accent lighting and light for areas such as walkways, driveways, and patios. Low-voltage outdoor lighting includes a transformer to step down receptacle voltages from 110 V to 12 V. Transformer enclosures can have a timer or be stand-alone. **See Figure 11-43.**

Some types of transformers can be powered by a light sensor time switch, which switches lights ON once a certain level of darkness is reached and OFF once it is daylight. Wiring the transformer requires attaching the low-voltage conductors to screw terminals. Powering the system only requires plugging the prewired transformer plug into an energized receptacle. If the timer is a separate unit from the

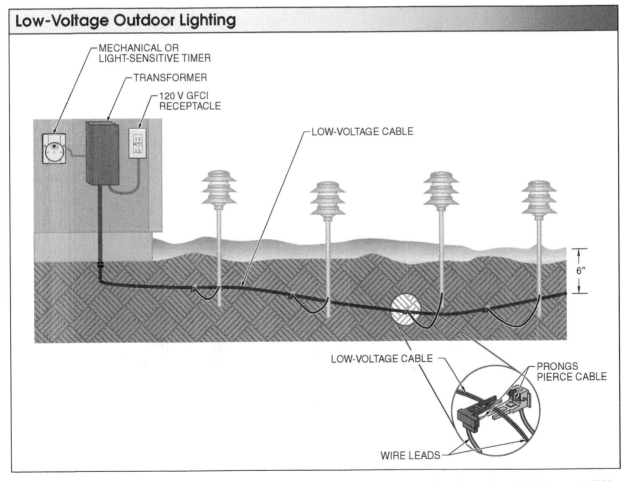

Low-Voltage Outdoor Lighting

MECHANICAL OR LIGHT-SENSITIVE TIMER

TRANSFORMER

120 V GFCI RECEPTACLE

LOW-VOLTAGE CABLE

6"

LOW-VOLTAGE CABLE

PRONGS PIERCE CABLE

WIRE LEADS

Figure 11-43. *Low-voltage outdoor lighting includes a transformer to step the receptacle voltages from 110 V down to 12 V.*

transformer, additional connections must be made. The manufacturer instructional manuals should always be referred to for the completion of all tasks.

If necessary, the low-voltage light fixtures for attachment to low-voltage cables should be assembled and laid out. Because it is low voltage (12 VDC), the cable only needs to be buried in a shallow trench, 4´ to 6´ deep.

PROPERLY INSTALLING AND TESTING RECEPTACLES

To be properly installed, receptacles must be tested. Receptacles are tested for proper polarization, grounding, and GFCI operation.

Polarization

In the early stages of residential wiring, receptacles were considered properly installed if the receptacle was polarized. Polarized plugs and receptacles keep the external surfaces of appliances such as toasters on the "ground" side of any current flow. The white wire (neutral) is connected to the long slot of a receptacle and the black wire (hot) is connected to the shorter slot for proper polarization. **See Figure 11-44.**

For ease of installation, low-voltage outdoor lighting may be solar powered and cordless.

Grounding

Grounding is recognized as being required to protect homeowners from residential electrical system problems. A properly grounded receptacle has a ground wire or metallic sheathing that is tightly connected at all points leading back to the ground bar of the service panel. **See Figure 11-45.**

A grounded receptacle provides an unbroken path to earth for current flow. Current flow will travel to earth ground and not through the body when proper grounding is in place. The third prong of a grounded plug fits into an arched hole in a grounded receptacle. The arched hole is connected through a wire or metal sheathing to the ground bar of the service panel. The ground bar is connected to earth ground through an electrode. **See Figure 11-46.**

> ⚠ **DANGER**
>
> *Many homeowners simply replace old, ungrounded receptacles with new three-hole grounded receptacles. The newly installed receptacles look grounded but are not, and are dangerous.*

Polarized Receptacles

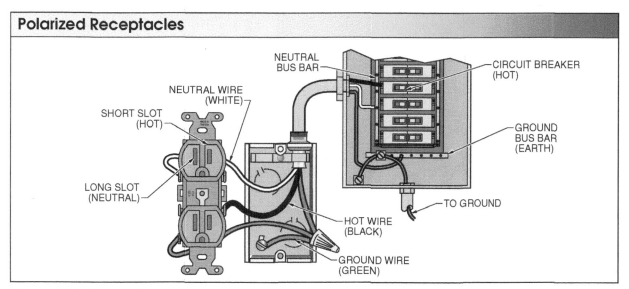

Figure 11-44. *A polarized receptacle has a white wire (neutral) connected to the long slot of the receptacle and a black wire (hot) connected to the short slot of the receptacle.*

Properly Grounded Receptacles

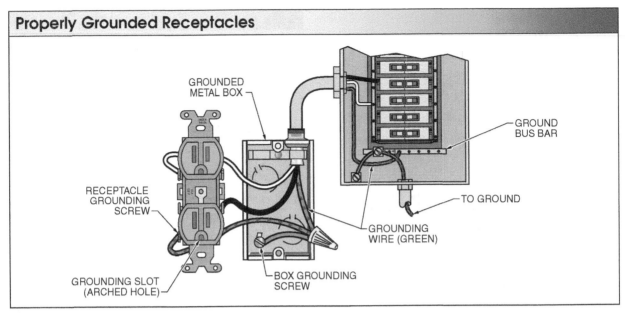

Figure 11-45. *A grounded receptacle provides an unbroken path to earth for current flow.*

Plug Grounding Path to Earth

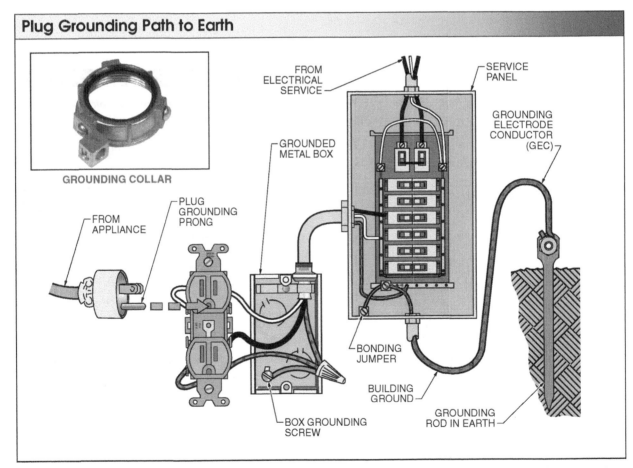

Figure 11-46. *The arched hole of a receptacle is connected through a wire or metal sheathing to the ground bar of the service panel and from there to a grounding electrode in the earth.*

GROUND FAULT CIRCUIT INTERRUPTERS

The latest device to provide protection from electrical shock in the home is the ground fault circuit interrupter (GFCI). **See Figure 11-47.** GFCIs can be permanent or plug-in. Check manufacturer specifications as to ratings and locations for proper GFCI installation instructions.

A GFCI that is being retrofitted into a box of an older home typically requires a box extender. **See Figure 11-48.** A box extender allows all connections to be made without stuffing the GFCI device into a cramped, wire-filled area. The pressure on the wires can cause connections to come apart or to overheat during operation.

Ground Fault Circuit Interrupters

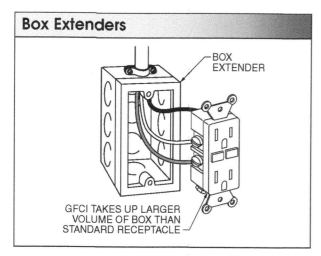

COMMERCIAL GRADE GFCI – 15 A

SWITCHED GFCI PROTECTED RECEPTACLE

Cooper Wiring Devices

Figure 11-47. *Ground fault circuit interrupters (GFCIs) can be of a permanently installed design or of a portable (plug-in) design.*

Box Extenders

BOX EXTENDER

GFCI TAKES UP LARGER VOLUME OF BOX THAN STANDARD RECEPTACLE

Figure 11-48. *GFCIs being retrofitted into a box of an older home typically require that a box extender be used.*

Plug-in GFCIs

Plug-in GFCIs are the most convenient ground fault circuit interrupter device to use. By simply plugging a GFCI module into any grounded receptacle, any tool plugged into the GFCI device and the operator are protected. Electricians typically keep a plug-in GFCI unit in a toolbox, and use the device whenever working with power tools outdoors, especially when conditions are wet. **See Figure 11-49.**

Plug-in GFCIs

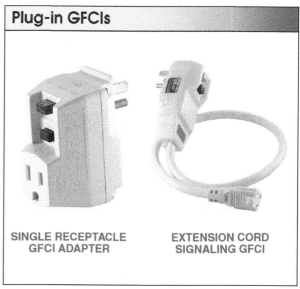

SINGLE RECEPTACLE GFCI ADAPTER

EXTENSION CORD SIGNALING GFCI

Cooper Wiring Devices

Figure 11-49. *By simply plugging a GFCI module into any grounded receptacle, the tool plugged into the GFCI device and the operator are protected.*

TESTING ELECTRICAL CIRCUITS

Electrical circuits can be tested safely and inexpensively with a neon tester. Most hardware stores sell simple test instruments such as neon testers. A neon tester is designed to light up in the presence of 115 V and 230 V circuits. A soft glow indicates a 115 V circuit; a brighter glow indicates a 230 V circuit. Many residential electrical system problems can be diagnosed by determining the presence or absence of voltage.

Testing for Defective Receptacles

One of the most common tests made with a neon tester is determining whether a receptacle is providing power. **See Figure 11-50.** The test leads of a neon tester or other test instrument must be firmly pressed into the receptacle slots to form a good electrical connection.

When voltage is present, the neon bulb will glow softly. When the tester does not light, the receptacle cover should be removed so that a second voltage check can be made at the terminals of the receptacle. **See Figure 11-51.** When voltage is present at the terminals of a receptacle but not at the slots, the receptacle is defective and must be replaced.

When voltage is not present at either the slots or terminals of a receptacle, the problem lies in the overload protection (fuses and circuit breakers) or in the electrical circuit (wires) leading to the troubled receptacle.

When the problem is in the electrical circuit leading to the receptacle, check each splice or each terminal point along the entire circuit for a break or loose connection.

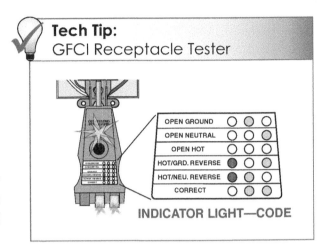

Tech Tip:
GFCI Receptacle Tester

OPEN GROUND	○	●	○
OPEN NEUTRAL	○	○	●
OPEN HOT	○	○	○
HOT/GRD. REVERSE	●	○	○
HOT/NEU. REVERSE	●	○	○
CORRECT	○	●	○

INDICATOR LIGHT—CODE

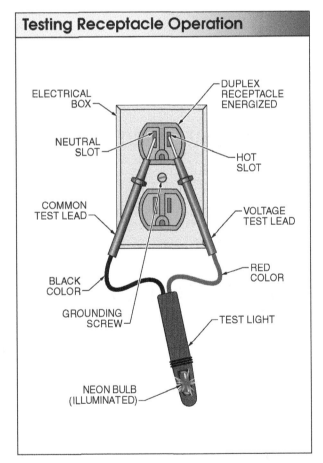

Testing Receptacle Operation

ELECTRICAL BOX

DUPLEX RECEPTACLE ENERGIZED

NEUTRAL SLOT

HOT SLOT

COMMON TEST LEAD

VOLTAGE TEST LEAD

BLACK COLOR

RED COLOR

GROUNDING SCREW

TEST LIGHT

NEON BULB (ILLUMINATED)

Figure 11-50. *The test leads of a test instrument must be firmly pressed into the slots of the receptacle to form a proper electrical connection.*

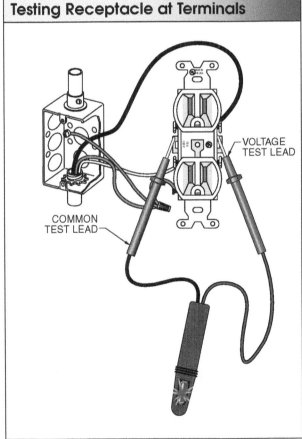

Testing Receptacle at Terminals

VOLTAGE TEST LEAD

COMMON TEST LEAD

Figure 11-51. *When voltage measurements are unsuccessful at the slots of a receptacle, the receptacle must be removed from the box and tested at the terminals.*

Testing for Defective Switches

Determining whether a switch is defective requires only a simple two-step procedure. The first step is to determine if voltage is reaching the switch. **See Figure 11-52.** The second step is to determine if voltage is passing through the switch. **See Figure 11-53.** A system that is properly grounded requires that the leads of the test instrument only touch the metal box and the terminals.

⚠️ **CAUTION**

In older homes without a ground, it is necessary to remove the wire nut from the neutral wire and use the neutral as the ground test point.

If voltage is not present at either switch terminal, the problem lies in the overload protection (fuses and circuit breakers) or in the electrical circuit leading to the troubled switch.

Testing Voltage from Switch

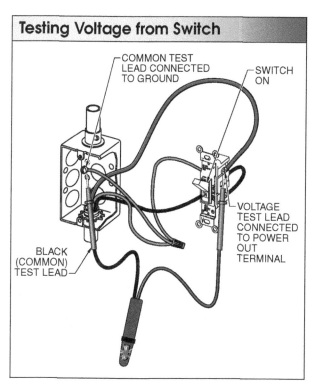

Figure 11-53. *When a switch has incoming voltage, the switch is turned ON and the switch is tested for voltage leaving the switch (top terminal).*

When the problem is in the electrical circuit leading to the switch, each splice or terminal point should be checked along the entire circuit for a break or loose connection.

⚠️ **DANGER**

Before starting any voltage test procedure, ensure the power to the suspected switch has been turned off. Remove the faceplate from the switch and unscrew the switch from the box. Pull the switch safely away from the metal box and position the switch so that no bare wires touch the box. When the switch is in a safe position, power is restored and the test procedures started.

Testing Voltage to Switch

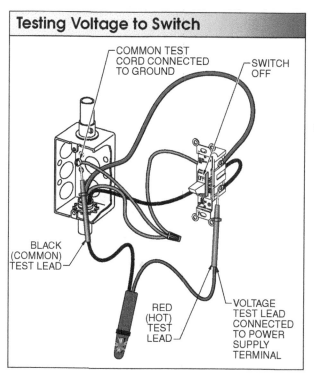

Figure 11-52. *A switch is tested by first measuring for incoming voltage to the switch (bottom terminal).*

Testing for Hot Wires

When remodeling, it is necessary to identify which wire(s) bring power to a circuit and which wire(s) feed other circuits. A neon tester can simplify this procedure by individually identifying hot wires. Any hot wire measured to common wire or ground by a neon tester will cause the neon tester to glow.

Grounded systems are the easiest to test for a hot wire because only the potential hot wires must be disconnected, separated, and tested. **See Figure 11-54.** The wire that causes a neon tester to glow is the hot lead.

An ungrounded system can be checked like a grounded system except any solderless connectors must be removed from the neutral wires. The neutral wires are used as the reference for ground in an ungrounded system. **See Figure 11-55.**

To determine if voltage is reaching a light fixture, turn the light fixture switch to the ON position. A properly installed circuit will cause the neon tester light to glow. **See Figure 11-56.**

Testing Receptacle Grounds

A simple test procedure is used to check the grounding of receptacles. **See Figure 11-57.** The test is used to ensure that each receptacle in a home is grounded.

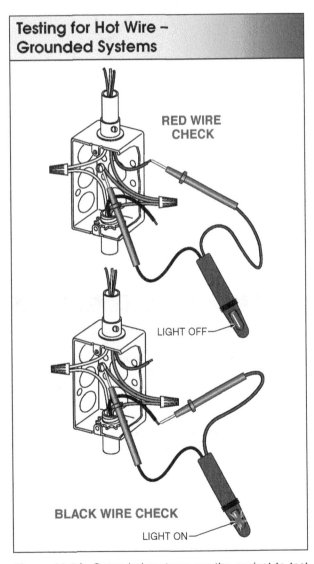

Figure 11-54. *Grounded systems are the easiest to test for a hot wire.*

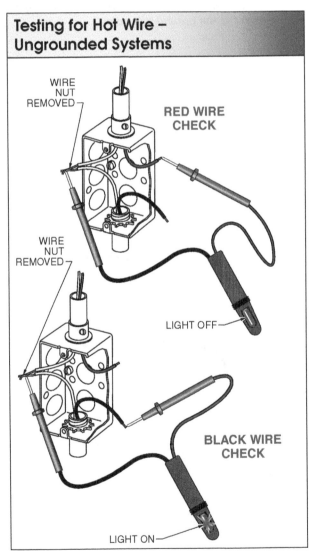

Figure 11-55. *The neutral wires are used as the reference for ground in ungrounded electrical systems.*

Testing for Voltage at Ceiling Fixture Box

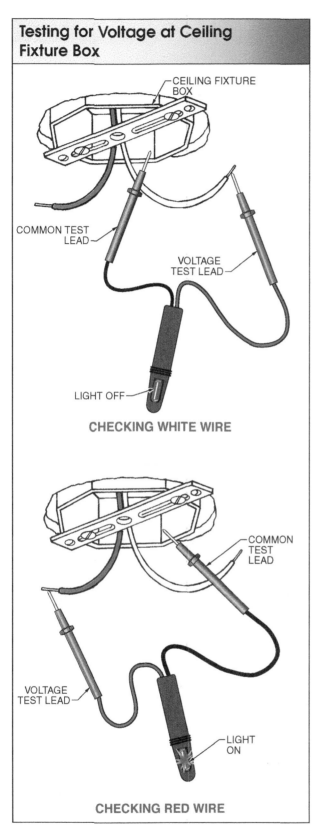

CHECKING WHITE WIRE

CHECKING RED WIRE

Figure 11-56. *The switch must be ON to a ceiling fixture box to test for voltage at the box.*

To test the grounding of receptacles, one lead (black) of the neon tester is held stationary on the ground terminal while the other lead is repositioned on each plug slot. If the receptacle is properly grounded, the neon light will only light when the test lead is placed in the hot slot. If the light does not glow with the test lead in either slot, the receptacle is not grounded.

Testing Circuit Breakers and Fuses

Circuit breakers can trip and fuses can blow, shutting off circuits in a residence. The ability to properly test circuit breakers and fuses is a must.

Circuit Breakers. A circuit breaker operates in much the same manner as a switch. A circuit breaker is either ON or OFF. To test a circuit breaker, the common test lead of a test instrument (voltage tester) is placed on the neutral bus bar and the positive test lead is placed on the screw terminal of the circuit breaker. **See Figure 11-58.** When the breaker is good, the voltage tester will indicate voltage when the breaker is in the ON position and will not indicate voltage when the breaker is in the "tripped" (OFF) position.

When a voltage tester indicates voltage in both circuit breaker positions, the breaker is shorted and must be replaced. When a voltage tester does not indicate voltage in either position, the circuit breaker is open and must be replaced.

Fuses. When a fuse is suspected of being defective, the fuse is checked with a multimeter set to measure resistance using the following procedure. **See Figure 11-59.**

1. Disconnect the power from the incoming lines to the fuses.
2. Remove the fuses from the fuse holder.
3. Measure the resistance of one of the removed fuses. The resistance of the fuse should be low (a few ohms).
4. Measure the resistance of the other removed fuse. The resistance of the fuse should be low (a few ohms) and within a measured ohm or two of the first measured fuse.
5. Reinstall tested fuses that are good or replace with new fuses.

Testing Receptacle Ground

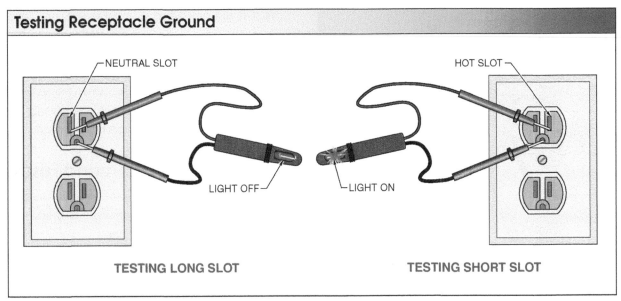

TESTING LONG SLOT   **TESTING SHORT SLOT**

Figure 11-57. *A properly wired and grounded receptacle should only turn on a neon light when the test light is connected from ground (arched hole) to the hot (short) slot.*

ADDING CIRCUIT PANELS (SUBPANELS)

Subpanels are added to existing service panels to obtain additional circuits. **See Figure 11-60.** The service panel must have breakers large enough to meet the demands of the subpanel—30 A, 50 A, or 60 A. The service panel circuit breakers protect the subpanel from overloading. Check with the specifications of the subpanel manufacturer to determine what size breakers to use in the service panel. Single-pole circuit breakers for residential installations are typically 15 A, 20 A, or 30 A.

An AC/DC voltage tester is different than an AC voltage tester as it includes a beeper or other indicator (solenoid that vibrates) when the tester is connected to AC voltage.

⚠ **CAUTION**

Do not replace fuses unless the fuses are de-energized and then only with the use of a fuse puller.

Testing Voltage at Circuit Breakers

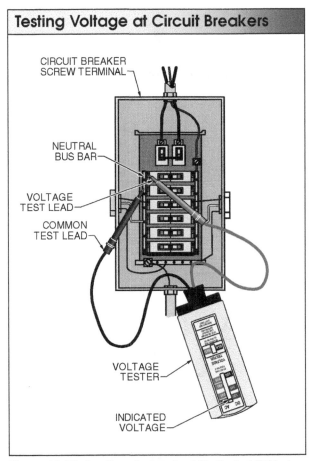

Figure 11-58. *A voltage tester should indicate 115 V when connected between the neutral bus bar and energized circuit breaker.*

Checking Fuses

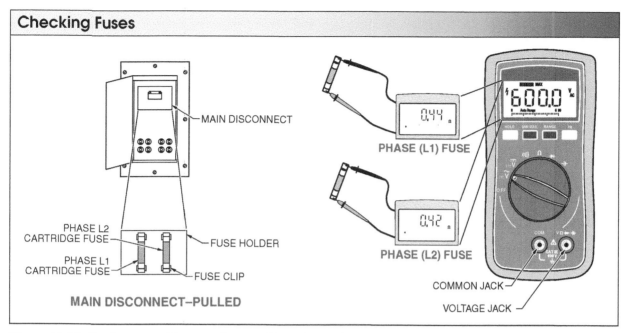

MAIN DISCONNECT

PHASE L2
CARTRIDGE FUSE

PHASE L1
CARTRIDGE FUSE

FUSE HOLDER

FUSE CLIP

MAIN DISCONNECT–PULLED

PHASE (L1) FUSE

PHASE (L2) FUSE

COMMON JACK

VOLTAGE JACK

Figure 11-59. *The fuses used in a service panel have low resistance and the resistance should be the same from fuse to fuse.*

Adding a Subpanel

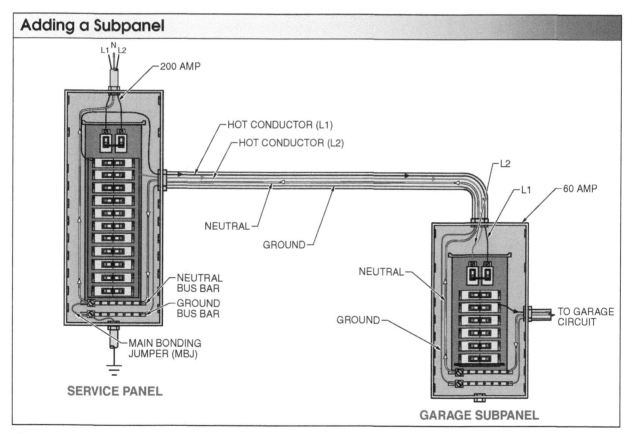

L1 N L2

200 AMP

HOT CONDUCTOR (L1)

HOT CONDUCTOR (L2)

NEUTRAL

GROUND

NEUTRAL
BUS BAR

GROUND
BUS BAR

MAIN BONDING
JUMPER (MBJ)

SERVICE PANEL

L2

L1

60 AMP

NEUTRAL

GROUND

TO GARAGE
CIRCUIT

GARAGE SUBPANEL

Figure 11-60. *Subpanels are added to residential electrical systems for garages and air conditioning condensers, or for additional circuit capacity.*

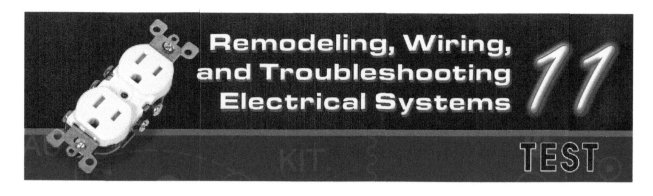

Name_____   Date _____

Remodeling, Wiring, and Troubleshooting Electrical Systems

_____ **1.** Studs and joists are typically laid out on ___″ centers.

       A. 16   **C.** 32

       B. 24   **D.** both A and B

_____ **2.** A(n) ___ knife can be used to cut small openings in drywall walls and ceilings.

_____ **3.** Metal ___ can be used to mount electrical boxes in existing walls.

_____ **4.** A(n) ___ tape is used to route cables through ceilings and walls.

    T   F   **5.** A soft glow on a neon tester indicates 230 V.

    T   F   **6.** When a light switch is operating properly and is in the ON position, voltage is present at the top terminal screw of the switch.

    T   F   **7.** A subpanel can be added to an existing service panel provided enough current is available.

    T   F   **8.** The hot wire of a grounded circuit is easier to check than the hot wire of an ungrounded circuit.

_____ **9.** A(n) ___ stud finder can be used to locate studs in existing walls.

_____ **10.** Folding ___ on each side of a box are used to keep the box from pulling back into the holes as the clamps are tightened.

_____ **11.** When voltage is present at the terminals but not at the receptacle slots, the receptacle is ___.

_____ **12.** To determine if a switch is good, ___ must reach and pass through the switch.

    T   F   **13.** Splices and terminal points in a circuit should be checked for breaks or loose connections.

T	F	**14.** A neon tester can be used to identify a hot wire.
T	F	**15.** To determine if a fuse is operating correctly, a multimeter set to measure resistance is used.
T	F	**16.** Fish tapes can be used to pull wires through back-to-back boxes.
T	F	**17.** Sawcuts should go completely through surface raceways when installing raceways around an outside corner.
_____		**18.** Receptacles in surface raceways should be at least ___″ from corners.
_____		**19.** A neon tester will illuminate when touched to a good circuit breaker and the ___ bar in the service panel.
_____		**20.** Horizontal ___ is used as a brace in stud walls.
_____		**21.** ___ boxes are typically very shallow.
T	F	**22.** Cables and wires can be routed around a door behind the trim.
T	F	**23.** Fish tape is easily used to fish wires through masonry walls.
T	F	**24.** A neon tester can be used to check splices for breaks or loose connections.
_____		**25.** When a switch is in the OFF position, voltage is present at the ___ terminal.

Exterior Wall Section

_____	**1.** Double top plate
_____	**2.** Backing
_____	**3.** Wall stud
_____	**4.** Gypsum board
_____	**5.** Frieze
_____	**6.** Rafter
_____	**7.** Sheathing
_____	**8.** Vent
_____	**9.** Plywood panel
_____	**10.** Roof joist

Testing Electrical Circuits

_____ 1. Voltage is reaching the switch.

_____ 2. The switch is good.

_____ 3. Testing for the hot wire in a grounded system.

_____ 4. Testing for the hot wire in an ungrounded system.

_____ 5. Voltage is not reaching the switch.

_____ 6. The switch is defective.

_____ 7. The receptacle is properly grounded.

_____ 8. The receptacle is not properly grounded.

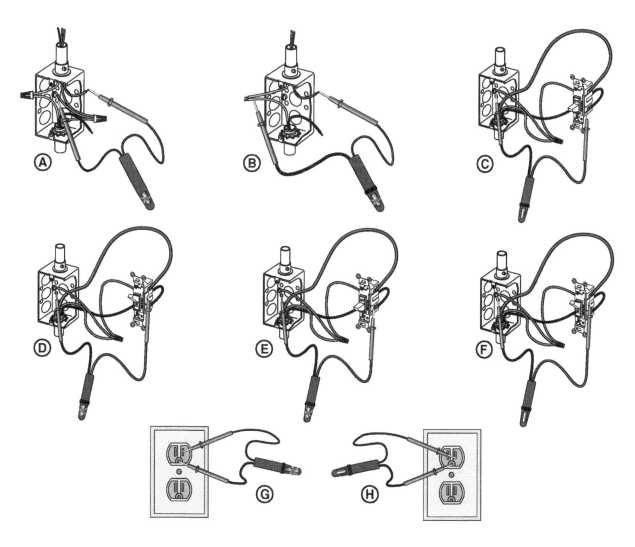

299

Residential Control Systems 12

Chapter 12 provides an overview of low-voltage residential electrical control systems and how the systems are installed and operated. Where appropriate, low voltage provides an economical and convenient way of signaling and controlling a variety of devices. Low voltage has become more popular because of the energy savings provided by low power consumption of the controller.

RESIDENTIAL CONTROL SYSTEMS

Residential control systems are considered low-voltage systems. Low-voltage systems in homes operate at less than 30 V. Typically, the range is between 12 V and 24 V. Doorbells, intercoms, security systems, and remote control wiring devices are examples of residential control (low-voltage) systems.

Low voltage is used with residential control systems because low voltage is safer and more economical than 115 V or 230 V. Safety is increased because the lower voltage reduces the hazard of electrical shock. Installation costs are reduced because low-voltage systems use smaller wires that

are less expensive. In addition, low-voltage systems are easy to install and require no elaborate protective measures. For additional information on the installation of low-voltage equipment, refer to NEC® Article 720, *Circuits and Equipment Operating at Less than 50 Volts.*

SIGNALING SYSTEMS

Signaling systems are used to notify the occupants of a visitor or of conditions existing in a specific room or area of a residence. Electromechanical systems include doorbell (bell/buzzer), chime, and radio/audio/intercom systems.

Doorbell Systems

The simplest low-voltage system typically found in a residence is a doorbell, or buzzer, system. A simple doorbell circuit consists of a step-down transformer, pushbutton switch, and a bell or buzzer. **See Figure 12-1.** A step-down transformer changes the 115 V to 12 V to safely power a doorbell or buzzer system.

Depressing a momentary contact pushbutton (PB) completes the circuit and allows current to flow through the doorbell as long as the pushbutton is held closed. Releasing the pushbutton opens the circuit.

Doorbell/Buzzer System

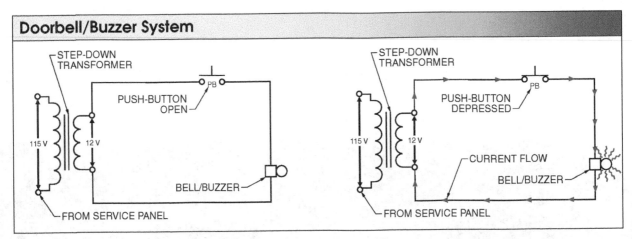

Figure 12-1. *A simple bell/buzzer circuit consists of a step-down transformer, a pushbutton, and a bell or buzzer.*

Multiple Location Activation. When a doorbell/buzzer is to be activated from two or more locations, a circuit with several pushbuttons is wired in parallel so that depressing any one of the pushbuttons activates the doorbell/buzzer. **See Figure 12-2.**

Doorbell/Buzzer System — Multiple Activation Locations

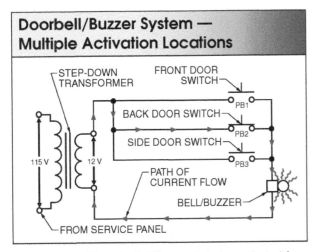

Figure 12-2. *When the bell/buzzer is to be activated from two or more locations, a circuit with several pushbuttons is wired in parallel so that depressing any one of them would activate the bell/buzzer.*

Return Call Systems. Return call systems provide a circuit for return signals to be generated, similar to an intercom system. **See Figure 12-3.** Depressing NO pushbutton 1 (PB1) completes a circuit, allowing current to flow through bell A and return to the transformer. Depressing NO pushbutton 2 (PB2) completes the opposite circuit, allowing current to flow through bell B and return to the transformer.

Return Call Systems

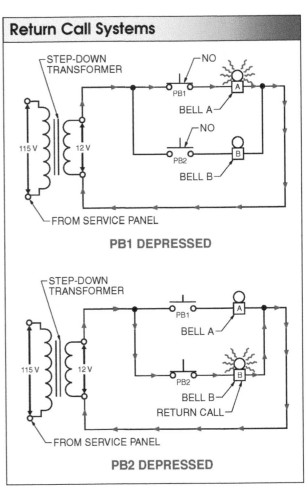

Figure 12-3. *Depressing pushbutton one (PB1) completes one circuit allowing current to flow through Bell A and return to the opposite side of the transformer. Depressing pushbutton two (PB2) completes the opposite circuit, allowing current to flow through Bell B and return to the opposite side of the transformer.*

The return call system can be simplified by the use of double-contact pushbuttons. Double-contact pushbuttons are arranged in series, rather than in parallel as with a single-contact pushbutton system. **See Figure 12-4.**

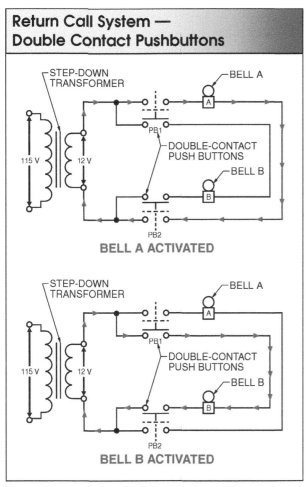

Figure 12-4. *The return call system can be simplified by the use of double-contact pushbuttons.*

Chime Systems

Chimes, although more expensive, have replaced doorbell/buzzer systems in many homes. **See Figure 12-5.** Chimes are available with simple musical notes or with very complex melodies. The complexity of the sound and the outward physical appearance dictate the price. The introduction of digital chimes allows the homeowner to program a wide range of sounds or melodies.

Chimes

Wilden Enterprises, Inc.

Figure 12-5. *Chimes, although more expensive, have replaced bell/buzzer systems in many homes.*

An intercom can be used with a doorbell/buzzer system and be connected to the same circuit.

Most magnetic chime systems have wiring similar to a doorbell circuit. **See Figure 12-6.** When pushbutton 1 (PB1) is depressed, chime A (front door) is activated. When pushbutton 2 (PB2) is depressed, chime B (back door) is activated. A common transformer is used to power both systems. Different tone sequences are used to differentiate between the front and back doors. For example, the front door chime may produce several tones while the back door chime produces only a single tone.

Magnetic Chime System

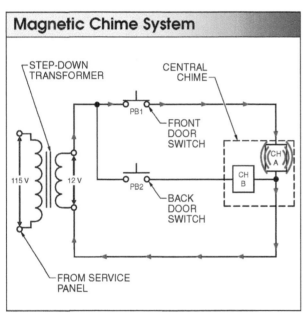

STEP-DOWN TRANSFORMER

CENTRAL CHIME

PB1

FRONT DOOR SWITCH

PB2

BACK DOOR SWITCH

CH A

CH B

115 V

12 V

FROM SERVICE PANEL

Figure 12-6. With a magnetic chime system, when pushbutton one (PB1) is depressed, the front door chime is activated. When pushbutton two (PB2) is depressed, the back door chime is activated.

To install a door chime, route a hot wire from each pushbutton (front and back door) and connect the hot wires to the chime. **See Figure 12-7.** Connect the chime common terminal to the transformer.

In multilevel homes, such as two-story homes or bi-levels, centrally located chimes can be impossible to hear. In multilevel homes, two chimes that are wired in parallel are typically used. **See Figure 12-8.** Both chime circuits are activated simultaneously, allowing the chimes to be heard anywhere in the house.

Wireless Doorbell/Chime Systems

When a doorbell or chime system malfunctions or fails, a wireless doorbell/chime can be used as a replacement. The doorbell/chime should be installed in a location where it can be heard most easily. The device must be plugged into a receptacle for power, however, the activation button (located at the door) is battery operated. The activation button sends a signal to the bell or chime when it is depressed and should be mounted directly on the house. **See Figure 12-9.**

There are advantages and disadvantages to using wireless doorbell/chime systems. The advantages of wireless doorbell/chime systems include ease of installation and no wires to route within the home. Disadvantages include battery replacement and limited installation locations (near receptacles).

Radio/Audio/Intercom

A radio/audio/intercom combines the functions of chimes, intercoms, and audio systems into one master unit. The master unit is used to control the various systems. **See Figure 12-10.** With radio/audio/intercom systems, a piece of equipment can be used to serve more than one function. Each manufacturer has several models and types ranging from economy to deluxe, including one or several of the following features:

- Room-to-room intercom that allows for calling any or all stations at one time from any station
- Audio monitoring of specific stations such as a child's bedroom
- Remote station privacy settings to shut off specific rooms
- Hands-free answering of calls
- Music muting or intercom override
- FM-AM radio
- CD and other audio inputs

Low-voltage wiring for doorbell circuits must be separate from power and lighting circuits and not occupy the same conduit or outlet boxes as these circuits unless a divider is provided to separate the systems.

Door Chime Connections

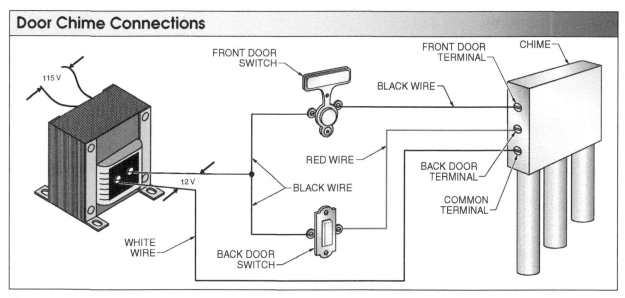

Figure 12-7. *To install a door chime, route two wires from each device and connect them at the chime.*

Multilevel Chime System

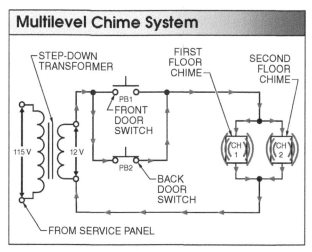

Figure 12-8. *In multilevel homes, two chimes wired in parallel are typically used. With this type of circuit both chimes are activated simultaneously, allowing them to be heard throughout the house.*

Wiring diagrams vary according to the manufacturer and should be followed closely when setting up a system. A properly installed radio/CD intercom can provide many hours of convenience and pleasure. A fire/security system may be part of most whole-house systems. Some individuals prefer stand-alone intercom and security systems because either the security or communication system still functions when the other has crashed.

Wireless Doorbell/Chime Systems

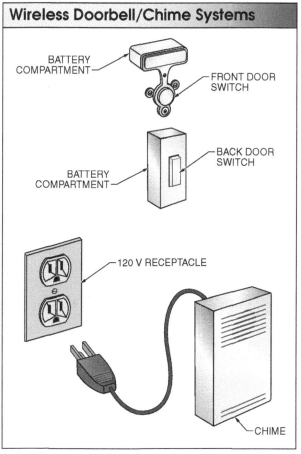

Figure 12-9. *A wireless doorbell or chime system can replace a hardwired system when the hardwired system fails.*

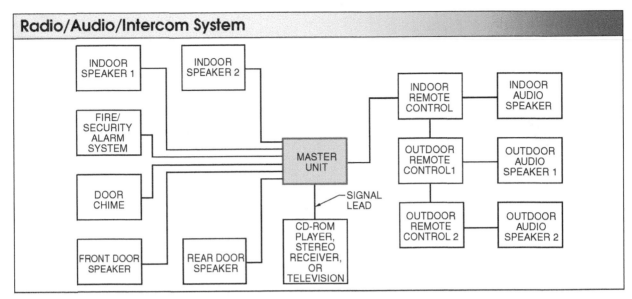

Radio/Audio/Intercom System

Figure 12-10. *A radio/audio/intercom combines the functions of chimes, intercoms, and audio systems into one master unit. The master unit is used to control the different systems.*

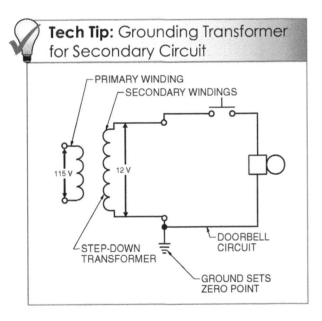

Tech Tip: Grounding Transformer for Secondary Circuit

SECURITY/FIRE SYSTEMS

The need for protection from intruders and the devastating effects of house fires have increased the demand for security/fire systems. A *security/fire system* is a home protection system composed of three elements: detectors, controls, and annunciators. **See Figure 12-11.** A *detector/sensor* is an electrical device used to detect noise, heat, gases, or movement. Detector/sensors are used to identify problems when the problem first exists and respond by opening or closing circuits to control devices. A *control* is an electronic device that accepts signals from detectors and turns annunciators ON and OFF. The control determines which detector is transmitting and uses that signal to trigger an annunciator. An annunciator is an electrical load that initiates an alarm. Annunciators respond to control signals by ringing a bell, flashing a light, or sounding a siren to alert individuals of danger. Many security/fire systems are equipped to automatically notify local authorities, such as fire and police departments, and activate sprinkler systems in the event of a fire.

Closed-Circuit Wiring (Security)

Security/fire systems operate with closed-circuit or open-circuit systems. Closed-circuit systems are typically used for the security portion of a security/ fire system. A closed-circuit system is used for security because the system is a series circuit, which remains continuous unless a detector is activated, or a line is cut, or power source is interrupted. **See Figure 12-12.** The detectors are in series with a normally closed relay. As long as the detectors remain in a closed position, the current to the detector circuit relay R1 will hold contacts C1 open in the circuit. If a detector circuit wire is cut or a detector is activated, the current to relay R1 stops because the series circuit is broken. Because relay R1 is de-energized, contacts C1 close and the response (bell) is activated. Although the circuit is quite simple, the circuit is difficult for an intruder to override. Modern closed-circuit systems are wireless.

Open-Circuit Wiring

Open-circuit wiring systems are typically used on the gas and smoke (fire) detection portions of a security/fire system. An open-circuit system consists of several open contacts wired in parallel with a normally open relay. An open-circuit system can be used on fire systems because the circuit is tamperproof. The principle behind an open-circuit system is that, as long as the sensors remain in an open position, no current will reach the detector circuit relay R1 and the annunciator circuit will remain inoperative.

See Figure 12-13. If any one of the sensors is activated, current passes through relay R1, and contacts C1 in the annunciator circuit close. With C1 closed, the bell, siren, or light turns on. The relay is used in this circuit so that the detectors can operate on low voltage while the annunciator operates on standard voltage.

Sensor Types

Sensors are the sensing portion in both fire and security systems. A sensor determines when a problem (smoke, fire, intruder) exists, and sends a signal to the controller indicating the problem. Sensor types include plunger button, magnetic, infrared, wireless control, heat, electronic-activated, and smoke/combustion gas detectors.

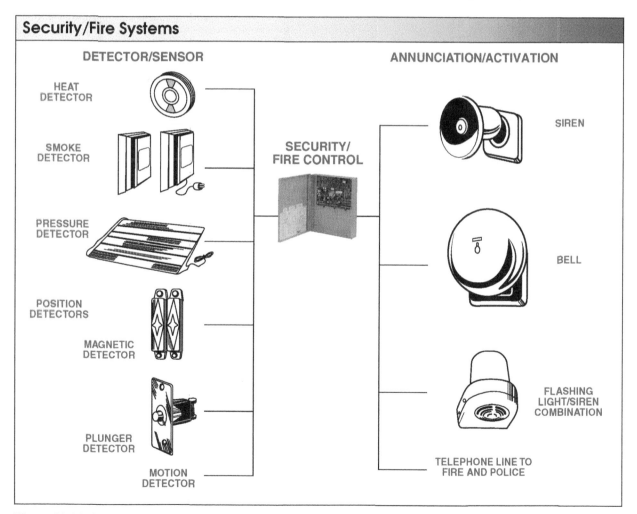

Security/Fire Systems

DETECTOR/SENSOR

ANNUNCIATION/ACTIVATION

HEAT DETECTOR

SMOKE DETECTOR

PRESSURE DETECTOR

POSITION DETECTORS

MAGNETIC DETECTOR

PLUNGER DETECTOR

MOTION DETECTOR

SECURITY/ FIRE CONTROL

SIREN

BELL

FLASHING LIGHT/SIREN COMBINATION

TELEPHONE LINE TO FIRE AND POLICE

Figure 12-11. *Security/fire systems are composed of three types of elements: detection, control, and annunciation.*

Closed Circuit Security System

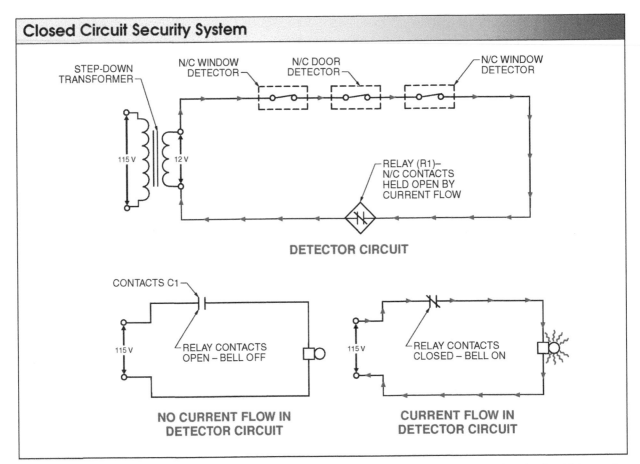

Figure 12-12. *A closed-circuit system is typically used with security systems because it is a series circuit, which remains continuous unless one of the detectors is activated, or the line is cut.*

Open Circuit Systems

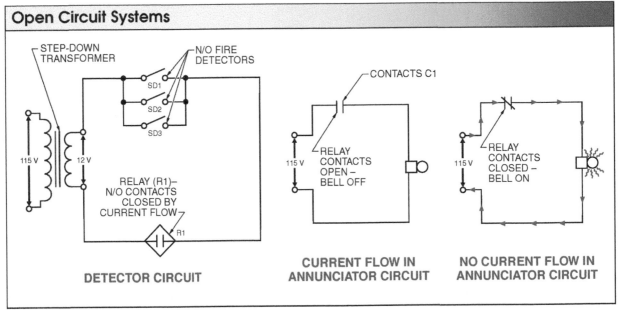

Figure 12-13. *Open-circuit wiring systems are typically used on the smoke (fire) detection portion of a security/fire system. An open circuit system consists of several open contacts wired in parallel with a normally open relay.*

Plunger Button Detectors. A *plunger button detector* is an electrical sensing device used on windows and doors to determine if the window or door is open or closed. **See Figure 12-14.** A plunger button detector is typically used in closed-circuit systems where a door or window holds the contacts of the detector in the closed position. Opening a door or window activates the detector. Plunger button detectors are easier to install on new construction than during remodeling.

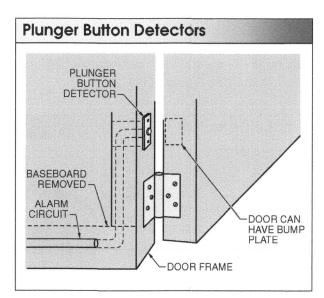

Plunger Button Detectors

Figure 12-14. *Plunger button detectors are typically used on windows and doors where they can be installed in a concealed location.*

Magnetic Detectors. A *two-piece magnetic detector* is an electrical sensing device used on windows and doors to determine if the window or door is open or closed. A magnetic detector is composed of a permanent magnet in one section and a sensor (switch) in the other section. As long as the two sections are close to each other, the switch remains closed. Movement of the door or window removes the switch from the magnet and the circuit is opened, activating a closed-circuit system. **See Figure 12-15.** The switch is permanently mounted with the magnet as the movable section. On existing construction, two-piece magnetic detectors are easier to mount than plunger button detectors.

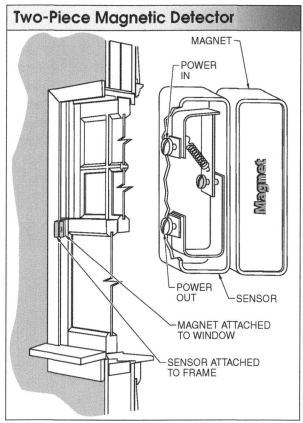

Two-Piece Magnetic Detector

Figure 12-15. *Movement of the door or window removes the switch from the magnet, activating the sensor and the circuit is opened, activating a closed circuit system.*

Infrared Detectors. An *infrared detector* is an electrical sensing device used for security purposes to determine if movement exists in a room or area. Infrared detectors are a combination transmitter-receiver recessed into the wall. The transmitter projects an invisible pulsating beam at a specially designed bounce-back reflector on the opposite wall. An interruption of the beam signals the receiver to activate the system alarm. **See Figure 12-16.** Because infrared is virtually invisible, infrared detectors have become more readily accepted than photoelectric detectors operating on a visible light beam.

Heat Detectors. A *heat detector* is an electrical sensing device used to detect excess heat levels within a structure. Heat detectors are often used in conjunction with smoke detectors. **See Figure 12-17.** Heat detectors warn of fire when the temperature in the area near the detector reaches a certain level. Heat detectors do not detect smoke. A heat detector is often used in areas such as kitchens and attics, where smoke detectors are not practical. Heat detectors are not recommended for use in bedrooms or sleeping areas. Heat detectors are typically classified as fixed temperature or rate of rise temperature detectors. A *fixed temperature detector* is a heat detector designed to respond when a room reaches a specific temperature. Once the specific temperature is reached, the contacts in the detector close and the alarm is activated.

Infrared Detectors

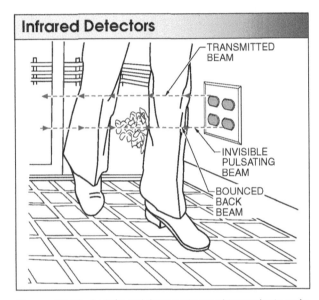

Figure 12-16. *An infrared detector transmitter projects an invisible pulsating beam at a specially designed bounce-back reflector on the opposite wall. An interruption of the beam signals the receiver to activate the system alarm.*

Heat Detectors

Nutone, Division of Scovill

Figure 12-17. *Heat detectors are used to detect excess heat levels within a structure and are often used in conjunction with smoke detectors.*

A *rate of rise detector* is a heat detector designed for flash fires or slow-burning fires. The elements of a rate of rise heat detector sense any rapid change in temperature, such as a 100°F temperature change in 30 seconds, to trigger an alarm. As added protection, most rate of rise units have fusible alloy elements which, when heated to a certain temperature, melt and cause the contacts of the detector to close.

Smoke Detectors. A *smoke detector* is an electrical device used to sense products of combustion (smoke) and activate an alarm when smoke is detected. Smoke detectors are required for home safety because smoke detectors provide an early warning to evacuate a home before a fire becomes out of control. **See Figure 12-18.**

Smoke Detectors

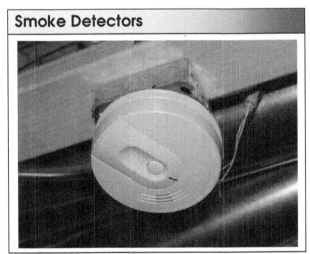

Figure 12-18. *Smoke detectors are essential to home safety because they provide the early warning necessary to help evacuate the home before the fire or amount of smoke is out of control.*

Smoke detectors use smoke or products of combustion to trigger the generation of a signal. Some units rely on the density of smoke to trigger an annunciator, while others are triggered by the presence of certain products of combustion. In either case, contacts are closed as with all detectors, and an alarm is sounded. Smoke detectors are available that operate on 115 VAC with a battery-power backup, while others operate only on battery power. A smoke detector that is installed in the center of a room will protect a 400 sq ft area.

National Fire Protection Association (NFPA) Standard 72 indicates the number and placement of smoke/heat detectors. **See Figure 12-19.** NFPA Standard 72 indicates that early-warning fire detection is best achieved by the installation of fire detection equipment in all rooms and areas of the house including the following:

- outside each separate sleeping area
- on each additional story of the family living unit
- basements
- living rooms
- dining rooms
- kitchens
- hallways
- attics
- furnace areas
- utility/storage rooms
- attached garages

When installing smoke detectors, it is important to follow local and national codes. For best performance, smoke detectors should be wired in parallel. **See Figure 12-20.** Size 14/2 wire is fed to the first detector, while size 14/3 wire is used for the connection of any additional units. The yellow lead from each smoke detector interconnects the entire system so that all detectors will sound when one is triggered. For maximum protection, it is best to hardwire smoke detectors with a back-up battery.

Smoke Detector Placement

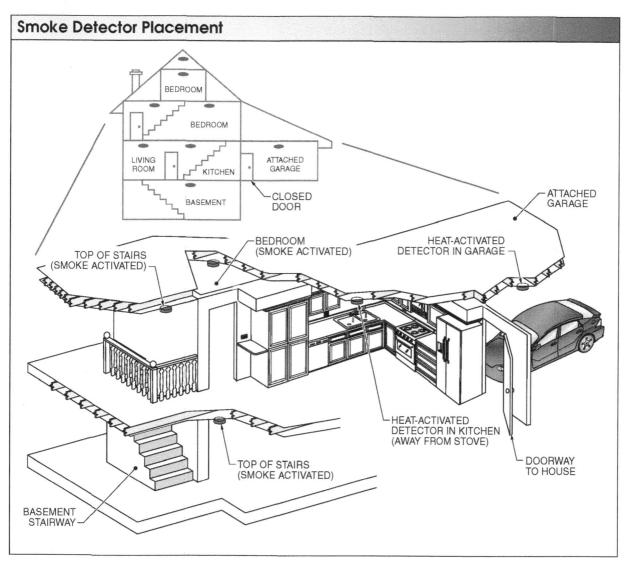

Figure 12-19. *The NFPA has several recommendations for the placement of smoke detectors in residential areas.*

Smoke Detector Installation

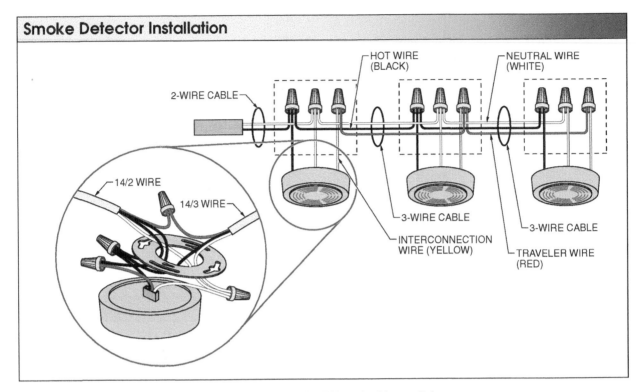

Figure 12-20. *For best performance, smoke detectors should be wired in parallel.*

A smoke detector system must be tested frequently to verify that it is in proper operating condition. A smoke detector system can be tested by depressing the test button located on each unit until the alarm sounds. Also, the alkaline batteries installed in smoke detectors should be replaced at least twice a year.

Wireless Control Detectors. A *wireless control detector* is a device used for security purposes in existing construction to eliminate hard wiring. Because running wires for long distances is impractical in an existing building, wireless control detectors are used. Each detector is attached to a transmitter that is capable of sending a signal to a radio receiver. When a detector senses a forced entry, the transmitted signal triggers the wireless receiver control unit, which sounds an alarm.

In addition to regular detectors, some wireless units provide a portable transmitter that can be hand-carried or placed on a piece of furniture.

Security/Fire Alarm System Controls

The control section of a security/fire alarm system receives signals from the detectors (sensors) and causes the appropriate annunciator to be activated. A control box is composed of a series of relays which open and close in response to the opening and closing of sensors. Each manufacturer labels terminals differently, but all manufacturers supply a complete set of step-by-step installation instructions. The choice of which unit to install typically depends on specific needs and cost considerations.

Annunciators

An *annunciator* is a signaling device that notifies individuals that one or more detectors have been activated. Annunciators, controls, and detectors are integrated into a system to fit specific needs. A complete security-alarm/fire-detection system includes various types of signaling devices (detectors) controlled by a master panel. Depending on the application, the system can be as simple or as complex as required. Manufacturers of annunciators and related equipment try to design systems that fit the requirements of the user. **See Figure 12-21.**

Security — Alarm/Fire System

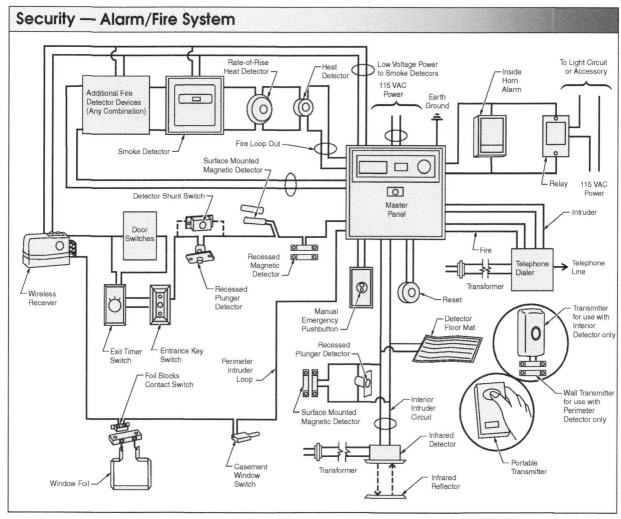

Figure 12-21. *Specific manufacturers of annunciators and related equipment can help design a system that fits the requirements of the user.*

REMOTE CONTROL WIRING

Remote control wiring is a method of controlling standard-voltage devices through the use of low-voltage relays and low-voltage switches. A basic remote control wiring system is used to control a light. A remote control system can be composed of a transformer, switch, and magnetic relay. **See Figure 12-22.** Remote control wiring systems were once popular because the transformer, switch, and magnetic relay are interconnected by low-voltage wire. The smaller the wire, the less the cost and the greater the ease of routing the wire. Due to wireless controls, these types of systems are becoming obsolete.

Remote Control Systems

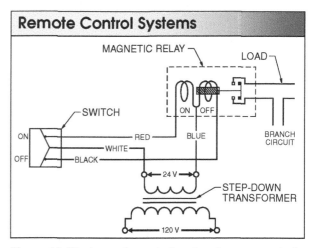

Figure 12-22. *A remote control system is composed of a transformer, switch, and magnetic relay.*

Relay Operation

The main component of the remote control wiring system is the relay. A remote control relay differs from those used in security-alarm/fire-detection systems. Remote control relays have an ON coil and an OFF coil with a common center plunger. **See Figure 12-23.** The use of two coils allows the movable plunger to be moved positively in either direction without the use of springs.

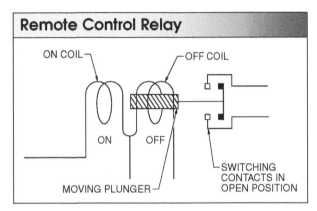

Figure 12-23. *Remote control relays have an ON coil and an OFF coil with a common center plunger.*

When switch SW1 is pressed, current will pass through the ON coil and cause the movable plunger to shift into the ON position. **See Figure 12-24.** As the movable plunger moves to the ON position, contacts C1 close and the load circuit (light) goes on.

The relay will move to and maintain the ON position even when switch SW1 is closed only momentarily. Other types of relays spring back from the closed position once power is cut off. When pushbutton PB1 is released, current will pass through the OFF coil, causing the movable plunger to shift to the OFF position. As the plunger moves, contacts C1 are opened and power to the load (light) is turned off. The relay maintains the OFF position until the switch energizes the ON coil again.

Relays (mounted to junction box knockouts) for low-voltage systems are capable of controlling 20 A of tungsten, fluorescent, or inductive loads at 115 VAC or 277 VAC. Low-voltage relays are also available for ½ HP 115 VAC and 1½ HP 230 VAC operation.

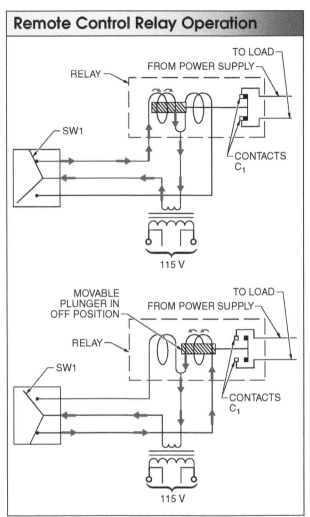

Figure 12-24. *The use of two coils allows the movable plunger to be moved positively in either direction without the use of springs.*

STRUCTURED WIRING SYSTEMS

Homeowners use structured wiring systems for various applications. The main purpose of a structured wiring system is to centralize the control of electrical and electronic devices and appliances in the home. Structured wiring systems use electrical and electronic control devices, such as stand-alone controllers, mini-controllers, and touch-panel controllers, to allow homeowners to control devices directly, indirectly, and automatically, based on need. **See Figure 12-25.**

Electrical and Electronic Control Devices

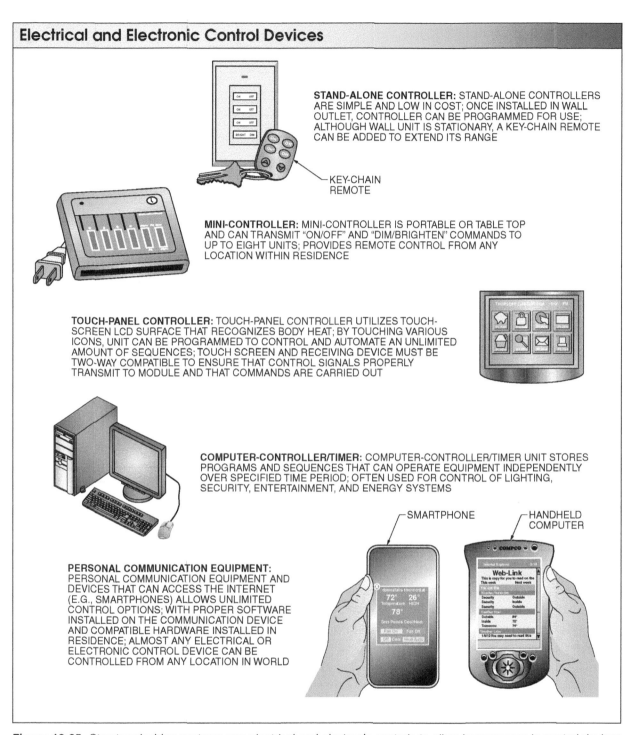

STAND-ALONE CONTROLLER: STAND-ALONE CONTROLLERS ARE SIMPLE AND LOW IN COST; ONCE INSTALLED IN WALL OUTLET, CONTROLLER CAN BE PROGRAMMED FOR USE; ALTHOUGH WALL UNIT IS STATIONARY, A KEY-CHAIN REMOTE CAN BE ADDED TO EXTEND ITS RANGE

KEY-CHAIN REMOTE

MINI-CONTROLLER: MINI-CONTROLLER IS PORTABLE OR TABLE TOP AND CAN TRANSMIT "ON/OFF" AND "DIM/BRIGHTEN" COMMANDS TO UP TO EIGHT UNITS; PROVIDES REMOTE CONTROL FROM ANY LOCATION WITHIN RESIDENCE

TOUCH-PANEL CONTROLLER: TOUCH-PANEL CONTROLLER UTILIZES TOUCH-SCREEN LCD SURFACE THAT RECOGNIZES BODY HEAT; BY TOUCHING VARIOUS ICONS, UNIT CAN BE PROGRAMMED TO CONTROL AND AUTOMATE AN UNLIMITED AMOUNT OF SEQUENCES; TOUCH SCREEN AND RECEIVING DEVICE MUST BE TWO-WAY COMPATIBLE TO ENSURE THAT CONTROL SIGNALS PROPERLY TRANSMIT TO MODULE AND THAT COMMANDS ARE CARRIED OUT

COMPUTER-CONTROLLER/TIMER: COMPUTER-CONTROLLER/TIMER UNIT STORES PROGRAMS AND SEQUENCES THAT CAN OPERATE EQUIPMENT INDEPENDENTLY OVER SPECIFIED TIME PERIOD; OFTEN USED FOR CONTROL OF LIGHTING, SECURITY, ENTERTAINMENT, AND ENERGY SYSTEMS

SMARTPHONE

HANDHELD COMPUTER

PERSONAL COMMUNICATION EQUIPMENT: PERSONAL COMMUNICATION EQUIPMENT AND DEVICES THAT CAN ACCESS THE INTERNET (E.G., SMARTPHONES) ALLOWS UNLIMITED CONTROL OPTIONS; WITH PROPER SOFTWARE INSTALLED ON THE COMMUNICATION DEVICE AND COMPATIBLE HARDWARE INSTALLED IN RESIDENCE; ALMOST ANY ELECTRICAL OR ELECTRONIC CONTROL DEVICE CAN BE CONTROLLED FROM ANY LOCATION IN WORLD

Figure 12-25. *Structured wiring systems use electrical and electronic controls to allow homeowners to control devices directly, indirectly, and automatically, based on need.*

For example, with a structured wiring system, homeowners can directly adjust the temperature thermostat when at home or set the thermostat to automatic control when away from home or asleep. Homeowners may even remotely adjust the thermostat with the proper software, control devices, and Internet access. Each type of control system has different advantages, disadvantages, and applications. **See Figure 12-26.**

Control System Comparisons

Technology Type	Advantages	Disadvantages	Applications
Remote technology	Speed and reliability	Costly to install	Video cameras and monitors
Hardwired	Reliability	Difficult to retrofit	Network systems
Infrared	Affordability	Line of sight limitations	TV/Consumer electronics
X-10	Whole-house system compatibility	Can require noise filters	Lighting, appliances, and security systems
Radio frequency (RF) wireless	Whole-house system compatibility	Short-term compatibility issues with existing software	Lighting, appliances, and security systems

Figure 12-26. *Each type of control system has different advantages, disadvantages, and applications.*

Structured wiring systems are used in homes and businesses for the routing of low-voltage electrical systems such as telephone, voice-data-video, and home security systems. Typical structured wiring applications use copper connectivity, but with recent technological advances, fiber optic connectivity is used more frequently to outfit a home or business. Fiber optic technology offers several advantages over copper systems such as a smaller space requirement, the ability to deliver more data, higher security, and lower attenuation rate.

Distribution Systems

Structured wiring is used to distribute a variety of audio, video, or data signals throughout a home or business. Examples are data lines, telephone lines, and audio/video information. **See Figure 12-27.** Structured wiring may also include security systems and computer networking. The challenge in creating an effective system is to create an infrastructure that meets current needs while establishing a system for the home or business of the future.

A distribution panel is the main part of a structured wiring system. Distribution panels are where the majority of cables are connected to distribute data and control signals to subsystem components throughout the home. A distribution panel is the common point of wire origination, distribution, and termination. Distribution panels contain a series of modules designed to distribute and control telephone,

data, and entertainment signals originating from within the home or from outside the home. Distribution panels are designed to allow changes to be made to the configuration of the low-voltage wiring network either by rearranging the appropriate patch cables or by adding additional modules.

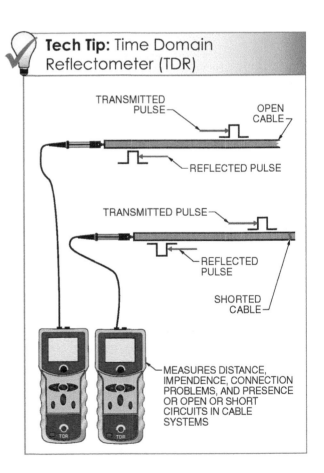

Tech Tip: Time Domain Reflectometer (TDR)

TRANSMITTED PULSE

OPEN CABLE

REFLECTED PULSE

TRANSMITTED PULSE

REFLECTED PULSE

SHORTED CABLE

MEASURES DISTANCE, IMPENDENCE, CONNECTION PROBLEMS, AND PRESENCE OR OPEN OR SHORT CIRCUITS IN CABLE SYSTEMS

Structured Wiring Systems

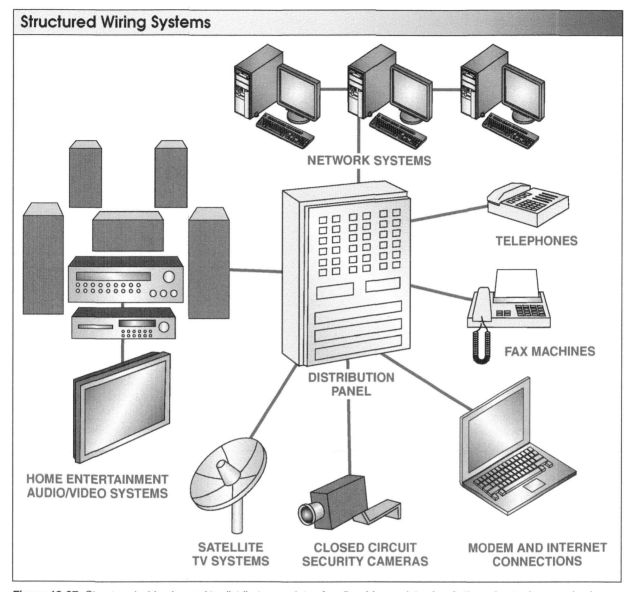

NETWORK SYSTEMS

TELEPHONES

FAX MACHINES

DISTRIBUTION PANEL

HOME ENTERTAINMENT AUDIO/VIDEO SYSTEMS

SATELLITE TV SYSTEMS

CLOSED CIRCUIT SECURITY CAMERAS

MODEM AND INTERNET CONNECTIONS

Figure 12-27. *Structured wiring is used to distribute a variety of audio, video or data signals throughout a home or business.*

Cabling Systems

Integral to the distribution system is the transmission cabling installed in each room. Cabling is the pipeline that delivers the signals and data throughout a home to specific locations. Typical transmission cable types include wire, communication cable, cable, and cord. **See Figure 12-28.** Communication cable includes unshielded twisted pair (UTP), coaxial cable, shielded twisted pair (STP), screened twisted pair (ScTP), and fiber optic cable. UTP, STP, and ScTP type cables are also identified as "copper cable." Copper cabling is more commonly used when routing a home network because of its low cost and ease of termination. Manufacturers provide a variety of cabling options based on the need to transfer power signals or data.

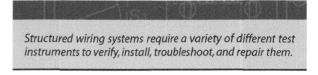

Structured wiring systems require a variety of different test instruments to verify, install, troubleshoot, and repair them.

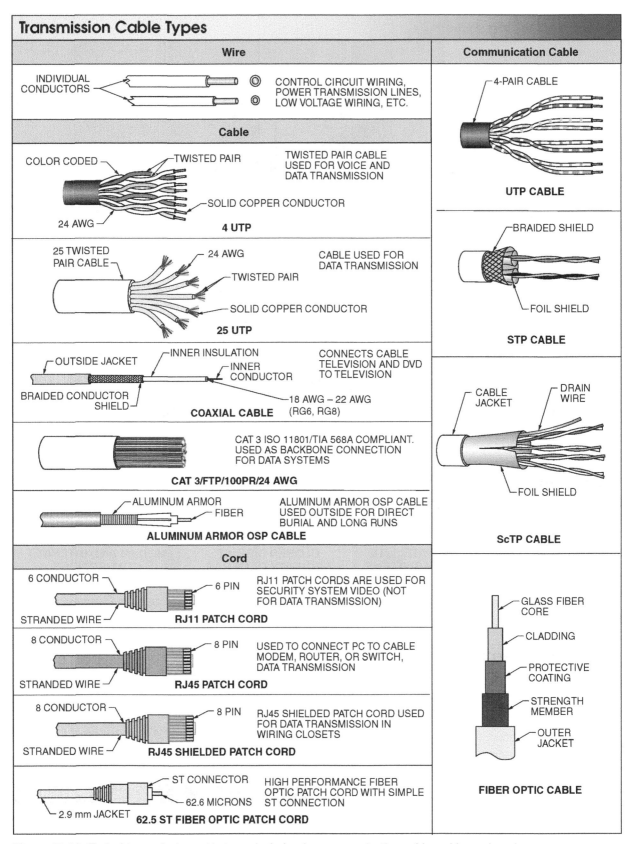

Figure 12-28. *Typical transmission cable types include wire, communication cable, cable, and cord.*

Cabling can be very simple or very complex, depending on the application. To make installation easier, cabling can be purchased packaged in bundles to accommodate many of the common uses required in a specific location. **See Figure 12-29.** Manufactured bundling makes it easier to pull cables through walls when compared to routing six separate cables.

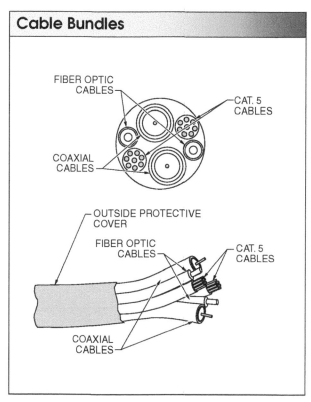

Figure 12-29. *To make installation easier, cabling can be purchased packaged in bundles to accommodate many of the common uses required in a specific location.*

Understanding the terminology associated with communication cables is required when installing the cables in a home. For example, copper communication cables are classified into categories designated Category 1 through Category 8. **See Figure 12-30.** A typical analog telephone line in an older existing structure uses Category 1 cabling. Category 5 and Category 5e copper cabling and above is recommended for all new construction.

Cable ratings are based on the final application and the transmission capacity of the cable. The greater a cable's transmission capacity, the greater the cost. **See Figure 12-31.**

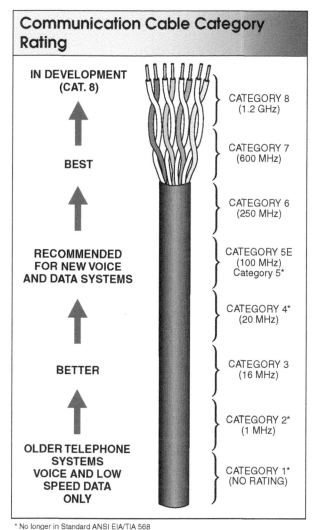

* No longer in Standard ANSI EIA/TIA 568

Figure 12-30. *Copper communication cables are classified from Category 1 through Category 8.*

FIBER OPTIC CABLE

As the need for data transfer and streaming video has increased, use of fiber optic technology for telecommunications cabling systems has increased. Currently, most cable bundles include at least one fiber optic cable as part of the bundle. Fiber optic cables must be protected from physical damage and moisture. To protect the glass fibers, protective cladding and coatings are used along with strength members. **See Figure 12-32.** *Cladding* is a layer of glass or other transparent material surrounding the fiber core of fiber optic cable that causes light to be reflected back to the central core for the purpose of maintaining signal strength over long distances.

Cable Ratings and Applications

Cable Rating	Typical Applications
Category 1 (CAT 1)	Older construction, used with Plain Old Telephone Service (POTS) and with frequencies less than 1 MHz
Category 2 (CAT 2)	Indoor wire and cable, voice and data transmission of up to 4 MHz
Category 3 (CAT 3)	Indoor wire and cable, voice and data transmission of up to 16 MHz
Category 4 (CAT 4)	Indoor wire and cable, voice and data transmission of up to 20 MHz
Category 5 (CAT 5)	Indoor wire and cable, voice and data transmission of up to 100 MHz
Category 5e (CAT 5e)	Indoor wire and cable, voice and data transmission of up to 100 MHz Used with longer cable runs because wire is packaged tighter.
Category 6 (CAT 6)	Indoor wire and cable, voice and data transmission of up to 250 MHz
Category 7 (CAT 7)	Indoor wire and cable, voice and data transmission of up to 600 MHz
Category 8 (CAT 8)	Indoor wire and cable, voice and data transmission of up to 1.2 GHz

Figure 12-31. Cable ratings are based on the final applications and the cost of the cable. The greater the transmission capacity of the cable, the greater the cost of the cable.

Fiber Optic Cable

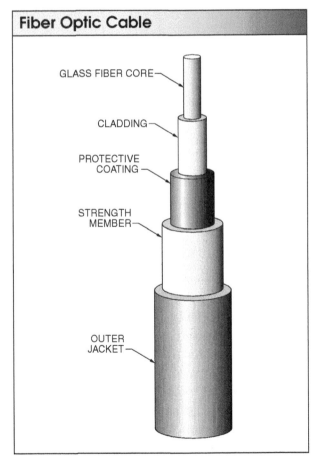

GLASS FIBER CORE

CLADDING

PROTECTIVE COATING

STRENGTH MEMBER

OUTER JACKET

Figure 12-32. A standard fiber optic cable is composed of a glass fiber core, cladding, protective coating, strength member, and an outer jacket.

Fiber Optic Operating Principles

Fiber optic technology is typically used as a transmission link. As a transmission link, fiber optics connect two electronic circuits, a transmitter and a receiver. Fiber optics uses a thin flexible glass or plastic optical fiber to transmit light. **See Figure 12-33.**

Fiber Optic Signal Transmission

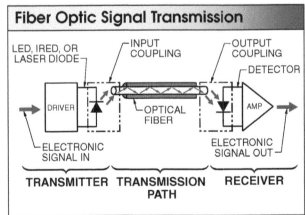

LED, IRED, OR LASER DIODE

INPUT COUPLING

OUTPUT COUPLING

DETECTOR

DRIVER

OPTICAL FIBER

AMP

ELECTRONIC SIGNAL IN

ELECTRONIC SIGNAL OUT

TRANSMITTER **TRANSMISSION PATH** **RECEIVER**

Figure 12-33. Fiber optics uses a thin flexible glass or plastic optical fiber to transmit light.

Fiber optic cable is classified as single-mode (yellow covering) or multimode (orange covering).

The central part of the transmitter is the light source. The source consists of a light emitting diode (LED), infrared emitting diode (IRED), or laser diode that changes electrical signals into light signals. A receiver typically contains a photodiode that converts light back into electrical signals. The output circuit of a receiver also amplifies the electric signal to produce the desired results, such as voice transmission or video signals. Advantages of fiber optic cables over copper cables include large bandwidth, low cost, low power consumption, low signal loss (attenuation), shielding from electromagnetic interference (EMI), small size, lightweight construction, and security (the inability to be tapped by unauthorized users).

Light Source

The light source feeding a fiber optic cable must be properly matched to the light-activated device for a fiber optic cable to operate effectively. A light source must also be of sufficient intensity to drive the light-activated device.

Laser Diodes. A *laser diode* is a fiber optic light source (similar to a light emitting diode) with an optical cavity that is used to produce laser light (coherent light). **See Figure 12-34.** An optical cavity is formed by coating opposite sides of a chip to create two highly reflective surfaces.

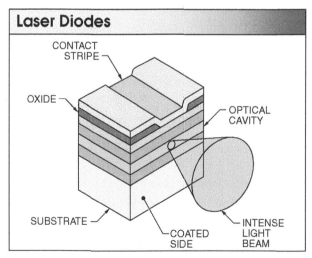

Figure 12-34. *A laser diode is a diode similar to a light emitting diode (LED) but with an optical cavity, which is required for laser (emitting coherent light) production.*

Fiber Connectors

The ideal interconnection of one fiber to another is an interconnection that has two fibers that are optically and physically identical. A *fiber connector* is a device that splices two fiber cables by squarely aligning the center axes of both cables and holding the two fiber cables together. The joining of the fibers is completed with an index-matching gel so that the interface between the cables has no influence on light propagation. A perfect connection is limited by variations in fibers and the need for high tolerances in the connector or splice.

Fiber-Connecting Hardware

Splices and fiber interconnections are often a negative factor due to alignment problems that can arise. Alignment problems can be eliminated through proper installation of fiber splices, connectors, and couplers. Recent advances in fiber optic technology allow for improved termination of connectors to cables through the use of crimp-style, prepolished connectors. Formerly, fiber connectors had to be factory polished, or hand polished in the field, resulting in lower quality connections. Poor connections result in signal loss. Common causes of signal loss with fiber optic cable include axial and angular misalignment, excessive end separation, and rough (unpolished) ends. **See Figure 12-35.**

Applications for Fiber Optic Cables

Because of the increased data transfer capabilities of fiber optic cable over other types of cabling, fiber optic cable is used extensively in high-speed computer networking, voice and data transmission, and security camera systems. As methods for installing fiber optic cable are improved and new devices are created, technologies such as digital satellite television, cable television, Internet, and telephone communications are improved by adding higher speed data transfer at a lower cost.

Wall Plates and Inserts

The type of wall plate and insert are chosen to suit the room of the house. The type of wall plate chosen must accommodate the type of service routed to a particular box. Wall plates are designed to accommodate different cable jacks (inserts) for single or multiple voice, data, and video connections. **See Figure 12-36.** Wall plates are decorated to lend particular designs to a room. The two types of wall-mounted service jack arrangements used to connect equipment in each room are interactive outlets and service outlets. An *interactive jack* is a connecting device such as a computer Ethernet connection that allows signals to be sent and received. A *service jack* (outlet) is a connecting device that provides connection to a service such as television reception through a coaxial cable.

Wallplates and Cable Jacks

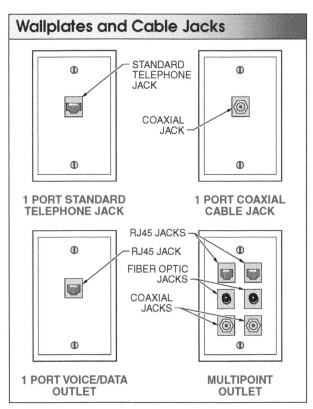

Figure 12-36. *Wall plates are designed to accommodate different cable jacks (inserts) for single or multiple voice, data, and video connections.*

STRUCTURED CABLING SYSTEM APPLICATIONS

Structured cabling system applications include computers and related networks, telephone systems, home entertainment systems, home security systems, and video monitoring systems. Systems can be routed so that moves, adds, or changes to the system can be performed with little or no difficulty.

Computer Network Systems

Personal computers (PCs), scanners, and printers are a sizable investment for a home. With a structured cabling system, multiple PCs can share a single Internet connection. Also, network system files that are located on any network PC's hard drive can be shared. Network systems allow for sharing devices such as printers and scanners from any location within the network. **See Figure 12-37.** With a network system, multiple users work on various projects simultaneously.

Causes of Signal Loss

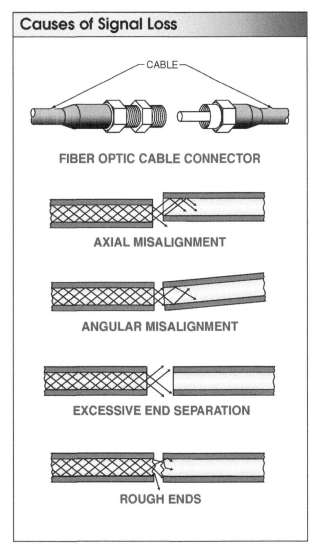

Figure 12-35. *Common causes of signal loss with fiber optic cable include axial and angular misalignment, excessive end separation, and rough (unpolished) ends.*

Computer Network Systems

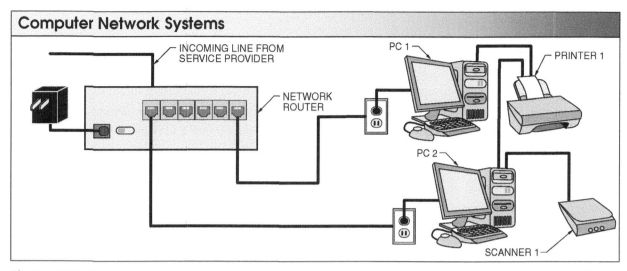

Figure 12-37. *Network systems allow for sharing devices such as printers and scanners from any location within the network.*

Home Entertainment

As with computer systems, a home entertainment system is a sizable investment. A home that has a structured cabling system can transmit audio and video signals to multiple rooms and locations throughout the home. **See Figure 12-38.**

Telephone System

Telephone systems can be hardwired, wireless, or part of a structured cabling system. The main advantage of a hardwired system is that a hardwired system has a separate power source and continues to operate during a power failure without a backup generator. Wireless systems allow roaming throughout a home or within 200′ of the telephone base unit without being tied to a permanent location. Structured cabling offers the greatest ability to manage and change the distribution of telephone lines to wall plate outlet locations in various rooms throughout a house.

While a digital video disc (DVD) is the same diameter as a compact disc read-only memory (CD-ROM), a DVD disc has smaller physical pits beneath the disc surface, allowing it to store more information than an ordinary CD-ROM.

Home Entertainment System

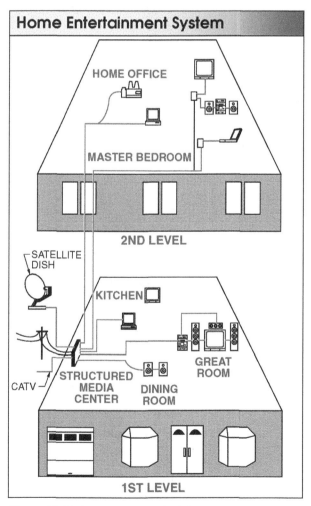

Figure 12-38. *With a structured cabling system, a home entertainment system audio and video signals can be routed throughout the home into a variety of rooms and locations.*

A structured cabling system allows control of which telephone lines will be dedicated for voice and which will be dedicated for data, such as for a computer. By accessing the distribution panel, cables can be moved, added, or changed to redirect telephone lines.

SECURITY SYSTEMS

Along with home entertainment, home security has become a requirement in home construction and re-modeling. The most common accessory added for home security through a structured cabling system is a video monitoring system.

Video Monitoring

Indoor and outdoor video monitoring allows complete visual monitoring anywhere in or near a house by the use of video cameras. **See Figure 12-39.** Video monitoring systems allow areas inside and outside of a home to be viewed from any television or monitor in the house. For example, to monitor a child's bedroom, an indoor camera can be installed in the bedroom. Typical video monitoring camera types include wall mount, dome (ceiling) mount, and standard mount. **See Figure 12-40.**

Through the use of a modulator, a preset channel can be selected with a remote control that will display live pictures from any room on any television or monitor in the home tuned to the preset channel. By adding a video sequencer to the distribution panel, any room can be watched by viewing pictures from each installed camera at regular intervals.

STRUCTURED CABLING ROUGH-IN

When preparing a home for a structured cabling system, proper rough-in techniques must be used. The structured cabling system should be designed so as to locate the distribution panel in an accessible location so that cabling, patch cords, and other equipment can be easily installed or changed. The type of wall material (drywall or plaster) must be taken into consideration when roughing in a structured cabling system.

Originally designed to monitor safety concerns, video security is continually evolving to serve many business and home uses. Businesses use video security systems to monitor employees, to trigger alerts in case of fire or motion, and to watch over entrances to buildings.

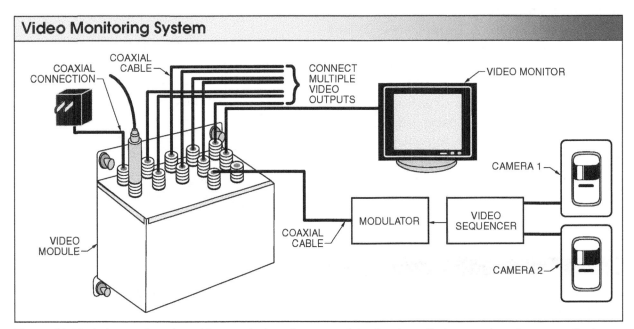

Video Monitoring System

Figure 12-39. *Indoor and outdoor video monitoring allows complete visual monitoring anywhere in or near the house through installation of video cameras.*

Video Monitoring Camera Types

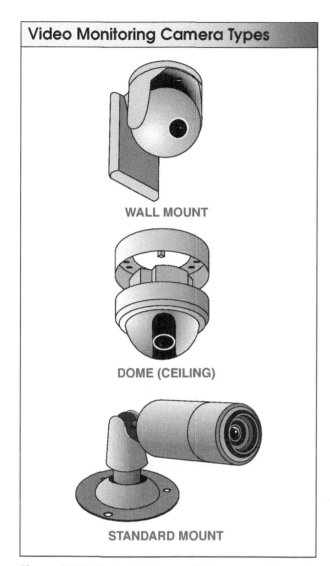

WALL MOUNT

DOME (CEILING)

STANDARD MOUNT

Figure 12-40. *Typical video monitoring camera types include wall mount, dome (ceiling) mount, and standard mount.*

Locating the Distribution Panel

The rough-in process refers to work required to install a structured cabling system prior to drywall installation. Although the first step in the rough-in process is to mount the structured cabling distribution panel, the process can vary according to the equipment manufacturer. Refer to the specification manual for the equipment when installing any distribution panel. The location of the panel determines the installation procedures.

Running Cables

Cables and cable bundles are typically routed through rafters and stud walls. **See Figure 12-41.** The drilling of holes and the routing of cables and bundles are accomplished with typical remodeling methods. Routing structured cabling requires that care be taken to avoid damaging the cable and to minimize electrical interference. Steel nailing plates are required to cover areas where a cable has been routed through a notch or hole in a stud. **See Figure 12-42.** When hole openings are less than 1¼″ from the edge of the stud, the area must be protected by a steel plate to prevent damage from drywall fasteners. To avoid electrical noise, a structured cabling box should not be placed near a power box or share drilled stud and rafter holes with power feed conductors or other conductors. Power feed cables and structured wiring cables should only cross each other at 90° angles.

Routing Cables

Figure 12-41. *Cables and cable bundles are routed through rafters and stud walls.*

Cable Routing Requirements

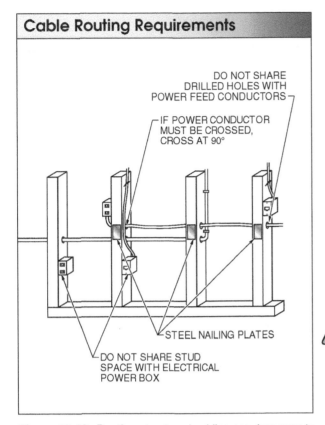

DO NOT SHARE DRILLED HOLES WITH POWER FEED CONDUCTORS

IF POWER CONDUCTOR MUST BE CROSSED, CROSS AT 90°

STEEL NAILING PLATES

DO NOT SHARE STUD SPACE WITH ELECTRICAL POWER BOX

Figure 12-42. *Routing structured cabling requires care to avoid damage to the cable and to minimize electrical interference.*

When a voice-data-video (VDV) cable such as a CAT 5E copper cable or single-mode fiber optic cable is installed and certified, continued cable testing must be part of a preventative maintenance program. As loads are added to a system and new cables installed, cables that appear to be working properly can be damaged to the point that the cable is unusable. Problems never occur at a time when lost data is not a problem. Routine testing of a VDV cable will indicate signal losses (attenuation) and other potential problems.

Securing Cables

Cables are secured in place by several different methods. Staple guns are a quick method of securing cables. Staples must be of sufficient size as to not pinch or penetrate the cable. The same precautions must be followed when hammering staples manually, as with a staple gun. For multiple cables and some cable bundles, J-hooks or similar devices

are used to support the cables. Multiple low-voltage cables can be bundled together through the use of cable ties or a cable tie gun.

A *cable tie gun* is a handheld device that is used to hold and tighten cables together with plastic or steel ties. **See Figure 12-43.** Bundles of low-voltage cable can be quickly tied together in multiple locations by squeezing the trigger handle of the cable tie gun at different points along the wires to be bundled. Cable tie guns automatically tension and cut off the excess tie material. Cable tie guns are designed to reduce fatigue and increase the productivity of the individual. Cable ties should only be used on low-voltage telecommunications cable.

⚠️ **WARNING**

Never bundle standard power conductors together as the heat from the current flow could damage insulation and possible cause a fire or electric shock.

Cable Tie Guns

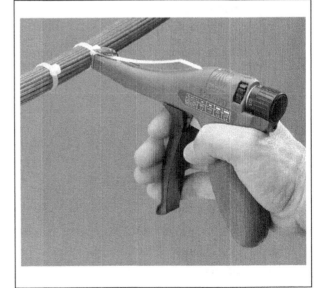

Panduit Corp.

Figure 12-43. *Multiple low-voltage cables may be quickly and efficiently bundled together through the use of cable ties and a cable tie gun.*

Over 50% of signal failure and transmission loss is caused by cable being installed incorrectly. Cable not properly installed changes the impedance or the ability of the cable to carry the signal at the required operating speed. Common installation problems include cables bent tighter than the minimum bend radius, cables overstretched through a cable support or overstretched throughout a run, staples driven too tightly, over tightened cable ties, and twists or knots in the cable. **See Figure 12-44.**

Plaster/Drywall Rings

Cables are typically terminated to single-gang plaster or drywall rings. Single-gang plaster or drywall rings are mounted either horizontally or vertically, depending on the location and application. Typically, at least 12″ of cable should extend beyond the ring for termination purposes. **See Figure 12-45.** Refer to individual manufacturer's guides as to the tools and techniques necessary for final terminations to outlet plates.

X10 TECHNOLOGY

X10 technology is a communications protocol that allows AC control devices to "talk" on existing household electrical wiring. X10 technology communications uses a low-level signal intermixed with current flow to transmit instructions to turn appliances or lights ON or OFF as programmed.

For example, if two dimmer switches are plugged into different receptacles, X10 technology uses the wires that connect all of the house receptacles to the service panel as a circuit to communicate. X10 uses a language that allows compatible X10 devices to talk to each other using the existing wiring in the home. X10 transmitting devices send a coded low-voltage signal that is superimposed on the 115 VAC current. An X10 receiver device connected to the house electrical system receives the coded signals. The receiver may respond to 256 different addresses that are programmed into the device. When more than one device must respond to the same signal (address), all affected devices are programmed with the same address.

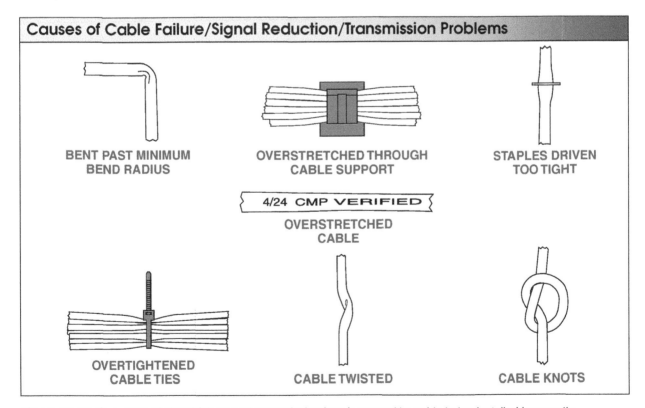

Figure 12-44. *Over 50% of signal failure and transmission loss is caused by cable being installed incorrectly.*

Plaster/Drywall Rings

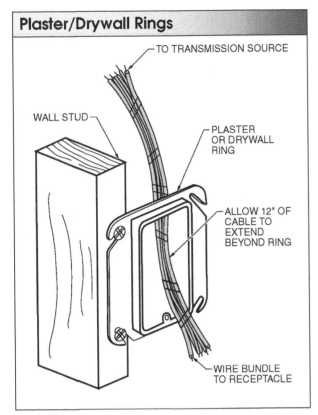

TO TRANSMISSION SOURCE

WALL STUD

PLASTER OR DRYWALL RING

ALLOW 12" OF CABLE TO EXTEND BEYOND RING

WIRE BUNDLE TO RECEPTACLE

Figure 12-45. *Cables are typically terminated to single-gang plaster or drywall rings and at least 12" of cable should extend past the plaster or drywall ring.*

This communication protocol originally was referred to as *power line carrier* (PLC) but is now commonly referred to as X10 communication. Many different companies manufacture X10-compatible switches, receptacles, and control devices.

X10 Programming

Typically, X10 programming consists of 16 unique house codes and 16 unique unit codes. The X10 house and unit codes are always used together (house codes are A–P and unit codes are 1–16) to send status or control instructions to various devices within a home such as A-1, A-15, and B-2. The instruction or status is encoded and transmitted on the power line in an order that is predefined. Because there are 16 house codes and 16 unit codes, up to 256 separate electrical devices can be connected to a network system to communicate within the home. Any instruction to be transmitted is first converted into a binary code at the transmitting device and then imposed on the power line in the form of a low-voltage signal. The binary coding is required because each X10 device receives every X10 command and the coded address lets the specific device respond accordingly.

X10 Applications

X10 applications include lamp/lighting control, dimmer switch control, control of 3-way switches, control of energy-saving devices, motion detectors for security purposes, and leak detectors to help quickly locate water leaks and prevent structural damage.

X10 Lamp Control. The most common X10 modules are lighting control devices. Lighting control devices simply plug into a standard 115 VAC receptacle, between the lamp and the receptacle. X10 lighting control devices are used to control incandescent bulbs used in typical table lamps unless wired into a home during construction. **See Figure 12-46.**

X10-Compatible Dimmer Switch Control. An X10 transmitter controls X10 dimmer switches directly at the dimmer switch or remotely. An X10 dimmer switch can be programmed to turn lights ON to a level of brightness ranging from 3% to 100% of maximum output. Once the program is set, there is no need to readjust the program each time the program is used. The unit has full manual control to make changes as needed. Switches with dimmer functions must be used only with incandescent lights. Fluorescent lighting systems cannot use a dimmer switch.

X10 3-Way Switches. When a light must be controlled from two or more locations, a 3-way ready switch and additional multi-way companion switch are used for communication purposes. A 3-way switch can be manually or electronically switched. When a 3-way switch is manually activated, the X10 controller has no way of taking commands and will not operate properly. By using an X10 switch that communicates with the controller, the controller will always be able to read whether the switch is "ON" or "OFF."

X10 Lamp Control

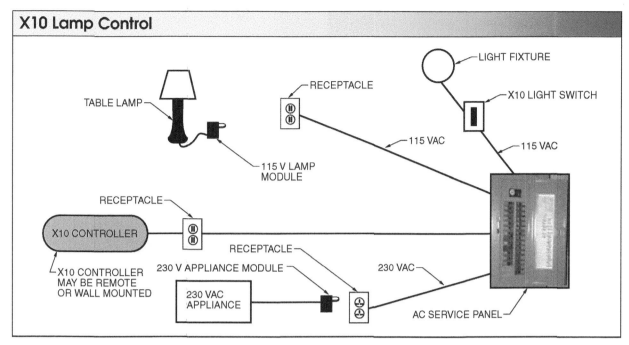

Figure 12-46. *X10 lighting control devices are used to control incandescent bulbs usually used in table lamps.*

Neutral Wire. Some switches require a connector that includes a neutral wire. When a neutral wire is not present in a switch box, a switch must be used that does not require a neutral. Another, more costly, option is to have an electrician install a neutral wire to the switch location.

X10 Energy-Saving Devices. As the cost of energy continues to climb, various energy cost-cutting methods are used. One of the most common methods of cutting energy costs is to install a programmable thermostat. By using a programmable thermostat, changes in the heating cycle of a dwelling can be made based on the time of day and the day of the week. Other methods for cutting energy costs include controlling natural gas usage, controlling appliances, and motorizing window coverings. X10 technology can control any X10 devices from any location with telephone or computer access. With the proper interface, programs can be accessed through the Internet to turn on or off appliances such as hot water heaters, air conditioners, or furnaces. Window coverings can be programmed to open or close, depending upon the angle of the sun or time of day.

Units are available for up to 7800 W and must be hardwired between the overcurrent protection (service panel) and the appliance. These types of units require a 230 VAC device rated for that appliance as well as special heavy-duty switches because they require a large amount of electrical current to operate.

X10 Motion-Activated Floodlights. Motion-activated floodlights not only turn ON when motion is detected but can also be used to turn lights ON inside a home. Because motion-activated floodlights can interact with other X10 controlled devices, motion-activated floodlights can be used to turn ON bedroom lights, kitchen lights, or activate an alarm. X10 motion sensor units are adjustable to activate from 6 sec to 30 min after the last motion was detected.

Water Damage Protection. Using a wireless leak detector that automatically shuts OFF the water source when a leak is detected from an appliance or plumbing fixture prevents water damage. By placing water sensors in an area where leaks can occur, any sign of moisture sends a wireless signal to the shutoff valve that immediately shuts OFF the main water line before any damage occurs. **See Figure 12-47.**

Wireless Leak Detectors

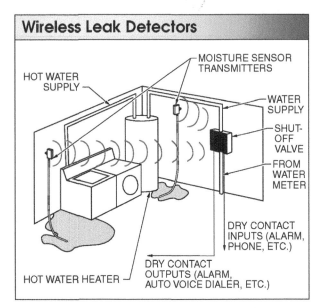

Figure 12-47. *By placing water sensors in an area where leaks may occur, any sign of moisture sends a wireless signal to the shut-off valve that immediately shuts OFF the main water line before any damage occurs.*

Motion Detectors. Motion detectors are used indoors and outdoors to monitor the possibility of an intruder. Indoor motion detectors are typically used in combination with other devices such as foils and magnetic switches to provide a more comprehensive security system.

The three types of motion detectors include passive infrared (PIR), microwave, and combination (PIR/microwave devices). When used outdoors, PIR devices detect large objects moving in the area by detecting the heat given off by the object. A PIR device is combined with a light source that is turned ON when motion is detected. This type of sensor is typically designed to eliminate triggering by small animals or wind.

Microwave devices can be installed either indoors or outdoors. Indoor motion sensors are typically mounted to the ceiling or angled down from the top of a wall to scan a particular area. Outdoor sensors are generally located near the doors and windows of a building. Combination devices use passive infrared and microwave technology to detect motion in a given area.

X10 Controlled Receptacles. X10 receptacles are installed to control lights and appliances from remote locations. **See Figure 12-48.** The X10 receptacle control unit fits into normal receptacle boxes and can replace an existing receptacle. X10 receptacles can be programmed to have one outlet controlled by an X10 controller, while keeping the other outlet constantly powered. Dual receptacles are also available that allow for two X10 controlled receptacles.

X10 Controlled Receptacles

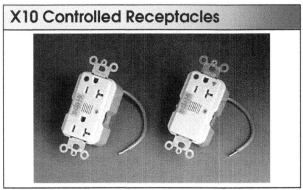

Leviton Manufacturing Co.

Figure 12-48. *A passive infrared (PIR) device is combined with a light source that is turned on when motion is detected.*

X10 Controllers. An X10 controller can be a simple wall-mount controller, a tabletop mini-controller, or a full-scale PC-based system. For remodeling projects, manufacturers recommend that mini-controllers be used. For new construction, a variety of controllers can be installed along with PC-based systems. As more complex controllers are used for a system, the number of devices controlled and the range of activities that can be controlled in the system also increase.

One of the major advantages of X10 technology is compatibility with other forms of remote control devices, such as hardwired systems found in structured cabling. X10 systems can link up with infrared, radio frequency, and Internet controls to allow control of systems from any remote location.

Low-Voltage Symbols

Symbols are used to provide information about low-voltage systems. Signaling, fire, communication, and X10 low-voltage systems all use symbols that are specific to the field. **See Figure 12-49.**

Suggested Symbols for Low Voltage Systems

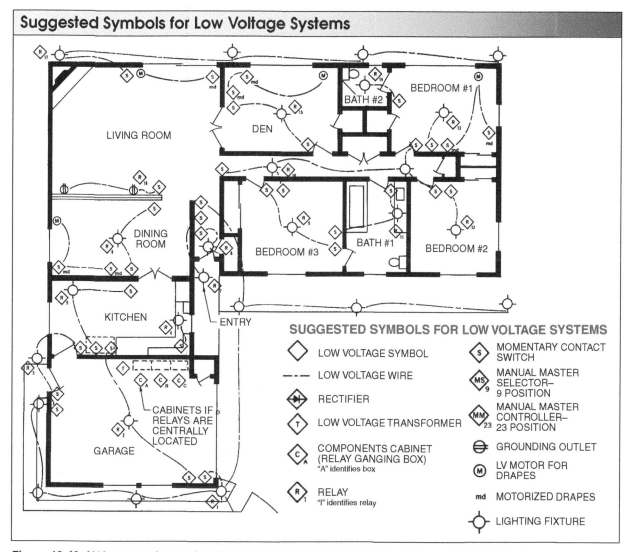

Figure 12-49. *X10 receptacles are installed to control lights and appliances from remote locations.*

NEXT GENERATION WIRELESS SYSTEMS

Next generation wireless technology is a modern wireless access control system that allows a homeowner and the electrical/electronic systems in the home to communicate with each other through remote control. Next generation wireless technology uses low-frequency radio waves that can travel easily through walls, floors, ceilings, and windows. This technology works to unify the electrical/electronic systems in the home into an integrated wireless system.

Next generation wireless technology is designed to work with many of the devices that are controlled remotely from personal computers through the Internet. The system is also compatible with cellular telephones (smartphones). Wireless systems communicate between devices with radio frequency (RF) signals, which require little physical wiring. This makes them faster to install and highly flexible with placement. Wireless systems are also particularly useful for retrofit and upgrade applications.

One disadvantage of wireless systems is that some depend on batteries, which must be replaced periodically. Another disadvantage is that they have a limited range of transmission. Some building materials, such as concrete and metal, affect the range of wireless transmission. In concrete and metal buildings, regenerative repeaters (regenerators) may be required for the operation of wireless systems. A *regenerative repeater (regenerator)* is a device inserted into a digital circuit to regenerate a transmitted circuit.

Next generation wireless control systems are compatible with X-10 technology, provided an X-10 bridge is installed. A *bridge* is a device that connects local area networks (LANs) with different hardware and protocols to one another.

ENERGY USAGE

Electrical energy is converted into motion, light, heat, sound, and visual outputs. **See Figure 12-50.** Approximately 62% of all electricity is converted into rotary motion by motors. Although three-phase (3φ) motors use the largest amount of electricity of all motors, they are the most energy efficient. Unfortunately, most residential structures do not have 3φ power lines. However, 220 V, single-phase (1φ) power is available and used for equipment that requires high voltage to operate such as HVAC systems, electric ovens, and electric clothes dryers. To effectively and efficiently use energy, electrical equipment must be designed and controlled properly for the specific application.

Energy Star is a U.S. government program that mandates limits on the power consumption of electric equipment. Energy Star approval ratings are used to help compare the operating efficiency of electrical appliances and components.

Energy Efficiency

Using advanced sensors and controls along with high-efficiency motors can significantly increase energy efficiency. *Energy efficiency* is the ratio of output energy (energy used by a device) to input energy (energy created to run a device) and is expressed as a percentage. Energy efficiency cannot exceed 100% in any system. A device that is 100% efficient converts all supplied energy to output power. However,

an electrical device with 100% efficiency does not exist because all electrical devices consume energy in order to produce work.

Energy-Efficient Lamps. Incandescent lamps are the most common type of lamp used in homes. However, incandescent lamps are inefficient and, over their lifetime, can result in energy usage costs five to ten times higher than their initial cost. To increase energy efficiency, incandescent lamps can be replaced with compact fluorescent lamps (CFLs).

On average, CFLs are four times more efficient and can last five times longer than incandescent lamps. CFLs are more efficient than incandescent lamps because they require less electricity to produce the same amount of light. The lesser the electricity generated, the lesser the amount of carbon dioxide (CO_2), sulfur oxide, and nuclear waste that is produced.

Light-emitting diode (LED) lamps have become an even more efficient means of producing light than CFLs for residential applications. LED lamps are normally grouped together to produce a high light output. LED lamps commonly last approximately 125 times longer than incandescent lamps and approximately 10 times longer than CFLs. In addition, they use approximately 85% less energy than incandescent lamps. Although it has a higher initial cost than other lamp types, an LED lamp saves energy over its lifetime.

Lighting-system control saves energy and can substantially reduce operating costs. For example, the lighting in a specific area of a home can be turned on automatically when an individual enters the area. Lights can also be set to turn off automatically after a predetermined time period of not sensing motion or body heat.

According to the U.S. Department of Energy, the energy consumed for lighting purposes accounts for approximately 22% of the total electricity generated in the United States. More than half of the energy used for lighting is consumed by the commercial sector. Residential consumers are responsible for the remaining lighting energy consumption. In total, billions of dollars are spent each year on lighting.

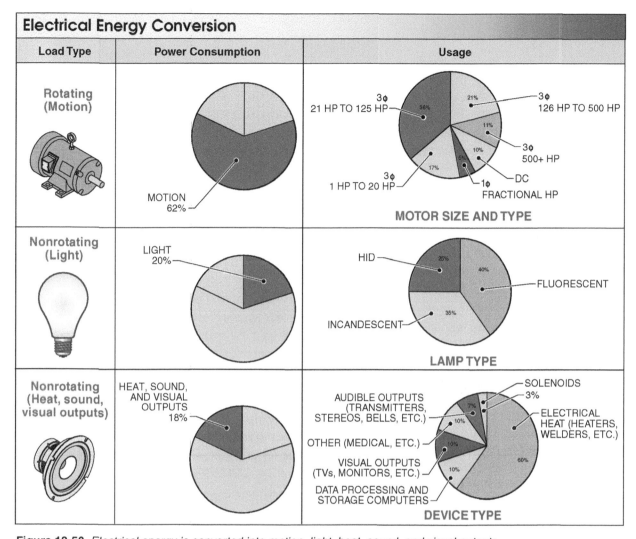

Electrical Energy Conversion		
Load Type	**Power Consumption**	**Usage**
Rotating (Motion)	MOTION 62%	3φ 21 HP TO 125 HP — 36% — 21% — 3φ 126 HP TO 500 HP — 11% — 10% — 3φ 500+ HP — 17% — DC — 1φ — FRACTIONAL HP — 3φ 1 HP TO 20 HP — **MOTOR SIZE AND TYPE**
Nonrotating (Light)	LIGHT 20%	HID — 25% — 40% — FLUORESCENT — 35% — INCANDESCENT — **LAMP TYPE**
Nonrotating (Heat, sound, visual outputs)	HEAT, SOUND, AND VISUAL OUTPUTS 18%	SOLENOIDS 3% — AUDIBLE OUTPUTS (TRANSMITTERS, STEREOS, BELLS, ETC.) — 7% — ELECTRICAL HEAT (HEATERS, WELDERS, ETC.) — 10% — OTHER (MEDICAL, ETC.) — 10% — 60% — VISUAL OUTPUTS (TVs, MONITORS, ETC.) — 10% — DATA PROCESSING AND STORAGE COMPUTERS — **DEVICE TYPE**

Figure 12-50. *Electrical energy is converted into motion, light, heat, sound, and visual outputs.*

Time-Based Control. A *time-based control system* is an automatic control system that uses the time of day to determine the desired operation of energy-consuming loads in a home. Time-based control requires loads to be turned on and off at specific times. Time-based control strategies consist of adjusting the time schedules of loads in the various areas of a home. Strategies include seven-day programming, daily multiple-time-period scheduling, and timed overrides.

Seven-day programming allows homeowners to individually program on and off time functions for each day of the week. HVAC, lighting, and other loads can be scheduled to operate independently during multiple time periods. A timed override allows the homeowner to change a zone from an unoccupied mode to an occupied mode for temporary use. A timed override can be activated by a switch that is configured as a digital input and tied to the control system. It can also be activated by a personal computer.

RECYCLING ELECTRICAL AND ELECTRONIC EQUIPMENT

Electrical and electronic equipment may contain materials that can be reused or recycled to produce useful byproducts. Some materials may be hazardous. Recycling helps reduce the harmful impact of electrical and electronic equipment on the environment.

Federal government directives have been established for the collection, recycling, and recovery of electrical equipment. These directives target the original equipment manufacturers (OEMs) of consumer electronic equipment. They are charged with the responsibility of making it easier to collect, recycle, and recover electrical equipment. Key objectives of these directives include the following:

• reducing electrical and electronic equipment waste
• increasing recycling and recovery
• improving the environmental disposal of nonrecyclable materials

Making electrical and electronic equipment manufacturers responsible for the reuse, recycling, and recovery of equipment provides an incentive for manufacturers to design efficient products. Consumers should also assume the responsibility of working within the guidelines.

RENEWABLE ENERGY SYSTEMS

A *renewable energy system* is an energy system that uses natural resources to produce heat and electric power. Solar energy, geothermal energy, and wind power are common types of renewable energy sources. Solar energy can be converted to thermal or electrical energy. Geothermal heating systems use heat pumps to pull heat up from the earth to heat or cool residences. Wind power systems use wind turbines to convert wind into usable electric power.

Solar Energy Systems

Solar energy is energy recovered from the sun in the form of sunlight. A *solar thermal energy system* is a renewable energy system that collects and stores solar energy and is used to heat air and water in a residential structure. Well-designed solar thermal energy systems maximize the amount of thermal energy received. **See Figure 12-51.** Focus is placed on the locations of windows and skylights, which can increase direct solar heat in living spaces and have an impact on the amount of illumination required by lighting systems.

Solar energy can be converted to electrical energy for use in a residence through photovoltaic technology. *Photovoltaic technology* is solar energy technology that uses the unique properties of semiconductors to directly convert solar radiation into electricity. A group of wafers, or cells, are combined to form a larger arrangement known as a module. A group of modules, known as a photovoltaic (PV) array, produces an appreciable amount of electrical power with no moving parts, noise, or emissions. The PV array is usually mounted on a roof or nearby on the ground and may include battery storage.

A *photovoltaic system* is an electrical system consisting of a PV array and other electrical components needed to convert solar energy into electricity usable by loads. The most common residential photovoltaic system is a utility-connected system that operates a PV array in parallel with and connected to an electric utility grid. **See Figure 12-52.**

Geothermal Energy Systems

Geothermal energy is derived from heat contained within the earth. Heat contained within the earth is maintained at a constant temperature (50°F to 70°F) in the soil and rocks below the frost line. With geothermal heating systems, ground-source heat pumps pull heat energy from the earth and convey it through a residence to provide heat during cold weather and cooling during warm weather. **See Figure 12-53.**

Sharp Electronics Corp.
Solar thermal energy is applied to residential structures through the use of roof panels attached directly to the building.

Solar Thermal Energy Systems

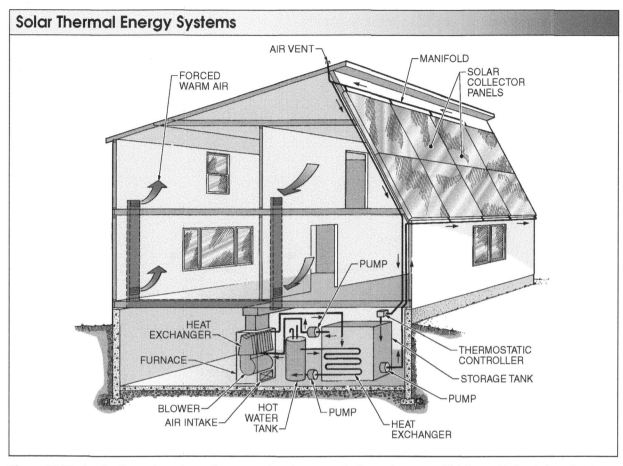

Figure 12-51. *A solar thermal energy system converts solar energy to thermal energy, which is used to heat air and water.*

Residential Photovoltaic Systems

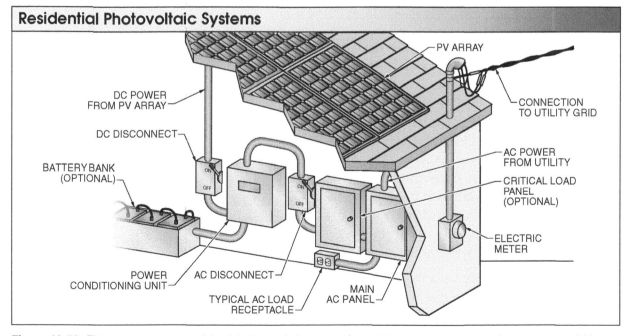

Figure 12-52. *The most common residential photovoltaic system is a utility-connected system that operates a PV array in parallel with and connected to the electric utility grid.*

Geothermal Heating and Cooling Systems

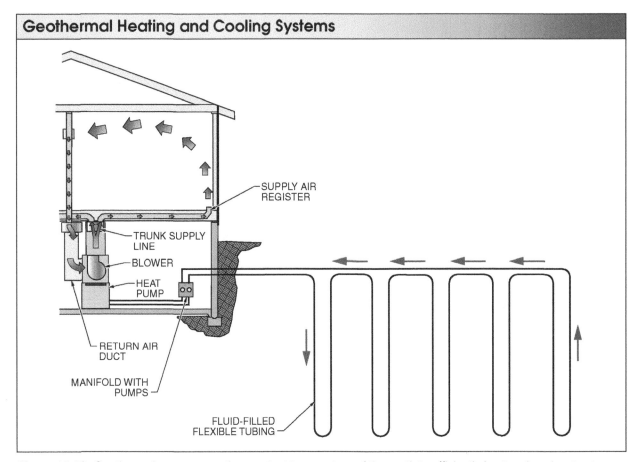

Figure 12-53. *Geothermal systems use the constant temperature of the earth to efficiently heat and cool a structure.*

A typical ground-source heat pump uses flexible tubing filled with water, refrigerant, or an antifreeze solution. As the fluid flows through the tubing into the ground, it absorbs heat energy from the earth. The warmed fluid is pumped into the living space to provide heat. After the energy is transferred to the living space, the fluid flows through return tubing into the earth to repeat the heating process. Geothermal systems may be used in combination with other piped heating systems to augment their heating capacity and minimize the amount of fuel or electricity required to bring the circulating fluid up to the necessary temperature prior to recirculation.

Heat Pumps. A *heat pump* is a mechanical compression refrigeration system that contains devices and controls that reverse the flow of refrigerant to move heat from one area to another. Reversing the flow of refrigerant switches the relative position of the evaporator and condenser in a heat pump system. A reversing valve is used to reverse the flow of refrigerant through the system.

Heat pumps can absorb heat from inside a building and reject the heat outdoors and absorb heat from the outdoors and reject it inside. A heat pump in the cooling mode moves heat from inside a building to the air outside a building. When a heat pump is in the cooling mode, the indoor unit is the evaporator and the outdoor unit is the condenser. A heat pump in the heating mode moves heat from outside a building to the air inside a building. When a heat pump is in the heating mode, the indoor unit is the condenser and the outdoor unit is the evaporator.

During heat pump operation, refrigerant flows from the compressor to a reversing valve. The refrigerant flows from the reversing valve to either the coil in the indoor unit or the coil in the outdoor unit. The direction of refrigerant flow depends on whether the system is in the cooling or heating mode.

The most common type of heat pump used in residential applications is an air-to-air heat pump. An air-to-air heat pump uses outdoor air as heat sink material when the system operates in the heating mode. Return air from the residence is blown across the indoor coil to absorb heat for heating living spaces. Air-to-air heat pumps are available with cooling capacities from 1 ton to 5 tons of cooling. They are available as split systems and as packaged systems. **See Figure 12-54.**

Wind Power Systems

A *wind power system* is a renewable energy system that converts wind energy into usable electric power. Wind power systems use wind turbines to help convert wind into usable power. A *wind turbine* is a machine that converts the energy in wind into mechanical energy. If the mechanical energy is converted to electricity, the machine is called a wind generator or wind turbine. Although wind turbines are old technology, the adaptation of this technology with modern equipment has produced a dramatic increase in the output of the wind turbine.

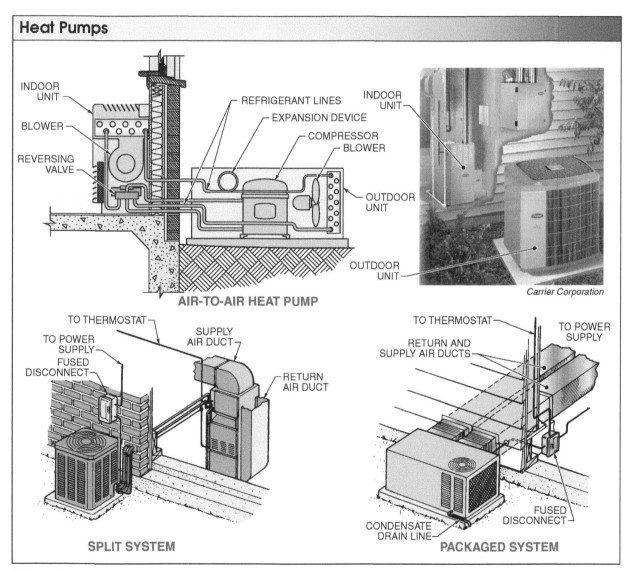

Heat Pumps

AIR-TO-AIR HEAT PUMP

Carrier Corporation

SPLIT SYSTEM

PACKAGED SYSTEM

Figure 12-54. *The most common type of heat pump used in residential applications is an air-to-air heat pump, which is available as a split system or packaged system.*

Wind turbine systems are usually composed of six major subsystems. The six subsystems of wind turbines are the following:
- turbine blade
- gear box transmission
- generator
- tower or mounting
- battery
- electronic control

Note: In some small systems, a tail is used to direct the blades into the wind.

Wind turbine manufacturers provide information on the maximum wind speed at which the turbine is designed to operate safely. Most wind turbines have automatic overspeed governors to keep the rotor from spinning out of control in high winds. A wind turbine can be connected to an electric utility distribution system (grid) or stand alone.

Grid-Connected Wind Power Systems. A *grid-connected wind power system* is a wind power system that is connected to an electric utility distribution system, or grid. **See Figure 12-55.** A grid-connected wind turbine can reduce the consumption of utility-supplied electricity for lighting, appliances, and electric heaters. When the wind turbine cannot deliver the amount of energy needed, the utility makes up the difference. When the wind system produces more electricity than required, the excess electricity is sold back to the utility.

With grid-connected wind power systems, the wind turbine operates only when the utility grid is available. During power outages, the wind turbine is required to shut down due to safety concerns. Federal regulations in the Public Utility Regulatory Policies Act (PURPA) require utilities to connect with and purchase power from energy-efficient wind power systems. To address any power quality and safety concerns, the utility must be contacted before connecting to its distribution lines. The utility usually provides a list of requirements for connecting the system to the grid.

Stand-Alone Wind Power Systems. A *stand-alone (off-grid) wind power system* is a wind power system that is not connected to an electric utility distribution system, or grid. The storage capacity of these systems must be large enough to supply electrical needs during noncharging periods. Stand-alone wind power systems usually provide power to a single residence, structure, or compound. **See Figure 12-56.**

Battery banks are typically sized to supply an electric load for one to three days. A wind turbine with an output capacity of 900 W to 1200 W can be sufficient for the operation of basic equipment. If more power is required, a wind turbine with an output capacity of 10,000 W to 15,000 W should be considered. It is a good practice to inventory all appliances and equipment that may run at the same time to determine the maximum load requirement for this type of system.

The capacity of utility-scale turbines ranges from 100 kW to several megawatts. Single small turbines, below 100 kW, are used for homes. Small turbines are sometimes used in connection with diesel generators, batteries, and photovoltaic systems. These systems are known as hybrid wind power systems and are typically used in remote off-grid locations, where connection to a utility grid is not available.

Grid-connected wind power systems are typically connected to a wind farm consisting of many wind turbines.

Electronic Controllers. Wind turbines are monitored and operated by electronic controllers. An electronic controller monitors the voltage of the batteries in the wind power system. The controller sends power from the wind turbine into the batteries to recharge them.

It also redirects power from the wind turbine into a secondary load if the batteries are fully charged. A secondary load can be another battery bank or a heating element, such as in a residential water heater. When batteries are at full capacity, the electricity of the wind turbine goes to the power grid.

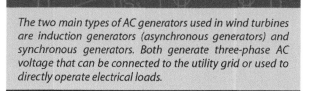

The two main types of AC generators used in wind turbines are induction generators (asynchronous generators) and synchronous generators. Both generate three-phase AC voltage that can be connected to the utility grid or used to directly operate electrical loads.

Grid-Connected Wind Power Systems

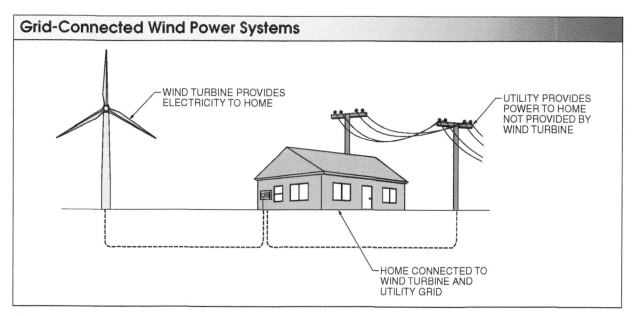

Figure 12-55. *A grid-connected wind power system is connected to an electric utility distribution system (grid).*

Stand-Alone Wind Power Systems

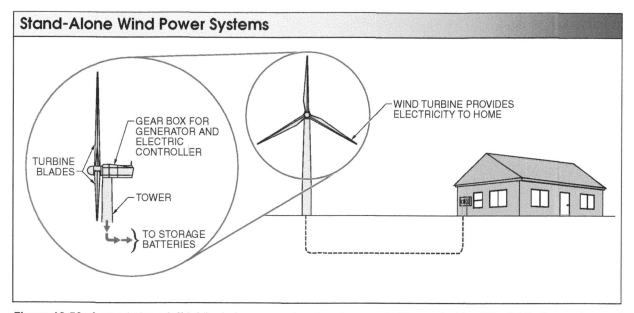

Figure 12-56. *A stand-alone (off-grid) wind power system is not connected to an electric utility distribution system and usually provides power to a single residence, structure, or compound.*

Name_____ **Date** _____

Residential Control Systems

      T   F   **1.** Residential low-voltage systems operate at less than 30 V.

_____ **2.** Low-voltage systems are safer than standard-voltage systems because the hazard of ___ is reduced.

_____ **3.** Low-voltage systems are more economical than standard-voltage systems because smaller ___ are used.

_____ **4.** The most simple low-voltage system is the ___ system.

_____ **5.** ___ open or close a circuit when a problem develops in a low-voltage system.

_____ **6.** The ___ of an X10 system can communicate with each other.

_____ **7.** Closed-circuit wiring is typically used for wiring ___ systems.

      T   F   **8.** Open-circuit wiring is typically used for fire detection systems.

_____ **9.** ___ allow detectors to operate on low voltage while the annunciators operate on standard voltage.

_____ **10.** Plunger button detectors are typically used in closed-circuit systems where a door or window holds the contacts in a(n) ___ position.

_____ **11.** In a two-piece magnetic detector system, the ___.
           A. switch is permanently mounted   **C.** both A and B
           B. magnet is allowed to move   **D.** none of the above

_____ **12.** A(n) ___ detector projects an invisible, pulsating beam at a bounce-back deflector.

      T   F   **13.** Annunciators, when triggered, ring a bell, flash a light, or sound an alarm.

_____ **14.** Closed-circuit wiring is wired as a(n) ___ circuit.

_____ **15.** Open-circuit wiring is wired as a(n) ___ circuit.

T   F   **16.** Controls determine which detector is responding and uses that signal to trigger an annunciator.

_____   **17.** A(n) ___ heat detector responds when a specific temperature is reached.

_____   **18.** Heat detectors are typically used in conjunction with ___ detectors.

_____   **19.** Relays for low-voltage systems are capable of controlling ___ A of tungsten, fluorescent, or inductive loads at 115 VAC and 277 VAC.

_____   **20.** Transformers in a remote control system reduce 120 V line voltage to ___ V.

T   F   **21.** Typically, transformers are mounted in 2″ x 2″ x 4″ boxes.

T   F   **22.** Transformers are sized according to the environment in which they are installed.

T   F   **23.** Relays may be mounted in junction box knockouts.

_____   **24.** A(n) ___ heat detector is designed for use with flash fires or slow-burning fires.

T   F   **25.** X10 technology operates on a separately run low-voltage wiring system.

_____   **26.** Low-voltage wire is available as three-, four-, and ___ -conductor cables.

_____   **27.** Smoke/gas detectors are available to operate on 120 V or on ___.

_____   **28.** The three basic parts of a security-fire system are ___.
        A. wiring, detection, and control
        B. wiring, control, and annunciation
        C. wiring, detection, and annunciation
        D. none of the above

_____   **29.** A(n) ___-controlled detector can be used when running wires for long distances is impractical.

_____   **30.** When two or more switches are wired in ___, each is capable of controlling relays.

Low-Voltage Symbols

_____ **1.** Relay

_____ **2.** Motor

_____ **3.** Low-voltage transformer

_____ **4.** Lighting fixture

_____ **5.** Low-voltage wire

_____ **6.** Rectifier

_____ **7.** Low-voltage symbol

_____ **8.** Motor master controller

_____ **9.** Momentary contact switch

_____ **10.** Grounding outlet

(A) (B) (C) (D) (E) (F) (G) (H) (I) (J)

Low-Voltage Circuits

_____ **1.** Bell ___ will sound when PB_1 is depressed.

_____ **2.** Bell D will sound when ___ is depressed.

_____ **3.** Bell C will sound when ___ is depressed.

_____ **4.** When PB_6 is depressed, ___.

         A. Chime F sounds   **B.** Chime G sounds
         C. both A and B   **D.** none of the above

_____ **5.** Bell B will sound when ___ is depressed.

_____ **6.** When PB_7 is depressed, ___.

         A. Chime F sounds   **B.** Chime G sounds
         C. both A and B   **D.** none of the above

_____ **7.** To sound Bell E, depress ___.

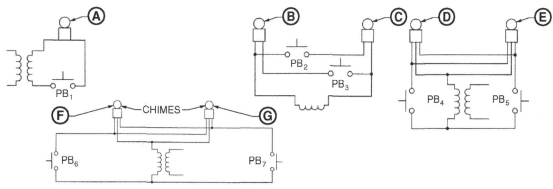

One-Family Dwelling

_____   **1.** The motor in Bedroom #1 operates motorized ___.

_____   **2.** Outside lighting at the rear of the house is controlled by a(n) ___ switch from the Living Room.

_____   **3.** The lighting fixture in Bath #1 may be controlled by a switch near the bathroom door and in Bedroom #___.

_____   **4.** Outside lighting on the Garage and front of the house may be controlled by switches located in the Garage and ___.

_____   **5.** ___ wire is used to connect all switches and relays.

_____   **6.** Grounding outlets are installed in the ___.

_____   **7.** The low-voltage transformer is located in the ___.

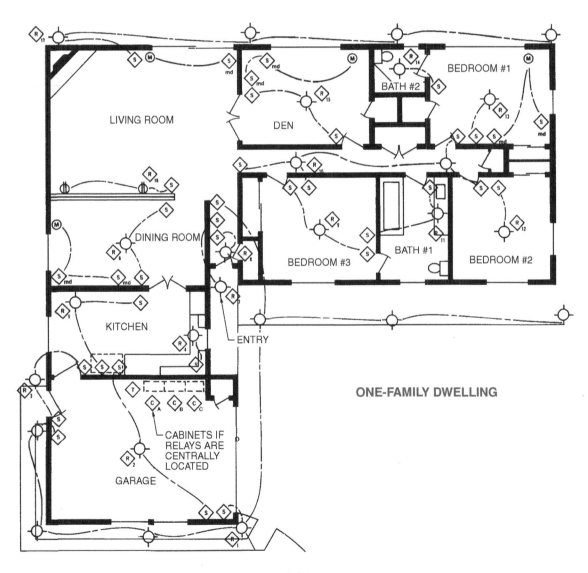

ONE-FAMILY DWELLING

Power Formulas - 1φ, 3φ

Phase	To Find	Use Formula	Example			
			Given	Find	Solution	
1φ	I	$I = \dfrac{VA}{V}$	32,000 VA, 240 V	I	$I = \dfrac{VA}{V}$ $\quad I = \dfrac{32{,}000\ VA}{240\ V}$	$I = $ **133 A**
1φ	VA	$VA = I \times V$	100 A, 240 V	VA	$VA = I \times V \quad VA = 100\ A \times 240\ V$	$VA = $ **24,000 VA**
1φ	V	$V = \dfrac{VA}{I}$	42,000 VA, 350 A	V	$V = \dfrac{VA}{I}$ $\quad V = \dfrac{42{,}000\ VA}{350\ A}$	$V = $ **120 V**
3φ	I	$I = \dfrac{VA}{V \times \sqrt{3}}$	72,000 VA, 208 V	I	$I = \dfrac{VA}{V \times \sqrt{3}}$ $\quad I = \dfrac{72{,}000\ VA}{360\ V}$	$I = $ **200 A**
3φ	VA	$VA = I \times V \times \sqrt{3}$	2 A, 240 V	VA	$VA = I \times V \times \sqrt{3} \quad VA = 2 \times 416$	$VA = $ **832 VA**

Electrical Symbols . . .

Lighting Outlets		Convenience Outlets		Switch Outlets	
OUTLET BOX AND INCANDESCENT LIGHTING FIXTURE	CEILING WALL	SINGLE RECEPTACLE OUTLET		SINGLE-POLE SWITCH	S
INCANDESCENT TRACK LIGHTING		DUPLEX RECEPTACLE OUTLET–120 V		DOUBLE-POLE SWITCH	S_2
BLANKED OUTLET	B B	TRIPLEX RECEPTACLE OUTLET–240 V		THREE-WAY SWITCH	S_3
		SPLIT-WIRED DUPLEX RECEPTACLE OUTLET		FOUR-WAY SWITCH	S_4
DROP CORD	D	SPLIT-WIRED TRIPLEX RECEPTACLE OUTLET			
EXIT LIGHT AND OUTLET BOX, SHADED AREAS DENOTE FACES.		SINGLE SPECIAL-PURPOSE RECEPTACLE OUTLET		AUTOMATIC DOOR SWITCH	S_D
		DUPLEX SPECIAL-PURPOSE RECEPTACLE OUTLET		KEY-OPERATED SWITCH	S_K
OUTDOOR POLE-MOUNTED FIXTURES		RANGE OUTLET	R	CIRCUIT BREAKER	S_{CB}
JUNCTION BOX	J J	SPECIAL-PURPOSE CONNECTION	DW	WEATHERPROOF CIRCUIT BREAKER	S_{WCB}
LAMP HOLDER WITH PULL SWITCH	L$_{PS}$ L$_{PS}$	CLOSED-CIRCUIT TELEVISION CAMERA		DIMMER	S_{DM}
MULTIPLE FLOODLIGHT ASSEMBLY		CLOCK HANGER RECEPTACLE	C	REMOTE CONTROL SWITCH	S_{RC}
		FAN HANGER RECEPTACLE	F		
EMERGENCY BATTERY PACK WITH CHARGER		FLOOR SINGLE RECEPTACLE OUTLET		WEATHERPROOF SWITCH	S_{WP}
INDIVIDUAL FLUORESCENT FIXTURE		FLOOR DUPLEX RECEPTACLE OUTLET		FUSED SWITCH	S_F
OUTLET BOX AND FLUORESCENT LIGHTING TRACK FIXTURE		FLOOR SPECIAL-PURPOSE OUTLET		WEATHERPROOF FUSED SWITCH	S_{WF}
CONTINUOUS FLUORESCENT FIXTURE		UNDERFLOOR DUCT AND JUNCTION BOX FOR TRIPLE, DOUBLE, OR SINGLE DUCT SYSTEM AS INDICATED BY NUMBER OF PARALLEL LINES		TIME SWITCH	S_T
SURFACE-MOUNTED FLUORESCENT FIXTURE				CEILING PULL SWITCH	S
Panel Boards		**Busducts and Wireways**		SWITCH AND SINGLE RECEPTACLE	$_S$
FLUSH-MOUNTED PANELBOARD AND CABINET		SERVICE FEEDER, OR PLUG-IN BUSWAY	B B B	SWITCH AND DOUBLE RECEPTACLE	$_S$
SURFACE-MOUNTED PANELBOARD AND CABINET		CABLE THROUGH LADDER OR CHANNEL	C C C	A STANDARD SYMBOL WITH AN ADDED LOWERCASE SUBSCRIPT LETTER IS USED TO DESIGNATE A VARIATION IN STANDARD EQUIPMENT	$O_{a,b}$ $_{a,b}$ $S_{a,b}$
		WIREWAY	W W W		

. . . Electrical Symbols

Commercial and Industrial Systems	Underground Electrical Distribution or Electrical Lighting Systems	Panel Circuits and Miscellaneous
PAGING SYSTEM DEVICE	MANHOLE — M	LIGHTING PANEL
FIRE ALARM SYSTEM DEVICE	HANDHOLE — H	POWER PANEL
COMPUTER DATA SYSTEM DEVICE	TRANSFORMER- MANHOLE OR VAULT — TM	WIRING – CONCEALED IN CEILING OR WALL
PRIVATE TELEPHONE SYSTEM DEVICE	TRANSFORMER PAD — TP	WIRING – CONCEALED IN FLOOR
SOUND SYSTEM	UNDERGROUND DIRECT BURIAL CABLE	WIRING EXPOSED
FIRE ALARM CONTROL PANEL — FACP	UNDERGROUND DUCT LINE	HOME RUN TO PANELBOARD Indicate number of circuits by number of arrows. Any circuit without such designation indicates a two-wire circuit. For a greater number of wires indicate as follows: —///— (3 wires) —////— (4 wires), etc.
Signaling System Outlets for Residential Systems	STREET LIGHT STANDARD FED FROM UNDERGROUND CIRCUIT	
	Above-Ground Electrical Distribution or Lighting Systems	FEEDERS Use heavy lines and designate by number corresponding to listing in feeder schedule
PUSHBUTTON		
BUZZER	POLE	WIRING TURNED UP
BELL	STREET LIGHT AND BRACKET	WIRING TURNED DOWN
BELL AND BUZZER COMBINATION	PRIMARY CIRCUIT	GENERATOR — G
COMPUTER DATA OUTLET	SECONDARY CIRCUIT	MOTOR — M
BELL RINGING TRANSFORMER — BT		INSTRUMENT (SPECIFY) — I
ELECTRIC DOOR OPENER — D	DOWN GUY	TRANSFORMER — T
CHIME — CH	HEAD GUY	CONTROLLER
TELEVISION OUTLET — TV	SIDEWALK GUY	EXTERNALLY-OPERATED DISCONNECT SWITCH
THERMOSTAT — T	SERVICE WEATHERHEAD	PULL BOX

Receptacle Connection Slot Configurations

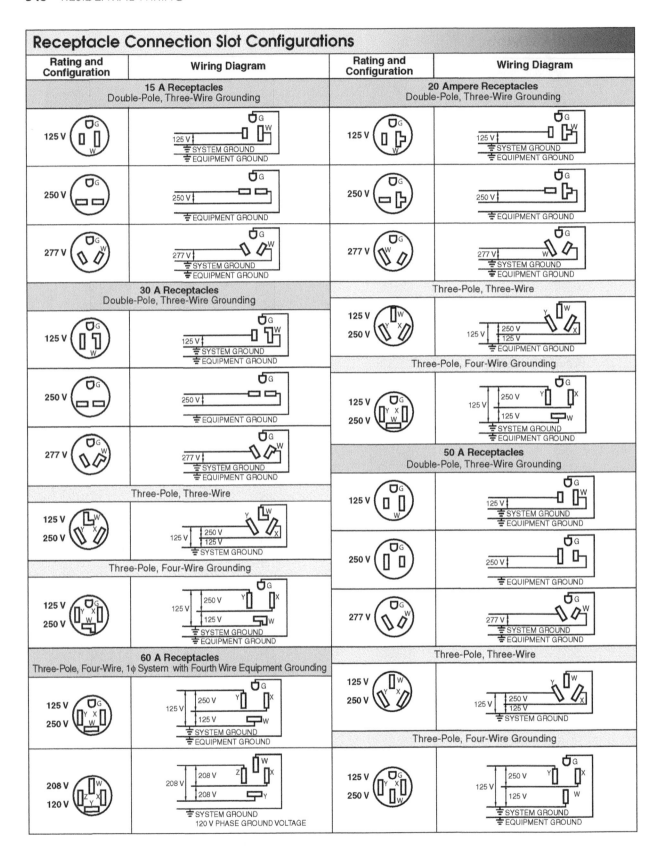

Component Plan Schedules . . .

Area	Convenience Receptacles	Special-Purpose Outlets	General Lighting	Major Appliances	General Switching for All Areas
Bedrooms	No space along a wall should be more than 6' from a receptacle outlet. Any wall space 2' or larger should have a minimum of 1 receptacle. NEC® 210.52 (A).	TV outlet, intercom, speakers (music), and telephone jack	Ceiling, wall or valence light, lamp switched at receptacle	Room air conditioner, electric baseboard heat	Switches are typically placed opposite the hinged side of door. When there are two or more entrances to a room, multiple switching should be used.
Living Room	No space along a wall should be more than 6' from a receptacle outlet. Any wall space 2' or larger should have a minimum of 1 receptacle. NEC® 210.52 (A).	TV outlet, intercom, speakers (music)	Ceiling fixture, recessed lighting, valence light, lamp switched at receptacle, possible dimmer	Room air conditioner, built-in stereo system, electric baseboard heat	Door switch can be used for closet light so that the switch turns the light ON and OFF as the door opens and closes.
Family Room	No space along a wall should be more than 6' from a receptacle outlet. Any wall space 2' or larger should have a minimum of 1 receptacle. NEC® 210.52 (A).	TV outlet, intercom, speakers (music), bar area (ice maker, blenders, small refrigerator, hot plate), telephone jack, thermostat	Ceiling fixture, recessed lighting, valence lights, studio spot lights, lamp switched at receptacle, fluorescent light, possible dimmer	Room air conditioner, built-in stereo system, electric baseboard heat	Switches with pilot lights should be used on lights, fans, and other electrical devices which are in locations not readily observable.
Dining Room	No space along a wall should be more than 6' from a receptacle outlet. Any wall space 2' or larger should have a minimum of 1 receptacle. NEC® 210.52 (A).	Elevated receptacles (48") for buffet tables, speakers (music)	Ceiling chandelier, recessed lighting, valence lighting, lighting for china hutch	Room air conditioner, electric baseboard heat	Combination switch/ receptacle may be used where space is limited. Fused switch may be used as a shut off for service and fuse protection as on a furnace.
Kitchen	Receptacle for each 4' of work space. Double duplex (Quad) receptacles in food preparation area. Requires two 20 A small appliance circuits. GFCIs to serve counter tops. NEC® 210.8 (A)(6), 210.52 (B)(3).	Clock outlet, TV outlet, intercom, speakers (music), exhaust fan, telephone jack	Light over sink, light over desk, recessed lights over work areas, illuminated ceiling, central room lighting, light under cabinets	Refrigerator, microwave oven, dishwasher, garbage disposal, oven, range top, trash compactor, electric baseboard heat	Dimmer switches should be used where accent lighting is appropriate.
Laundry		Exhaust fan, intercom, speakers (music), telephone jack	Central light, light over work area	Washer, dryer, hot water heater, water softener, electric baseboard heat	Timer switch may be installed to turn lights or appliances ON and OFF at specific times. Photoswitch may be used to control equipment by light changes.
Garage	Elevated receptacles (48") placed for convenience. GFCI receptacles. NEC® 210.8 (A)(2), 210.52 (G).	Garage door openers, work bench outlets, freezer, intercom, speakers (music), telephone jack	Ceiling fixtures located in front of garage door when open. Outside wall mount light to illuminate drive. Light beside outside door	Shop tools, welder, electric fan, forced air heater	
Bathrooms	Duplex receptacles on GFCI near sink and mirror. NEC® 210.8 (A)(1), 210.52 (D).	Ventilating fan, intercom, telephone jack,	Light on either side of mirror (wall-mount or hanging), overhead general light, night light, illuminated ceiling light in enclosed shower, heat/light on timer	Electric fan, forced air heater	Switches are usually placed opposite the hinge side of the door. When there are two or more entrances to a room, multiple switching should be used.

... Component Plan Schedules

Area	Convenience Receptacles	Special-Purpose Outlets	General Lighting	Major Appliances	General Switching for All Areas
Hallways	One receptacle every 10' of hallway. NEC® 210.52 (H).	Thermostat	Recessed ceiling lights, wall-mounted lights, ceiling mounted lights	Electric baseboard heat	Door switch may be used on closet so that it turns ON and OFF as the door opens and closes.

Switches with pilot lights should be used on lights, fans, and other electrical devices which are in locations not readily observable.

Combination switch/receptacle may be used where space is limited. Fused switch may be used as a shut off for service and fuse protection as on a furnace.

Dimmer switches should be used where accent lighting is appropriate.

Timer switch may be installed to turn lights or appliances ON and OFF at specific times. Photoswitch may be used to control equipment by light changes. |
Stairways	One receptacle may be necessary for vacuuming.		Ceiling light at top and bottom of stairs. Wall mount light (must be accessable to change bulb).		
Closets			Recessed light, fluorescent light, ceiling-mounted light		
Foyers or Entrances	Receptacles for convenience in cleaning. Weatherproof receptacles if outside.		Light inside entrance. Wall-mounted light outside entrance, recessed lights	Electric baseboard heat	
Attic	Receptacle for general use.	Extra circuits tied off for future expansion	Lights as necessary for area used	Attic fan, power exhaust vents	
Basement or Utility Area	(Unfinished) Receptacles for convenience. (Finished) Use guidelines for family room, living room, and bathroom. NEC® 210.8 (A)(5) and (6), 210.52 (G).	TV outlet, intercom, telephone jack, speakers (music)	Workbench light, light near furnace, light at stairs. If finished for recreation refer to family room, living room, and bathroom.	Hot water heater, furnace, central air conditioning, soft water system, electronic air cleaner	
Outside House	Weatherproof GFCI receptacles. NEC® 210.8 (A)(3), 210.52 (E).	Intercom/public address speakers (music), telephone jack	Lighting at entrance, spot lights, decorative low voltage lighting yard light	Connection for central air conditioning	

Electrical/Electronic Abbreviations/Acronyms

Abbr/ Acronym	Meaning	Abbr/ Acronym	Meaning	Abbr/ Acronym	Meaning
A	Ammeter; Ampere; Anode; Armature	FU	Fuse	PNP	Positive-Negative-Positive
AC	Alternating Current	FWD	Forward	POS	Positive
AC/DC	Alternating Current; Direct Current	G	Gate; Giga; Green; Conductance	POT.	Potentiometer
A/D	Analog to Digital	GEN	Generator	P-P	Peak-to-Peak
AF	Audio Frequency	GRD	Ground	PRI	Primary Switch
AFC	Automatic Frequency Control	GY	Gray	PS	Pressure Switch
Ag	Silver	H	Henry; High Side of Transformer; Magnetic Flux	PSI	Pounds Per Square Inch
ALM	Alarm			PUT	Pull-Up Torque
AM	Ammeter; Amplitude Modulation	HF	High Frequency	Q	Transistor
AM/FM	Amplitude Modulation; Frequency Modulation	HP	Horsepower	R	Radius; Red; Resistance; Reverse
		Hz	Hertz	RAM	Random-Access Memory
ARM.	Armature	I	Current	RC	Resistance-Capacitance
Au	Gold	IC	Integrated Circuit	RCL	Resistance-Inductance-Capacitance
AU	Automatic	INT	Intermediate; Interrupt	REC	Rectifier
AVC	Automatic Volume Control	INTLK	Interlock	RES	Resistor
AWG	American Wire Gauge	IOL	Instantaneous Overload	REV	Reverse
BAT.	Battery (electric)	IR	Infrared	RF	Radio Frequency
BCD	Binary Coded Decimal	ITB	Inverse Time Breaker	RH	Rheostat
BJT	Bipolar Junction Transistor	ITCB	Instantaneous Trip Circuit Breaker	rms	Root Mean Square
BK	Black	JB	Junction Box	ROM	Read-Only Memory
BL	Blue	JFET	Junction Field-Effect Transistor	rpm	Revolutions Per Minute
BR	Brake Relay; Brown	K	Kilo; Cathode	RPS	Revolutions Per Second
C	Celsius; Capacitance; Capacitor	L	Line; Load; Coil; Inductance	S	Series; Slow; South; Switch
CAP.	Capacitor	LB-FT	Pounds Per Foot	SCR	Silicon Controlled Rectifier
CB	Circuit Breaker; Citizen's Band	LB-IN.	Pounds Per Inch	SEC	Secondary
CC	Common-Collector Configuration	LC	Inductance-Capacitance	SF	Service Factor
CCW	Counterclockwise	LCD	Liquid Crystal Display	1 PH; 1φ	Single-Phase
CE	Common-Emitter Configuration	LCR	Inductance-Capacitance-Resistance	SOC	Socket
CEMF	Counter Electromotive Force	LED	Light Emitting Diode	SOL	Solenoid
CKT	Circuit	LRC	Locked Rotor Current	SP	Single-Pole
CONT	Continuous; Control	LS	Limit Switch	SPDT	Single-Pole, Double-Throw
CPS	Cycles Per Second	LT	Lamp	SPST	Single-Pole, Single-Throw
CPU	Central Processing Unit	M	Motor; Motor Starter; Motor Starter Contacts	SS	Selector Switch
CR	Control Relay			SSW	Safety Switch
CRM	Control Relay Master	MAX.	Maximum	SW	Switch
CT	Current Transformer	MB	Magnetic Brake	T	Tera; Terminal; Torque; Transformer
CW	Clockwise	MCS	Motor Circuit Switch	TB	Terminal Board
D	Diameter; Diode; Down	MEM	Memory	3 PH; 3φ	Three-Phase
D/A	Digital to Analog	MED	Medium	TD	Time Delay
DB	Dynamic Braking Contactor; Relay	MIN	Minimum	TDF	Time Delay Fuse
DC	Direct Current	MN	Manual	TEMP	Temperature
DIO	Diode	MOS	Metal-Oxide Semiconductor	THS	Thermostat Switch
DISC.	Disconnect Switch	MOSFET	Metal-Oxide Semiconductor Field-Effect Transistor	TR	Time Delay Relay
DMM	Digital Multimeter			TTL	Transistor-Transistor Logic
DP	Double-Pole	MTR	Motor	U	Up
DPDT	Double-Pole, Double-Throw	N; NEG	North; Negative	UCL	Unclamp
DPST	Double-Pole, Single-Throw	NC	Normally Closed	UHF	Ultrahigh Frequency
DS	Drum Switch	NEUT	Neutral	UJT	Unijunction Transistor
DT	Double-Throw	NO	Normally Open	UV	Ultraviolet; Undervoltage
DVM	Digital Voltmeter	NPN	Negative-Positive-Negative	V	Violet; Volt
EMF	Electromotive Force	NTDF	Nontime-Delay Fuse	VA	Volt Amp
F	Fahrenheit; Fast; Field; Forward; Fuse	O	Orange	VAC	Volts Alternating Current
FET	Field-Effect Transistor	OCPD	Overcurrent Protection Device	VDC	Volts Direct Current
FF	Flip-Flop	OHM	Ohmmeter	VHR	Very High Frequency
FLC	Full-Load Current	OL	Overload Relay	VLF	Very Low Frequency
FLS	Flow Switch	OZ/IN.	Ounces Per Inch	VOM	Volt-Ohm-Milliammeter
FLT	Full-Load Torque	P	Peak; Positive; Power; Power Consumed	W	Watt; White
FM	Frequency Modulation	PB	Pushbutton	w/	With
FREQ	Frequency	PCB	Printed Circuit Board	X	Low Side of Transformer
FS	Float Switch	PH; φ	Phase	Y	Yellow
FTS	Foot Switch	PLS	Plugging Switch	Z	Impedance

Metric Prefixes

Multiples and Submultiples	Prefixes	Symbols	Meaning
$1,000,000,000,000 = 10^{12}$	tera	T	trillion
$1,000,000,000 = 10^{9}$	giga	G	billion
$1,000,000 = 10^{6}$	mega	M	million
$1000 = 10^{3}$	kilo	k	thousand
$100 = 10^{2}$	hecto	h	hundred
$10 = 10^{1}$	deka	da	ten
$1 = 10^{0}$			
$0.1 = 10^{-1}$	deci	d	tenth
$0.01 = 10^{-2}$	centi	c	hundredth
$0.001 = 10^{-3}$	milli	m	thousandth
$0.000001 = 10^{-6}$	micro	μ	millionth
$0.000000001 = 10^{-9}$	nano	n	billionth
$0.000000000001 = 10^{-12}$	pico	p	trillionth

Metric Conversions

Initial Units	Final Units											
	giga	mega	kilo	hecto	deca	base unit	deci	centi	milli	micro	nano	pico
giga		3R	6R	7R	8R	9R	10R	11R	12R	15R	18R	21R
mega	3L		3R	4R	5R	6R	7R	8R	9R	12R	15R	18R
kilo	6L	3L		1R	2R	3R	4R	5R	6R	9R	12R	15R
hecto	7L	4L	1L		1R	2R	3R	4R	5R	8R	11R	14R
dek	8L	5L	2L	1L		1R	2R	3R	4R	7R	10R	13R
base unit	9L	6L	3L	2L	1L		1R	2R	3R	6R	9R	12R
deci	10L	7L	4L	3L	2L	1L		1R	2R	5R	8R	11R
centi	11L	8L	5L	4L	3L	2L	1L		1R	4R	7R	10R
milli	12L	9L	6L	5L	4L	3L	2L	1L		3R	6R	9R
micro	15L	12L	9L	8L	7L	6L	5L	4L	3L		3R	6R
nano	18L	15L	12L	11L	10L	9L	8L	7L	6L	3L		3R
pico	21L	18L	15L	14L	13L	12L	11L	10L	9L	6L	3L	

Common Prefixes

Symbol	Prefix	Equivalent
G	giga	1,000,000,000
M	mega	1,000,000
k	kilo	1000
base unit	—	1
m	milli	0.001
μ	micro	0.000001
n	nano	0.000000001
p	pico	0.000000000001
Z	impedance	ohms—

Capacitors

Connected in Series		Connected in Parallel	Connected in Series/Parallel
Two Capacitors	**Three or More Capacitors**		
$C_T = \dfrac{C_1 \times C_2}{C_1 + C_2}$ where C_T = total capacitance (in μF) C_1 = capacitance of capacitor 1 (in μF) C_2 = capacitance of capacitor 2 (in μF)	$\dfrac{1}{C_T} = \dfrac{1}{C_1} + \dfrac{1}{C_2} + \ldots$	$C_T = C_1 + C_2 + \ldots$	1. Calculate the capacitance of the parallel branch. $\quad C_T = C_1 + C_2 + \ldots$ 2. Calculate the capacitance of the series combination. $\quad C_T = \dfrac{C_1 \times C_2}{C_1 + C_2}$

Temperature Conversions

Convert °C to °F	Convert °F to °C
$°F = (1.8 \times °C) + 32$	$°C = \dfrac{(°F - 32)}{1.8}$

Units of Power

Power	W	ft lb/s	HP	kW
Watt	1	0.7376	$.341 \times 10^{-3}$	0.001
Foot-pound/sec	1.356	1	$.818 \times 10^{-3}$	1.356×10^{-3}
Horsepower	745.7	550	1	0.7457
Kilowatt	1000	736.6	1.341	1

Branch Circuit Voltage Drop

$$\%V_D = \frac{V_{NL} - V_{FL}}{V_{FL}} \times 100$$

where
$\%V_D$ = percent voltage drop (in volts)
V_{NL} = no-load voltage (in volts)
V_{FL} = full-load voltage drop (in volts)
100 = constant

Standard Sizes of Fuses and CBs

NEC® 240-6(a) lists standard ampere ratings of fuses and fixed-trip CBs as follows:
15, 20, 25, 30, 35, 40, 45,
50, 60, 70, 80, 90, 100, 110,
125, 150, 175, 200, 225,
250, 300, 350, 400, 450,
500, 600, 700, 800,
1000, 1200, 1600,
2000, 2500, 3000, 4000, 5000, 6000

Voltage Conversions

To Convert	To	Multiply By
rms	Average	0.9
rms	Peak	1.414
Average	rms	1.111
Average	Peak	1.567
Peak	rms	0.707
Peak	Average	0.637
Peak	Peak-to-Peak	2

Raceways

EMT	Electrical Metallic Tubing
ENT	Electrical Nonmetallic Tubing
FMC	Flexible Metal Conduit
FMT	Flexible Metallic Tubing
IMC	Intermediate Metal Conduit
LFMC	Liquidtight Flexible Metal Conduit
LFNC	Liquidtight Flexible Nonmetallic Conduit
RMC	Rigid Metal Conduit
RNC	Rigid Nonmetallic Conduit

Cables

AC	Armored Cable
BX	Tradename for AC
FCC	Flat Conductor Cable
IGS	Integrated Gas Spacer Cable
MC	Metal-Clad Cable
MI	Mineral-Insulated, Metal Sheathed Cable
MV	Medium Voltage
NM	Nonmetallic-Sheathed Cable (dry)
NMC	Nonmetallic-Sheathed Cable (dry or damp)
NMS	Nonmetallic-Sheathed Cable (dry)
SE	Service-Entrance Cable
TC	Tray Cable
UF	Underground Feeder Cable
USE	Underground Service-Entrance Cable

Common Electrical Insulations

60°C 140°F	75°C 167°F	90°C 194°F
TW	FEPW	TBS
UF	RH	SA
	RHW	SIS
	THHW	FEP
	THW	FEPB
	THWN	MI
	XHHW	RHH
	USE	RHW-2
	ZW	THHN
		THHW
		THW-2
		THWN-2
		USE-2
		XHH
		XHHW
		XHHW-2
		ZW-2

INSULATION •TABLE 310-13

AMPACITY •TABLE 310-16

COPPER, ALUMINUM, OR COPPER-CLAD ALUMINUM

Fuses and ITCBs

Increase	Standard Ampere Ratings
5	15, 20, 25, 30, 35, 40, 45
10	50, 60, 70, 80, 90, 100, 110
25	125, 150, 175, 200, 225
50	250, 300, 350, 400, 450
100	500, 600, 700, 800
200	1000, 1200
400	1600, 2000
500	2500
1000	3000, 4000, 5000, 6000

1 A, 3 A, 6 A, 10 A, and 601 A are additional standard ratings for fuses.

Standard Calculation: One-Family Dwelling

1. GENERAL LIGHTING: *Table 220.12*

_____ sq ft × 3 VA =   _____ V.

Small appliances: *220.52(A)*

_____ VA × _____ circuits =   _____ VA

Laundry: *220.52(B)*

_____ VA × 1 =   _____ VA

_____ VA

Applying Demand Factors: *Table 220.42.*

First 3000 VA × 100% =   3000 VA

Next _____ VA × 35% =   _____ VA   **PHASES**   **NEUTRAL**

Remaining _____ VA × 25% =   _____ VA

Total   _____ VA   _____ VA   _____ VA

2. FIXED APPLIANCES: *220.53*

Dishwasher =   _____ VA

Disposer =   _____ VA

Compactor =   _____ VA

Water heater =   _____ VA

_____ = _____ VA

_____ = _____ VA

_____ = _____ VA   (120 V Loads × 75%)

Total   _____ VA × 75% = _____ VA   _____ VA   _____ VA

3. DRYER: *220.54; Table 220.54*

_____ VA × _____% =   _____ VA   _____ VA × 70% = _____ VA

4. COOKING EQUIPMENT: *Table 220.55; Notes*

Col A _____ VA × _____% =   _____ VA

Col B _____ VA × _____% =   _____ VA

Col C _____ VA × _____% =   _____ VA

Total   _____ VA   _____ VA × 70% = _____ VA

5. HEATING or A/C: *220.60*

Heating unit = _____ VA × 100% =   _____ VA

A/C unit = _____ VA × 100% =   _____ VA

Heat pump = _____ VA × 100% =   _____ VA

Largest Load   _____ VA   _____ VA   _____ VA

6. LARGEST MOTOR: *220.14(C)*

φ _____ VA × 25% =   _____ VA   _____ VA

N _____ VA × 25% =   _____ VA   _____ VA

1φ service: PHASES $I = \dfrac{VA}{V} =$ _____ A

NEUTRAL $I = \dfrac{VA}{V} =$ _____ A   _____ VA   _____ VA

220.61(B) First 200 A × 100% =   200 A

Remaining _____ A × 70% =   _____ A

Total   _____ A

Optional Calculation: One-Family Dwelling

1. HEATING or A/C: *220.82(C)(1 – 6)*

Heating units (3 or less) = _____ VA × 65% = _____ VA

Heating units (4 or more) = _____ VA × 40% = _____ VA

A/C unit = _____ VA × 100% = _____ VA

Heat pump = _____ VA × 100% = _____ VA   **PHASES**

Largest Load _____ VA

Total _____ VA   _____ VA

2. GENERAL LOADS: *220.82(B)(1 – 4)*

General lighting: *220.82(B)(1)*

_____ sq ft × 3 VA _____ VA

Small appliance and laundry loads: *220.82(B)(2)*

_____ VA × _____ circuits = _____ VA

Special loads: *220.82(B)(3 – 4)*

Dishwasher = _____ VA

Disposer = _____ VA

Compactor = _____ VA

Water heater = _____ VA

_____ = _____ VA

_____ = _____ VA

_____ = _____ VA

_____ = _____ VA

_____ = _____ VA

_____ VA   _____ VA

Total _____ VA

Applying Demand Factors: *220.82(B)*

First 10,000 VA × 100% = 10,000 VA

Remaining _____ VA × 40% = _____ VA

Total _____ VA   _____ VA

NEUTRAL (Loads from Standard Calculation)

1. General lighting = _____ VA
2. Fixed appliances = _____ VA
3. Dryer = _____ VA
4. Cooking equipment = _____ VA
5. Heating or A/C = _____ VA
6. Largest motor = _____ VA

Total _____ VA

1φ service: PHASES $I = \dfrac{VA}{V} =$ _____ A

NEUTRAL $I = \dfrac{VA}{V} =$ _____ A   _____ VA

Architectural Symbols . . .

Material	Elevation	Plan View	Section
EARTH			
BRICK	WITH NOTE INDICATING TYPE OF BRICK (COMMON, FACE, ETC.)	COMMON OR FACE / FIREBRICK	SAME AS PLAN VIEWS
CONCRETE		LIGHTWEIGHT / STRUCTURAL	SAME AS PLAN VIEWS
CONCRETE MASONRY UNITS		OR	OR
STONE	CUT STONE / RUBBLE	CUT STONE / RUBBLE / CAST STONE (CONCRETE)	CUT STONE / CAST STONE CONCRETE / RUBBLE OR CUT STONE
WOOD	SIDING / PANEL	WOOD STUD / REMODELING / DISPLAY	ROUGH MEMBERS / FINISHED MEMBERS / PLYWOOD
PLASTER		WOOD STUD, LATH, AND PLASTER / METAL LATH AND PLASTER / SOLID PLASTER	LATH AND PLASTER
ROOFING	SHINGLES	SAME AS ELEVATION	
GLASS	OR / GLASS BLOCK	GLASS / GLASS BLOCK	SMALL SCALE / LARGE SCALE

. . . Architectural Symbols

Material	Elevation	Plan View	Section
FACING TILE	CERAMIC TILE	FLOOR TILE	CERAMIC TILE LARGE SCALE — CERAMIC TILE SMALL SCALE
STRUCTURAL CLAY TILE			SAME AS PLAN VIEW
INSULATION		LOOSE FILL OR BATTS — RIGID — SPRAY FOAM	SAME AS PLAN VIEWS
SHEET METAL FLASHING		OCCASIONALLY INDICATED BY NOTE	
METALS OTHER THAN FLASHING	INDICATED BY NOTE OR DRAWN TO SCALE	SAME AS ELEVATION	SMALL SCALE — STEEL — CAST IRON — ALUMINUM — BRONZE OR BRASS
STRUCTURAL STEEL	INDICATED BY NOTE OR DRAWN TO SCALE	OR	REBARS — SMALL SCALE — LARGE SCALE — L-ANGLES, S-BEAMS, ETC.

Plot Plan Symbols

NORTH	FIRE HYDRANT	WALK	—E— OR ELECTRIC SERVICE
POINT OF BEGINNING (POB)	MAILBOX	IMPROVED ROAD	—G— OR NATURAL GAS LINE
UTILITY METER OR VALVE	MANHOLE	UNIMPROVED ROAD	—W— OR WATER LINE
POWER POLE AND GUY	TREE	BUILDING LINE	—T— OR TELEPHONE LINE
LIGHT STANDARD	BUSH	PROPERTY LINE	NATURAL GRADE
TRAFFIC SIGNAL	HEDGE ROW	PROPERTY LINE	FINISH GRADE
STREET SIGN	FENCE	TOWNSHIP LINE	+ XX.00′ EXISTING ELEVATION

A

abbreviation: A letter or combination of letters that represents a word.

AC voltage: Voltage that varies in strength at regular intervals.

alternating current (AC): Current that reverses its direction of flow at regular intervals.

ambient temperature: The temperature of air around a piece of equipment.

ampacity: The current that a conductor can carry continuously, under the conditions of use.

ampere (A): The number of electrons passing a given point in one second.

annunciator: A signaling device that notifies individuals that one or more detectors have been activated.

anti-short bushing: A plastic or heavy fiber paper device used to protect the conductors of armored cable.

appliance: Any utilization equipment which performs one or more functions, such as clothes washing, air conditioning, cooking, etc.

appliance branch circuit: A branch circuit that supplies energy to one or more outlets to which appliances are to be connected.

appliance plug: A plug used to power an appliance that produces heat, such as an electric grill, roaster, broiler, waffle iron, or large coffeemaker. Also known as a heater plug.

armature: The movable coil of wire in a generator that rotates through the magnetic field.

armored cable (AC): A factory assembly that contains the conductors within a jacket made of a spiral wrap of steel.

automatic transfer switch: An electrical device that transfers the load of a residence from public utility circuits to the output of a standby generator during a power failure.

audio system: An arrangement of electronic components that is designed to reproduce sound.

B

back-wired connector: A mechanical connection method used to secure wires to the backs of switches and receptacles. Also known as a quick connector.

bare conductor: A conductor with no insulation or covering of any type.

bathroom: An area with a basin and one or more of a toilet, tub, or shower.

battery: A DC voltage source that converts chemical energy to electrical energy.

bend: Any change in direction of a cable, conduit, or raceway.

blade cartridge fuse: A snap-in type electrical safety device that operates on the heating effect of an element.

bonding: Joining metal parts to form a continuous path to conduct any current safely that is commonly imposed.

bonding wire: An uninsulated conductor in armored cable that is used for grounding.

box: A metallic or nonmetallic electrical enclosure used for equipment, devices, and pulling or terminating conductors.

branch circuit: The portion of the electrical circuit between the last overcurrent device (fuse or CB) and the outlets (appliances, TVs, and lamps).

branch-circuit conductor: The circuit conductor(s) between the final overcurrent device protecting the circuit and the outlet(s).

branch-circuit rating: The ampere rating or setting of the overcurrent device (fuse or circuit breaker) protecting the conductors.

bridge: A device that connects local area networks (LANs) with different hardware and protocols to one another.

brush: The sliding contact that rides against the slip rings and is used to connect the armature to the external circuit (power grid).

bushing: A fitting placed on the end of a conduit to protect the conductor's insulation from abrasion.

C

cable: A factory assembly with two or more conductors and an overall covering.

cable assembly: A flexible assembly containing multi-conductors with a protective outer sheath.

cable tie gun: A handheld device that is used to hold and tighten cables together with plastic or steel ties.

Canadian Standard Association (CSA) label: A marking that indicates that extensive tests have been conducted on a device by the Canadian Standards Association.

cartridge fuse: A snap-in type electrical safety device that operates on the same basic heating principle as a plug fuse.

caution: A word used to indicate a potentially hazardous situation, which could result in minor or moderate injury.

circuit: A complete path (when ON) for current to take that includes electrical control devices, circuit protection, conductors, and load(s).

circuit breaker: A fixed electrical safety device designed to protect electrical devices and individuals from overcurrents.

circuit breaker fuse: A screw-in type electrical safety device that has the operating characteristics of a circuit breaker.

circular mil: A measurement used to determine the cross-sectional area of a conductor.

cladding: A layer of glass or other transparent material surrounding the fiber core of fiber optic cable that causes light to be reflected back to the central core for the purpose of maintaining signal strength over long distances.

clip grounding: A grounding method where a grounding clip is merely slipped over the grounding wire from the electrical device.

cold solder joint: A defective solder joint that results when the parts being joined do not exceed the liquid temperature of the solder.

color coding: A technique used to identify switch terminal screws through the use of various colors.

communication cable: Cable (usually copper or fiber optic) used to transmit data from one location (or device) within a system to another location.

compact disc (CD) player: An electric device that plays music stored on a compact disc (CD) by means of reflecting light from a laser beam.

complete path: A grouping of electrical devices and wires that create a path for current to take from the power source (service panel), through controls (switches), to the load (light fixtures and receptacles), and back to the power source.

component grounding: A grounding method where the ground wire is attached directly to an electrical component such as a receptacle.

component list: A list of electrical equipment indicating manufacturer, specifications, and how many of each electrical component are required for a room or area.

component plan: A group of schedules that state the required locations for receptacles, lights, and switches according to the NEC® and local codes.

compression connector: A type of box fitting that firmly secures conduit to a box by utilizing a nut that compresses a tapered metal ring (ferrule) into the conduit.

conductor: A slender rod or wire that is used to transmit the flow of electrons in an electrical circuit.

conductor symbol: An electrical symbol that represents copper and aluminum respectively.

conduit: A rugged protective tube (typically metal) through which wires are pulled.

conduit coupling: A type of fitting used to join one length of conduit to another length of conduit and still maintain a smooth inner surface.

conduit seal: A fitting which is inserted into runs of conduit to isolate certain electrical apparatus from atmospheric hazards.

contact: The conducting part of a switch that operates with another conducting part to make or break a circuit.

continuity tester: A test instrument that is used to test a circuit for a complete path for current to flow.

control: An electrical device that accepts signals from detectors and turns annunciators ON and OFF.

control switch: A switch that controls the flow of current in a circuit.

converter: An electronic device that changes AC voltage into DC voltage.

cord: Two or more flexible conductors grouped together and used to deliver power to a load by means of a plug.

covered conductor: A conductor not encased in a material recognized by the NEC®.

crimp-type solderless connector: An electrical device that is used to join wires together or to serve as terminal ends for screw connections.

current-limiting fuse: A fuse that opens a circuit in less than ½ of a cycle to protect circuit components from damage caused by short-circuit currents.

D

damp location: A partially protected area subject to some moisture.

danger: A word used to indicate an imminently hazardous situation, which could result in death or serious injury.

DC voltage: Voltage that flows in one direction at a constant strength.

dead front: A cover required for the operation of a plug or connector.

decibel (dB): A unit used to express the relative intensity (volume) of sound.

detector/sensor: An electrical device (sensor) used to detect noise, heat, gases, or movement.

device box: A box which houses an electrical device.

digital display: An electronic device that displays readings as numerical values.

digital multimeter: An electrical test instrument that can measure two or more electrical properties and display the measured properties as numerical values.

dimmer switch: A switch that changes the lamp brightness by changing the voltage supplied to a lamp.

disconnecting means: A device, or group of devices, by which circuit conductors are disconnected from their power source.

double-pole switch: A control device that is two switches in one for controlling two separate loads.

double offset (saddle): A common complex bend made in conduit to bypass obstructions.

downrod mount: A type of ceiling fan mounting used to adjust head clearance or for angle mounting to vaulted ceilings.

dry location: A location which is not normally damp or wet.

duplex receptacle: A receptacle that has two spaces for connecting two different plugs.

dwelling: A structure that contains eating, living, and sleeping space, and permanent provisions for cooking and sanitation.

E

Edison-base fuse: A plug fuse that incorporates a screw configuration which is interchangeable with fuses of other ampere ratings.

effectively grounded: Grounded with sufficient low impedance and current-carrying capacity to prevent hazardous voltage buildups.

electrical circuit: An assembly of conductors (wires), electrical devices (switches and receptacles), and components (lights and motors) through which current flows.

electrical layout: A drawing that indicates the connections of all devices and components in a residential electrical system.

electrical metallic tubing (EMT): A light-gauge electrical pipe often referred to as thin-wall conduit.

electrical noise: Any unwanted disturbance that interferes with an electrical signal.

electrical plan: A drawing and list that indicates what devices are to be used, where the electrical devices are to be placed, and how the devices are to be wired.

electrical tool pouch: A small, open tool container (pouch) for storing a few commonly used electrical tools.

electrical shock: The condition that occurs when an individual comes in contact with two conductors of a circuit or when the body of an individual becomes part of an electrical circuit.

electrical warning: A word used to indicate a high-voltage location and conditions that could result in death or serious injury from an electrical shock if proper precautions are not taken.

electric-discharge lighting fixture: A lighting fixture that utilizes a ballast for the operation of the lamp.

electricity: A form of energy where electrons jump from atom to atom through a conducting material.

electrode: A long metal rod used to make contact with the earth for grounding purposes.

electromagnetism: The magnetic field produced when electricity passes through a conductor.

enclosure: The case or housing of equipment or other apparatus which provides protection from live or energized parts.

energy: The capacity to do work.

energy efficiency: The ratio of output energy (energy used by a device) to input energy (energy created to run a device); expressed as a percentage.

equipment: Any device, fixture, apparatus, appliance, etc. used in conjunction with electrical installations.

equipment bonding jumper (EBJ): A conductor that connects two or more parts of the EGC.

equipment ground: A circuit designed to protect individual components connected to an electrical system.

equipment grounding conductor (EGC): An electrical conductor that provides a low-impedance path between electrical equipment and enclosures and the system grounded conductor.

eutectic alloy: An alloy that has one specific melting temperature with no intermediate stage.

explosion warning: A word used to indicate a location and conditions where exploding parts may cause death or serious personal injury if proper precautions and procedures are not followed.

exposed: As applied to wiring methods, is on a surface or behind panels with open access.

F

feeder: All circuit conductors between the branch-circuit overcurrent device or the source of a separately derived system and the final load (appliance or lighting).

fiber connector: A device that splices two fiber cables by squarely aligning the center axes of both cables and holding the two fiber cables together.

fishing: A term used for the process of pulling wires through conduit.

fish tape: A device used to pull wires through conduit.

fitting: An electrical system accessory that performs a mechanical function of connection.

fixed temperature detector: A heat detector designed to respond when a room reaches a specific temperature.

flat conductor cable (FCC): A cable with three or more flat copper conductors edge-to-edge and separated and enclosed by an insulating material.

flat-cord plug: An ungrounded plug that includes a core that snaps into a housing after the wires are connected.

flexible cable: An assembly of one or more insulated braided conductors contained within an overall outer covering and used for the connection of equipment to a power source.

flexible cord: An assembly of two or more insulated braided conductors contained within an overall outer covering and used for the connection of equipment to a power source.

flexible metallic conduit: A conduit that has no wires and can be bent by hand.

floodlight: A lamp that casts general light over a large area.

fluorescent lamp: A low-pressure discharge lamp in which ionization of mercury vapor transforms ultraviolet energy generated by the discharge into light.

flush mount: A type of ceiling fan mounting used to allow maximum head clearance.

footcandle (fc): The amount of light produced by a lamp (lumens) divided by the area that is illuminated.

four-way switch: A control device that is used in combination with two three-way switches to allow control of a load from three locations.

full-range speaker: A speaker designed to adequately reproduce most of the audio spectrum.

fuse: An electrical device used to limit the rate of current flow in a circuit.

G

general-purpose branch circuit: A branch circuit that supplies a number of outlets for lighting and appliances.

generated electricity: The alternating current (AC) created by power plant generators.

generator: An electromechanical device that converts mechanical energy into electrical energy by means of electromagnetic induction.

grade: The level or elevation of the earth on a job site.

grid-connected wind power system: A wind power system that is connected to an electric utility distribution system, or grid.

ground: A conducting connection between electrical circuits or equipment and the earth.

grounded conductor: A conductor that has been intentionally grounded.

ground fault: Any current above the level that is required for a dangerous shock.

ground fault circuit interrupter (GFCI): A fast-acting electrical device that detects low levels of leakage current to ground and opens the circuit in response to the ground fault.

ground fault circuit interrupter (GFCI)-receptacle: A fast-acting receptacle that detects low levels of leakage current to ground and opens the circuit in response to the leakage (ground fault).

grounding electrode conductor (GEC): The conductor that connects the grounding electrode(s) to the grounded conductor and/or the EGC.

grounding receptacle: Receptacle that includes a grounding terminal connected to a grounding slot in the receptacle configuration.

H

hard-wiring: The physical connection between electrical components.

heat detector: An electrical sensing device used to detect excess heat levels within a structure.

heater plug: *See* appliance plug.

heating element: A conductor that offers enough resistance to produce heat when connected to an electrical power supply.

heating system: A complete system of components such as heating elements, fasteners, nonheating circuit wiring, leads, temperature controllers, safety signs, junction boxes, raceways, and fittings.

heat pump: A mechanical compression refrigeration system that contains devices and controls that reverse the flow of refrigerant to move heat from one area to another.

horsepower (HP): A unit of power equal to 746 W or 33,000 lb-ft per minute (550 lb-ft per second).

hydromassage bathtub: A permanently installed bathtub with a recirculating pump designed to accept, circulate, and discharge water for each use.

I

illumination: The effect that occurs when light falls on a surface.

immersion detection circuit interrupter (IDCI): Circuit interrupter designed to provide protection against shock when appliances fall into a sink or bathtub.

incandescent lamp: An electric lamp that produces light by the flow of current through a tungsten filament inside a gas-filled, sealed glass bulb.

indenter connector: A type of box fitting that secures conduit to a box with the use of a special indenting tool.

individual branch circuit: A branch circuit that supplies only one box or receptacle.

infrared detector: An electrical sensing device used for security purposes to determine if movement exist in a room or area.

instantaneous-trip CB (ITCB): CB with no delay between the fault or overload sensing element and the tripping action of the device.

insulated conductor: A conductor covered with a material classified as electrical insulation.

interactive jack: A connecting device such as a computer Ethernet connection that allows signals to be sent and received.

inverse-time CB (ITCBs): CB with an intentional delay between the time when the fault or overload is sensed and the time when the CB operates.

inverter: Equipment used to change voltage level or waveform, or both, of electrical energy.

isolated-ground receptacle: A special receptacle that minimizes electrical noise by providing a separate grounding path for each connected device.

isolating switch: A switch designed to isolate an electric circuit from the power source. It has no interrupting rating and is operated only after the circuit has been opened by other means.

J

junction box: A box in which splices, taps, or terminations are made.

K

kick: A single bend in a raceway.

knockout: A round indentation punched into the metal of a box and held in place by unpunched narrow strips of metal.

L

lamp: An output device (load) that converts electrical energy into light.

lampholder: A device designed to accommodate a lamp for the purpose of illumination.

laser diode: A fiber optic light source (similar to a light emitting diode) with an optical cavity that is used to produce laser light (coherent light).

lateral service: An electrical service in which service-entrance conductors are run underground from the utility service to the dwelling.

lateral service entrance (cabinet): A cabinet that encloses the service wires, meter socket, and meter.

lateral service entrance (conduit): A service entrance where all service wires are buried underground, creating a condition where the wires are subjected to less damage from the environment (weather) and allowing unsightly poles and service wires to be removed from streets and alleys.

light: That portion of the electromagnetic spectrum which produces radiant energy.

lighting box: A box intended for the direct connection of a lampholder, a lighting fixture, or pendant cord terminating in a lampholder.

lighting track: An assembly consisting of an energized track and lighting fixture heads which can be positioned in any location along the track.

liquidtight flexible nonmetallic conduit (LFNC): A raceway of circular cross-section with an outer jacket which is resistant to oil, water, sunlight, corrosion, etc. The inner core varies based on intended use.

lockout: The process of removing the source of electrical power and installing a lock which prevents the power from being turned ON.

lubricant: A wet or dry compound that is applied to the exterior of wires to allow the wires to slide better.

lumen (lm): The unit used to measure the total amount of light produced by a light source.

luminaire: A complete lighting fixture consisting of the lamp or lamps, reflector or other parts to distribute the light, lamp guards, and lamp power supply.

luminaire temperature: The temperature at which a lamp delivers its peak light output.

M

magnet: A device that attracts iron and steel because of the molecular alignment of its material.

magnetic circuit breaker: An electrical safety device that operates with miniature electromagnets.

magnetic detector: An electrical sensing device used on windows to determine if the window is open or closed.

magnetism: A force that interacts with other magnets and ferromagnetic materials.

main bonding jumper (MBJ): The connection at the service equipment that ties together the EGC, the grounded conductor, and the GEC.

manually controlled circuit: Any circuit that requires a person to initiate an action for the circuit to operate.

mean lumen: The average light produced after a lamp has operated for approximately 40% of its rated life.

mercury-vapor lamp: An HID lamp that produces light by an electrical discharge through mercury vapor.

metal-clad cable (MC): A factory assembly of one or more conductors with or without fiber-optic members, enclosed in a metallic armor.

metallic-sheathed cable: A type of cable that consists of two or more individually insulated wires protected by a flexible metal outer jacket.

mobile home: A transportable factory-assembled structure or structures constructed on a permanent chassis for use as a dwelling. A mobile home is not constructed on a permanent foundation but is connected to the required utilities. The term "mobile home" includes manufactured homes.

momentary power interruption: A decrease to 0 V on one or more power lines lasting from 0.5 cycles up to 3 sec.

motion sensor: An infrared sensing device that detects the movement of a temperature variance and automatically switches when the movement is detected.

motor: A machine that develops torque (rotating mechanical force) on a shaft which is used to produce work.

motor efficiency: The measure of the effectiveness with which a motor converts electrical energy to mechanical energy.

motor torque: The force that produces, or tends to produce rotation in a motor.

N

National Electrical Code® (NEC®): A book of electrical standards that indicate how electrical systems must be installed and how work must be performed.

National Fire Protection Association (NFPA): A national organization that provides guidance on safety and assessing the hazards of the products of combustion.

network: A system of computers, terminals, and databases connected by communication lines.

nipple: A short piece of conduit or tubing that does not exceed 24″ in length.

nonadjustable-trip CB (NATCB): A fixed CB designed without provisions for adjusting either the ampere trip setpoint or the time-trip setpoint.

nongrounding receptacle: A receptacle with two wiring slots for branch-circuit wiring systems that do not provide an equipment grounding conductor.

nonmetallic extension: An assembly of two insulated conductors in a nonmetallic jacket or an extruded thermoplastic covering. Nonmetallic extensions are surface extensions intended for mounting directly on the surface of walls or ceilings.

nonmetallic-sheathed cable: An electrical conductor (cable) that has a set of insulated electrical conductors held together and protected by a strong plastic jacket.

nonmetallic wireway: A flame-retardant nonmetallic trough with a removable cover that houses and protects wires and cables laid in place after the wireway has been installed.

non-time delay fuse (NTDF): A fuse that may detect an overcurrent and open the circuit almost instantly.

O

offset: A compound bend in conduit used to bypass many types of obstructions.

Ohm's law: The relationship between voltage, current, and resistance properties in an electrical circuit.

one-family dwelling: A dwelling with one housing unit.

open circuit transition switching: A process in which power is momentarily disconnected when switching a circuit from one voltage supply or level to another.

outlet box: An electrical device designed to house electrical components and protect wiring connections.

output: A device that is used to produce work, light, heat, sound, or display information.

overcurrent: Any current in excess of that for which the conductor or equipment is rated.

overcurrent protection device (OCPD): A disconnect switch with a circuit breaker (CB) or fuse added to provide overcurrent protection of the switched circuit.

overhead riser service entrance: A type of service entrance that has wires running from a utility pole to a service head, where the meter socket and riser are firmly secured to the exterior of a dwelling, while the service panel is placed inside.

overlamping: Installing a lamp of a higher wattage than for which a fixture is designed.

overload: A small-magnitude overcurrent, that over a period of time, leads to an overcurrent which may operate the overcurrent protection device (fuse or CB).

oxide: A thin, but highly resistive coating that forms on metal when exposed to air.

P

pad: A small round conductor to which a component lead is soldered.

panel: A collection of modules, secured together, wired, and designed to provide a field-installable unit.

panelboard: A single panel or group of assembled panels with buses and overcurrent devices, which may have switches to control light, heat, or power circuits.

parallel conductors: Two or more conductors that are electrically connected at both ends to form a single conductor.

patch cord: A cable that consists of two wires: a thin center conductor that carries the signal and a woven shield wire that encircles it.

pegboard: A thin constructed board typically available in 4′ x 8′ sheets that is perforated with equally spaced holes for accepting hooks.

pendant: A hanging light fixture that uses a flexible cord to support a lampholder.

permanently-connected appliance: A hard-wired appliance that is not cord-and-plug connected.

permanently installed swimming pool: A pool constructed in ground or partially above ground and designed to hold over 42″ of water, and all indoor pools regardless of depth.

permanent whole-house generator: A permanently installed standby generator system that can supply temporary power to an entire residence.

personal protective equipment (PPE): Clothing and/or equipment worn by individuals to reduce the possibility of injury in the work area.

photovoltaic system: An electrical system consisting of a PV array and other electrical components needed to convert solar energy into electricity usable by loads.

photovoltaic technology: Solar energy technology that uses the unique properties of semiconductors to directly convert solar radiation into electricity.

pictorial drawing: A drawing that shows the length, height, and depth of an object in one view.

pigtail grounding: A grounding method where two grounding wires are used to connect an electrical device to a grounding screw in a box and then to system ground.

pigtail splice: A splice that consists of twisting two or more wires together.

plug: A device at the end of a cord that connects the device to an electrical power supply by means of a receptacle.

plug fuse: A fuse that uses a metallic strip which melts when a predetermined amount of current flows through it.

plunger button detector: An electrical sensing device used on windows and doors to determine if the window or door is open or closed.

polarized receptacle: A receptacle where the size of the connection slots determines the plug connection.

pole: The number of completely isolated circuits that a relay can switch.

polyethylene fish tape: Wire-pulling device typically used to pull wire within conduit systems.

power line carrier (PLC): Now commonly referred to as X10 communication.

power source: The service panel in a residence.

primer: A chemical agent that cleans and softens a surface and allows solvent cement to penetrate more effectively into the surface.

pull box: A box used as a point to pull or feed electrical conductors into conduit or raceway systems.

pulling grip: A device that is attached to a fish tape to allow more leverage.

Q

quick connector: *See* back-wired connector.

R

raceway: An exposed metal or nonmetallic enclosed channel for conductors.

rated voltage: A voltage range that is typically within ±10% of ideal voltage.

rate of rise detector: A heat detector designed for flash fires or slow-burning fires.

receptacle: A contact device installed for the connection of plugs and flexible cords to supply current to portable electrical equipment.

receptacle box: An electrical device designed to house electrical components and protect wiring connections.

receptacle tester: A test instrument that is plugged into a standard receptacle to determine if the receptacle is properly wired and energized.

redundant grounding: Grounding with two separate grounding paths.

regenerative repeater (regenerator): A device inserted into a digital circuit to regenerate a transmitted circuit.

remote control wiring: A method of controlling standard-voltage devices through the use of low-voltage relays and low-voltage switches.

renewable energy system: An energy system that uses natural resources to produce heat and electric power.

resistance (R): 1. The opposition to the flow of electrons. **2.** Any force that tends to hinder the movement of an object.

resistivity: The resistance of a conductor having a specific length and cross-sectional area.

resistor: A device that limits the current flowing in an electronic circuit (security).

rigid fish tape: A wire-pulling device used for pulling wires through conduit and pulling wires through walls and ceilings during remodeling.

rigid metal conduit (RMC): A heavy-duty pipe that is threaded on the ends much like standard plumbing pipe.

ripcord: a cord included in a cable that aids in removing the outer jacket.

roughing-in: A phrase that refers to the placement of electrical boxes and wires before wall coverings and ceilings are installed.

S

safety glasses: An eye protection device with special impact-resistant glass or plastic lenses, reinforced frames, and side shields.

safety label: A label that indicates areas or tasks that can pose a hazard to personnel and/or equipment.

schematic diagram: A drawing that indicates the electrical connections and functions of a specific circuit arrangement using graphic symbols.

security/fire system: A home protection system composed of three elements: detectors, controls, and annunciators.

selector switch: A switch with an operator that is rotated (instead of pushed) to activate a specified electrical contact.

self-grounding receptacle: A grounding type receptacle that utilizes a pressure clip around the mounting screw to ensure good electrical contact between the receptacle yoke and the outlet box.

service: The electrical supply, in the form of conductors and equipment, that provides electrical power to a building or structure.

service conductor: The conductor from the service point or other source of power to the service panel.

service drop: A method of electrical service that runs wires from a utility pole to a service head on or above a residence.

service entrance: The connecting link between a residence and the power company.

service-entrance cable (SE): A multiconductor assembly with or without an overall covering.

service-entrance conductor — overhead system: Conductor that connects the service equipment for a building or structure with the electrical utility supply conductors.

service-entrance conductor — underground system: Conductor that connects the service equipment with a service lateral.

service equipment: All of the necessary equipment to control the supply of electrical power to a building or a structure.

service jack: A connecting device that provides connection to a service such as television reception through a coaxial cable.

service lateral: Any service to a residence that is achieved by burying the wires underground.

service mast: An assembly consisting of a service raceway, guy wires or braces, service head, and any fittings necessary for the support of service-drop conductors.

service panel: An electrical device containing fuses or circuit breakers for protecting the individual circuits of a residence and is a means of disconnecting the entire residence from the distribution system.

service point: The point of connection between the local electrical utility company and the premises wiring of the building or structure.

set screw connector: A type of box fitting that relies on the pressure of a screw against conduit to hold the conduit in place.

short circuit: A condition that occurs when two ungrounded conductors (hot wires), or an ungrounded and a grounded conductor of a single-phase (1ϕ) circuit, come in contact with each other.

shunt: A permanent conductor placed across a water meter to provide a continuous flow path for ground current.

signal level: The amplitude or strength of an electronic (audio) signal.

single-pole switch: An electrical control device used to turn lights or appliances ON and OFF from a single location.

single receptacle: A single contact device with no other contact device on the same yoke.

slip rings: Metallic rings connected to the ends of an armature and are used to connect the induced voltage of a generator to the electrical distribution system.

smoke detector: An electrical device used to sense products of combustion (smoke) and activate an alarm when smoke is detected.

solar energy: Energy recovered from the sun in the form of sunlight

solar thermal energy system: A renewable energy system that collects and stores solar energy and is used to heat air and water in a residential structure.

solder: An alloy of tin (Sn) and lead (Pb) used to make permanent electrical connections.

soldering: The process of making a sound electrical and mechanical joint between certain metals by joining them with solder.

soldered connection: A connection that joins conductors by heat to make a strong electrical and mechanical connection.

solderless connector: A device used to join wires firmly without the help of solder.

solvent cement: A chemical agent that penetrates and softens the surface of plastic pipe and fittings.

spa (hot tub): An indoor or outdoor hydromassage pool or tub that is not designed to have the water discharged after each use.

speaker: An electric device that converts electrical signals into sound waves.

splice: The joining of two or more electrical conductors by mechanically twisting the conductors together or by using a special splicing device.

split-bolt connector: A type of solderless mechanical connection used for joining large cables such as in service entrances.

split-wired receptacle: A standard receptacle that has had the tab between the two brass-colored (hot) terminal screws removed.

stand-alone (off-grid) wind power system: A wind power system that is not connected to an electric utility distribution system, or grid.

standard plug fuse: A screw-in type electrical safety device that contains a metal conducting element designed to melt when the current through the fuse exceeds the rated value.

static electricity: An electrical charge at rest.

stereo system: An audio system that uses two independent channels that are routed to a pair of speakers situated to the right and to the left of the listener.

storable swimming pool: A pool constructed on or above ground and designed to hold less than 42″ of water, or a pool with nonmetallic, molded polymeric walls or inflatable fabric walls regardless of size.

strip gauge: A short groove that indicates how much insulation must be removed from a wire so the wire can be properly inserted into a switch.

surface raceway: An enclosed channel for conductors which is attached to a surface.

surge suppressor: A receptacle that provides protection from high-level transients by limiting the level of voltage allowed downstream from the surge suppressor.

surround sound: An effect that provides a feeling of left/right and front/back sound movement as well as reverberated and reflected sound similar to a concert hall.

sustained power interruption: A decrease to 0 V on all power lines for a period of more than 1 min.

switch: An electrical device used to control loads in a residential electrical circuit.

switchboard: A single panel or group of assembled panels with buses, overcurrent devices, and instruments.

symbol: A graphic element that represents a component, device, or quantity.

system ground: A special circuit designed to protect the entire distribution system of a residence.

T

tap splice: A splice that connects two wires together when one wire is to remain unbroken.

temperature: The measurement of the intensity of heat.

temperature rise: The amount of heat that an electrical component produces above the ambient temperature.

temperature switch: A switch that responds to temperature changes.

temporary power interruption: A decrease to 0 V on one or more power lines lasting between 3 sec and 1 min.

test light (neon tester): A test instrument that is designed to illuminate in the presence of 115 V and 230 V circuits.

thermal circuit breaker: An electrical safety device that operates with a bimetallic strip that warps when overheated.

thermally protected: Designed with an internal thermal protective device which senses excessive operating temperatures and opens the supply circuit to the fixtures.

thermal-magnetic circuit breaker: An electrical safety device that combines the heating effect of a bimetallic strip with the pulling strength of a magnet to move a trip bar.

three-way switch: An electrical control device used in pairs to control a light or load from two locations.

throw: The number of closed contact positions per switch pole.

time-based control system: An automatic control system that uses the time of day to determine the desired operation of energy-consuming loads in a home.

time-delay plug fuse: A screw-in type electrical safety device with a dual element.

timer: A control device that uses a preset time period as part of a control function.

tinning: The process of applying solder to a clean soldering iron tip to prevent corrosion on the tip.

torque: The force that causes an object to rotate.

transfer switch: A switch that isolates a generator from a power grid.

transformer: An electric device that uses electromagnetism to change (step-up or step-down) AC voltage from one level to another.

transient: A temporary, unwanted voltage in an electrical circuit.

transistor: A three-terminal device that controls current through the device depending on the amount of voltage applied to the base.

T rating: Special switch information that indicates a switch is capable of handling the severe overloading created by a tungsten load as the switch is closed.

travel trailer: A vehicle that is mounted on wheels, has a trailer area less than 320 sq ft (excluding wardrobes, closets, cabinets, kitchen units, fixtures, etc.), is of such size and weight that a special highway use permit is not required, and is designed as temporary living quarters while camping or traveling.

tuner: An electric device designed to receive radio broadcast signals and convert them into electrical signals that can be amplified.

tweeter: A small speaker that is designed to reproduce high frequencies between 4 kHz to 22 kHz.

twisted conductors: Conductors that are intertwined at a constant rate.

two-piece magnetic detector: An electrical sensing device used on windows and doors to determine if the window or door is open or closed.

2-way switch: A single-pole, single-throw (SPST) switch.

type NM cable: A nonmetallic-sheathed cable that has the conductors enclosed within a nonmetallic jacket.

type NMC cable: A nonmetallic-sheathed cable that has the conductors enclosed within a corrosion-resistant, nonmetallic jacket.

type S plug fuse: A screw-in type electrical safety device that has all the operating characteristics of a time-delay plug fuse plus the added advantage of being non-tamperable.

U

UL label: A stamped or printed icon that indicates that a device or material has been approved for consumer use by Underwriters Laboratories Inc.®

underground (service lateral): A method of electrical service that buries the wires to a residence in the ground.

Underwriters Laboratories Inc.® (UL): An independent organization that tests equipment and products to see if they conform to national codes and standards.

unfinished basement: The portion of area of a basement which is not intended as a habitable room, but is limited to storage areas, work areas, etc.

ungrounded conductor: A current-carrying conductor that is connected to loads through fuses, circuit breakers, and switches.

uninterruptible power system (UPS): A power supply that provides constant on-line power when the primary power supply is interrupted.

utilization equipment: Any electrical equipment which uses electrical energy for electronic, electromechanical, chemical, heating, lighting, etc. purposes.

V

visible light: The portion of the electromagnetic spectrum to which the human eye responds.

voltage (E): The amount of electrical strength in a circuit.

voltage dip: A momentary low voltage.

voltage indicator: A test instrument that indicates the presence of voltage when the test tip touches, or is near, an energized hot conductor or energized metal part.

voltage spike: An increase in voltage (normally several thousand volts) that lasts for a very short time (microseconds to milliseconds).

voltage stabilizer (regulator): A device that provides precise voltage regulation to protect equipment from voltage dips and voltage surges.

voltage surge: A higher-than-normal voltage that temporarily exists on one or more power line.

voltage tester: An electrical test instrument that indicates the approximate amount of voltage and the type of voltage (AC or DC) in a circuit by the movement of a pointer (and vibration, on some models).

voltage-to-ground: The difference of potential between a given conductor and ground.

voltage variance: The difference in voltage between a voltage surge and a voltage dip.

W

warning: A word used to indicate a potentially hazardous situation, which could result in death or serious injury.

water pipe ground: A continuous underground metallic pipe that supplies a residence with water and is typically the best electrical ground for a residential electrical system.

watt (W): A unit of measure equal to the power produced by a current of 1 A across a potential difference of 1 V.

wattage: The number of watts a device uses to operate.

Western Union splice: A type of splice that is used when the connection must be strong enough to support long lengths of heavy wire.

wet location: Any location in which a conductor is subjected to excessive moisture or saturation from any type of liquids or water.

wind power system: A renewable energy system that converts wind energy into usable electric power.

wind turbine: A machine that converts the energy in wind into mechanical energy.

wire: Any individual conductor.

wireless control detector: A device used for security purposes in existing construction to eliminate hard wiring.

wire marker: A preprinted peel-off sticker designed to adhere to insulation when wrapped around a conductor.

wire nut: An electrical device that holds wires together and provides insulative protection.

wire stripper: A tool that is designed to properly remove insulation from small gauge (normally AWG sizes 10–22) wires.

wireway: A metallic or nonmetallic trough with a hinged or removable cover designed to house and protect conductors and cables.

wiring diagram: A drawing that indicates the connections of all devices and components in a residential electrical system.

wiring layout: A schematic that indicates how the component parts of a circuit will be connected to one another and where the wires will be run.

wiring plan: A floor plan drawing that indicates the placement of all electrical devices and the wiring required to connect all the devices into circuits.

woofer: A large speaker that is designed to reproduce low-frequency sounds.

X

X10 technology: A communications protocol that allows AC control devices to "talk" on existing household electrical wiring.

Index

Page numbers in italic refer to figures.